Introductory
Computer
Mathematics

Introductory
Computer
Mathematics

Second Edition

Nigel P. Cook

Upper Saddle River, New Jersey
Columbus, Ohio

Library of Congress Cataloging-in-Publication Data

Cook, Nigel P.,
  Introductory computer mathematics / Nigel P. Cook.--2nd ed.
    p. cm.
  ISBN 0-13-045289-0
  1. Mathematics.   2. Computer science--Mathematics   I. Title.

  QA39.3.C67 2003
  004'.01'51--dc21   2002074990

Editor in Chief: Stephen Helba
Executive Editor: Frank I. Mortimer, Jr.
Media Development Editor: Michelle Churma
Production Editor: Louise N. Sette
Production Supervision: Clarinda Publication Services
Design Coordinator: Diane Ernsberger
Cover Designer: Jason Moore
Cover art: Photonica
Production Manager: Brian Fox
Marketing Manager: Tim Peyton

This book was set by The Clarinda Company. It was printed and bound by Courier Kendallville, Inc.
The cover was printed by The Lehigh Press, Inc.

Pearson Education Ltd.
Pearson Education Australia Pty. Limited
Pearson Education Singapore Pte. Ltd.
Pearson Education North Asia Ltd.
Pearson Education Canada, Ltd.
Pearson Educación de Mexico, S.A. de C.V.
Pearson Education—Japan
Pearson Education Malaysia Pte. Ltd.
Pearson Education, *Upper Saddle River, New Jersey*

10 9 8 7 6 5 4
ISBN: 0-13-045289-0

To Dawn, Candy, and Jon

Books by Nigel P. Cook

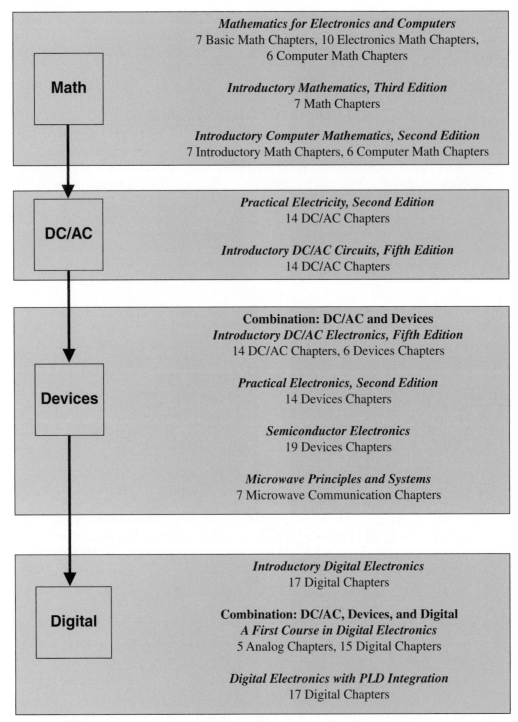

Math

Mathematics for Electronics and Computers
7 Basic Math Chapters, 10 Electronics Math Chapters,
6 Computer Math Chapters

Introductory Mathematics, Third Edition
7 Math Chapters

Introductory Computer Mathematics, Second Edition
7 Introductory Math Chapters, 6 Computer Math Chapters

DC/AC

Practical Electricity, Second Edition
14 DC/AC Chapters

Introductory DC/AC Circuits, Fifth Edition
14 DC/AC Chapters

Devices

Combination: DC/AC and Devices
Introductory DC/AC Electronics, Fifth Edition
14 DC/AC Chapters, 6 Devices Chapters

Practical Electronics, Second Edition
14 Devices Chapters

Semiconductor Electronics
19 Devices Chapters

Microwave Principles and Systems
7 Microwave Communication Chapters

Digital

Introductory Digital Electronics
17 Digital Chapters

Combination: DC/AC, Devices, and Digital
A First Course in Digital Electronics
5 Analog Chapters, 15 Digital Chapters

Digital Electronics with PLD Integration
17 Digital Chapters

For more information on any of the other textbooks by Nigel Cook, see his web page at www.prenhall.com/cook or ask your local Prentice Hall representative.

Preface

INTRODUCTION

Since World War II no branch of science has contributed more to the development of the modern world as electronics. It has stimulated dramatic advances in the fields of communication, computing, consumer products, industrial automation, test and measurement, and health care. It has now become the largest single industry in the world, exceeding the automobile and oil industries, with annual sales of electronic systems greater than $2 trillion. One of the most important trends in this huge industry has been a gradual shift from *analog electronics* to *digital electronics*. This movement began in the 1960s and is almost complete today. In fact, a recent statistic stated that, on average, 90% of the circuitry within electronic systems is now digital, and only 10% is analog. This digitalization of the electronics industry is merging sectors that were once separate. For example, two of the largest sectors or branches of electronics are *computing* and *communications*. Being able to communicate with each other using the common language of digital has enabled computers and communications to interlink, so that computers can now function within communication-based networks, and communications networks can now function through computer-based systems. Industry experts call this merging *convergence* and predict that digital electronics will continue to unite the industry and stimulate progress in practically every field of human endeavor.

Mathematics is interwoven into the very core of science, and therefore an understanding of mathematics is imperative for anyone pursuing a career in technology. As with any topic to be learned, the method of presentation can make a big difference between clear comprehension and complete confusion. Employing an "integrated math applications" approach, this text reinforces all math topics with extensive electronic and computer applications to show the student the value of math as a tool; therefore, if the need is instantly demonstrated, the tool is retained.

OUTLINE

To assist educators in curriculum chapter selection, the text has been divided into two parts to give it the versatility to adapt to a variety of course lengths.

Part A Basic Math

Part B Computer Math

SUPPLEMENTS

- Instructor's Solutions Manual (ISBN 0-13-045280-7)
- Companion Website: http://www.prenhall.com/cook   The companion website has an additional bank of questions that tests student understanding and e-mails the results to the instructor.

Whenever you have two different forms, like analog and digital, there is always a need to convert back and forth between the two. Computers are used at the heart of almost every electronic system today because of their ability to quickly process and store large amounts of data, make systems more versatile, and perform many functions. Many of the input signals applied to a computer for storage or processing are analog in nature, and therefore a data conversion circuit is needed to interface these analog inputs with a digital system. Similarly, a data conversion circuit may also be needed to interface the digital computer system with an analog output device.

Figure 8-5 shows a simplified block diagram of a computer. No matter what the application, a computer has three basic blocks: a microprocessor unit, a memory unit, and an input/output unit.

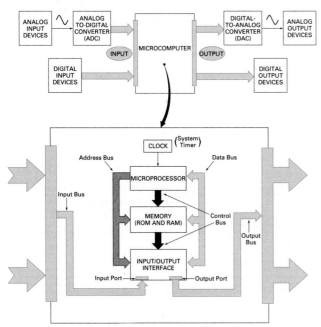

FIGURE 8-5   Connecting Analog and Digital Devices to a Computer.

NEW   Two new chapters have been added to expand on the mathematics needed for computer technology—Chapter 8, **Analog to Digital**, and Chapter 13, **Introduction to Computer Programming.**

Line No.	Instruction	Description
1	LET COUNT = 0	Set Count to 0
2	LET COUNT = COUNT + 1	Increment Count
3	IF COUNT = 10 THEN 1	Go to 1 if Count = 10
4	GO TO 2	Otherwise go to 2

(b)

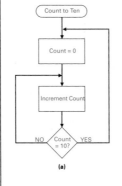

(a)

Label	Instruction	Comments
START:	MVI A,0	;Set A register to 0
LOOP:	INR A	;Increment A register
	CPI 10	;Compare A register to 10
	JZ START	;Go to beginning if A = 10
	JMP LOOP	;Repeat

(c)

Memory Address (Hex)	Memory Contents (Hex)	(Binary)
07F0	3E	00111110
07F1	00	00000000
07F2	3C	00111100
07F3	FE	11111110
07F4	0A	00001010
07F5	CA	11001010
07F6	F0	11110000
07F7	07	00000111
07F8	C3	11000011
07F9	F2	11110010
07FA	07	00000111

(d)

BASIC Language		8085 Assembly Language		8085 Machine Language		
Line No.	Instruction	Label	Instruction	Address	Contents	
1	LET COUNT = 0	START:	MVI A,0	07F0	3E	Opcode
				07F1	00	Data
2	LET COUNT = COUNT + 1	LOOP:	INR A	07F2	3C	Opcode
3	IF COUNT = 10 THEN 1		CPI 10_{10}	07F3	FE	Opcode
				07F4	0A	Data
			JZ START	07F5	CA	Opcode
				07F6	F0	Address
				07F7	07	
4	GO TO 2		JMP LOOP	07F8	C3	Opcode
				07F9	F2	Address
				07FA	07	

(e)

FIGURE 13-7   A Count-to-10 Programming Example. (a) Flowchart. (b) High-Level Language Program. (c) Assembly Language Program. (d) Machine Language Program. (e) Summary of Program Languages.

NEW The end-of-section **self-test evaluation points** list the objectives that should have been met up to that point, and a set of questions tests the student's level of comprehension.

Acute Angle
An angle that measures less than 90°.

Obtuse Angle
An angle that measures greater than 90°.

Figure 6-1(d) shows a few more geometric basics. A right angle is one of 90°, and can be symbolized with a small square in the corner. Two lines are perpendicular to one another when the point at which they intersect forms right angles. The angle measure of a straight angle is 180°. The terms **acute angle** and **obtuse angle** describe angles that are relative to a right angle (90°). An acute angle is one that is less than a right angle (<90°), while an obtuse angle is one that is greater than a right angle (>90°).

SELF-TEST EVALUATION POINT FOR SECTION 6-1

Now that you have completed this section, you should be able to:

■ *Objective 1.* *Describe the purpose of geometry, and define many of the basic geometric terms.*

■ *Objective 2.* *Demonstrate how to use a protractor to measure and draw an angle.*

Use the following questions to test your understanding of section 6-1:

1. Using a protractor, measure:

a. ∠AXB	**d.** ∠AXE	**g.** ∠AXH	**j.** ∠IXG
b. ∠AXC	**e.** ∠AXF	**h.** ∠AXI	**k.** ∠IXF
c. ∠AXD	**f.** ∠AXG	**i.** ∠IXH	**l.** ∠IXE

2. Which of the angles in the previous question are right angles, obtuse angles, and acute angles?

3. Draw:

a. A vertical line	**e.** An angle of 35°
b. A horizontal line	**f.** An angle of 172°
c. Three parallel lines	**g.** An obtuse angle
d. A right angle	**h.** An acute angle

6-2   PLANE FIGURES

A plane figure is a perfectly flat surface, such as a tabletop, floor, wall, or windowpane. In this section we will examine the details of several flat plane surfaces, which, as you will discover, have only two dimensions—length and width.

6-2-1   *Quadrilaterals*

Polygon
A plane geometry figure bounded by 3 or more line segments.

Parallelogram
A quadrilateral where both pairs of opposite sides are parallel.

A **polygon** is a plane geometric figure that is made up of three or more line segments. Quadrilaterals, such as the square, rectangle, rhombus, parallelogram, and trapezoid, are all examples of four-sided polygons.

The information given in Figure 6-2 summarizes the facts and formulas for the square, which has four right angles and four equal length sides, and is a **parallelogram** (which means it has two sets of parallel sides). The box in Figure 6-2 lists the formulas for calculating a square's perimeter, area and side length.

Figures 6-3, 6-4, 6-5 and 6-6 on pp.141–143 summarize the facts and formulas for other quadrilaterals. In addition, these diagrams include, when necessary, a visual breakdown of the formulas so that it is easier to see how the formula was derived.

6-2-2   *Right Angle Triangles*

The **right-angle triangle**, or right triangle shown in Figure 6-7(a) (p. 145), has three sides and three corners. Its distinguishing feature, however, is that two of the sides of this triangle are at right angles (at 90°) to one another. The small square box within the triangle is placed in the corner to show that sides *A* and *B* are *square* or at *right angles* to one another.

6-4   SOLID FIGURES

As mentioned previously, a plane figure is a flat surface that has only two dimensions—length and width. A **solid figure**, on the other hand, is a three-dimensional figure having length, width and height.

6-4-1   *Prisms*

A **prism** is a solid figure made up of plane figure faces, and has a least one pair of parallel surfaces. Figure 6-22 details the facts and formulas for the prism. The visual breakdown in Figure 6-22(a) shows how a cube is made up of an upper base, lower base, and a set of sides. The **lateral surface area (L)** of the cube is the area of the sides only, as shown in Figure 6-22(b). The **total surface area (S)** is the lateral surface area of the cube plus the area of the two bases, as shown in Figure 6-22(c). The holding capacity or **volume** of the cube measures the space inside the prism, and is described visually in Figure 6-22(d). Figure 6-22(e) shows examples of other prism types.

6-4-2   *Cylinders*

A **cylinder** is a solid figure with curved side walls extending between two identical circular bases. Figure 6-23 details the facts and formulas for the cylinder. As with prisms, plane figure formulas are used to calculate the cylinder's lateral surface area and base area, and as expected, the volume of a cylinder is the product of its base area and height.

6-4-3   *Spheres*

A **sphere** is a globular solid figure in which any point on the surface is at an equal distance from the center. Figure 6-24 lists the formulas for a sphere along with a visual breakdown for a sphere's surface area and volume.

6-4-4   *Pyramids and Cones*

Figure 6-25(a) (p. 164) lists the formulas for a **pyramid**, which has only one base and three or more lateral surfaces that taper up to a single point called an **apex**. Figure 6-25(a) lists the formulas for a **cone**, which has a circular base and smooth surface that extends up to an apex. Figures 6-25(b) and 6-26(b) (p. 165), list the formulas for the frustum of a pyramid and the frustum of a cone. A **frustum** is the base section of a solid pyramid or cone, and is formed by cutting off the top of the solid figure so that the upper base and lower base are parallel to one another.

Solid Figure
A 3-dimensional figure having length, width, and height.

Prism
A solid figure made up of plane figure faces; having at least one pair of parallel surfaces.

Lateral Surface Area
The side area of a solid figure.

Total Surface Area
The lateral surface area of a solid figure plus the area of its two bases.

Volume
The space within a solid figure, or its holding capacity.

Cylinder
A solid figure with curved side walls extending between two identical circular bases.

Sphere
A globular solid figure in which any point on the surface is at equal distance from the center.

Pyramid
A solid figure having only one base and 3 or more lateral surfaces that taper up to a single point.

Apex
The uppermost point or tip of a solid figure.

Cone
A solid figure having a circular base and smooth surface that extends up to an apex.

Frustum
The base section of a solid pyramid or cone.

NEW **Margin terms** highlight key terminology as it is introduced.

SELF-TEST EVALUATION POINT FOR SECTION 6-4

Now that you have completed this section, you should be able to:

■ *Objective 11.* *Define the term solid figure, and identify the different types.*

■ *Objective 12.* *List the formulas for determining the lateral surface area, total surface area, and volume of solid figures.*

Use the following questions to test your understanding of Section 6-4:

1. What is the difference between a plane figure and a solid figure?

2. What is the difference between lateral surface area and total surface area?

3. Sketch the following solid figures and list the formulas for lateral surface area, total surface area and volume.

a. Prism	**d.** Pyramid
b. Cylinder	**e.** Cone
c. Sphere	

4. What is a frustum?

In Figure 6-2(c), a 200-foot piece of string is stretched from the top of a 125-foot cliff to a point on the beach. What is the distance from this point to the cliff?

■ *Solution:*

In this example, B is unknown, and therefore

$$B = \sqrt{C^2 - A^2}$$
$$= \sqrt{200^2 - 125^2}$$
$$= \sqrt{40,000 - 15,625}$$
$$= \sqrt{24,375}$$
$$= 156 \text{ ft}$$

6-2-3 *Vectors and Vector Diagrams*

Vector or Phasor
A quantity that has magnitude and direction and that is commonly represented by a directed line segment whose length represents the magnitude and whose orientation in space represents the direction.

A **vector** or **phasor** is an arrow used to represent the magnitude and direction of a quantity. Vectors are generally used to represent a physical quantity that has two properties. For example, Figure 6-8(a) shows a motorboat heading north at 12 miles per hour. Figure 6-8(b) shows how vector **A** could be used to represent the vessel's direction and speed. The size of the vector represents the speed of 12 mph by being 12 cm long, and because we have made the top

FIGURE 6-8   **Vectors and Vector Diagrams.**

146   CHAPTER 6 / GEOMETRY AND TRIGONOMETRY

Once into the pages of this book, the reader will quickly be put at ease by the **student-friendly writing style** and the author's ability to simplify traditionally difficult subjects.

Each chapter opens with a **humanistic vignette**, a brief story that describes the history of a mathematical discovery, written in terms of the people who made these discoveries.

PART A   BASIC MATH

Fractions

There's No Sleeping When He's Around!

Carl Friedrich Gauss was born April 30, 1777, to poor, uneducated parents in Brunswick, Germany. He was a child of precocious abilities, particularly in mental computation. In elementary school he soon impressed his teachers, who said that mathematical ability came easier to Gauss than speech.

In secondary school he rapidly distinguished himself in ancient languages and mathematics. At 14, Gauss was presented to the court of the duke of Brunswick, where he displayed his computing skill. Until his death in 1806, the duke generously supported Gauss and his family, encouraging the boy with textbooks and a laboratory.

In the early years of the nineteenth century, Gauss's interest was in astronomy, and his accumulated work on celestial mechanics was published in 1809. In 1828, at a conference in Berlin, Gauss met physicist Wilhelm Weber, who would eventually become famous for his work on electricity. They worked together for many years and became close friends, investigating electromagnetism and the use of a magnetic needle for the measurement of current. In 1833 they constructed an electric telegraph system that could communicate across Göttingen from Gauss's observatory to Weber's physics laboratory. (This telegraph system of communication was later developed independently by U.S. inventor Samuel Morse.)

Gauss conceived almost all of his fundamental mathematical discoveries between the ages of 14 and 17. There are many stories of his genius in his early years, one of which involved a sarcastic teacher who liked giving his students long-winded problems and then resting, or on some occasions sleeping, in class. On his first day with Gauss, who was 8 years old, the teacher began, as usual, by telling the students to find the sum of all the numbers from 1 to 100. The teacher barely had a chance to sit down before Gauss raised his hand and said "5050." The dumbfounded teacher, who believed Gauss must have heard the problem before and memorized the answer, asked Gauss to explain how he had solved the problem. He replied: "The numbers 1, 2, 3, 4, 5, and so on to 100 can be paired as 1 and 100, 2 and 99, 3 and 98, and so on. Since each pair has a sum of 101, and there are 50 pairs, the total is 5050."

Basic theory is presented in intuitive form, augmented by numerous **worked-out examples** and **annotations** to clearly describe steps and procedures so the student can understand the mechanics of mathematics.

1-3 SUBTRACTING FRACTIONS

To subtract one fraction from another, we follow the same steps as in addition except that we subtract the numerators. Let us examine a few examples, starting with a subtraction that involves two fractions with the same denominator.

■ **EXAMPLE:**

What is $\dfrac{9}{16} - \dfrac{4}{16}$?

■ *Solution:*

$$\frac{9}{16} - \frac{4}{16} = \frac{9-4}{16} = \frac{5}{16}$$

If the two fractions involved in the subtraction are of different types, we will need to determine the lowest common denominator so that we are subtracting one type of fraction from the same type of fraction.

■ **EXAMPLE:**

What is the result of $\dfrac{6}{10} - \dfrac{4}{20}$?

■ *Solution:*

$$\frac{6}{10} - \frac{4}{20} = \frac{12-4}{20} = \frac{8}{20}$$
$$= \frac{8 \div 4}{20 \div 4} = \frac{2}{5}$$

Steps:
Find lowest common denominator (20).
10 into 20 = 2,
2 × 6 = 12.
20 into 20 = 1,
1 × 4 = 4.
12 − 4 = 8.
Reduce result.

1-3-1 *Subtracting Mixed Fractions*

As with addition, the way to subtract one mixed fraction from another is to deal with the whole numbers and fractions separately.

■ **EXAMPLE**

Subtract $2\dfrac{1}{3}$ from $5\dfrac{4}{5}$.

■ *Solution:*

$$= 3\frac{12-5}{15}$$
$$= 3\frac{7}{15}$$

Steps:
Subtract whole numbers: 5 − 2 = 3.
Determine the lowest common denominator.

3 | 5
5 | 5 ____ 1
   3 × 5 = 15
5 into 15 = 3, 3 × 4 = 12.
3 into 15 = 5, 5 × 1 = 5.
12 − 5 = 7.

FIGURE 5-3 **Home Electrical System.**

(Lamp Switch 1, Lamp Switch 2, 120 V Source, Lamp 1 (125 Ω), Lamp 1 (250 Ω), Outlet 1, Outlet 2, Hair Dryer (40 Ω), Space Heater (12 Ω))

Numerous general, electronic, and computer **integrated math applications** instantly demonstrate the value of math as a tool.

3. Power Formula Example

Another formula frequently used in electricity and electronics is the power formula. We will also study this formula so that we can practice formula transposition again.

Electric power is the rate at which electric energy is converted into some other form of energy. Power, which is symbolized by P, is measured in *watts* (symbolized by W) in honor of James Watt. We hear watt used frequently in connection with the brightness of a lightbulb: for instance, a 60 watt bulb, a 100 watt bulb, and so on. This designation is a power rating and describes how much electric energy is converted every second. Because a lightbulb generates both heat and light energy, we would describe a 100 watt lightbulb as a device that converts 100 watts of electric energy into heat and light energy every second. Referring to Figure 5-1(b), you can see a lightbulb connected in an electric circuit. The amount of electric power supplied to that lightbulb is dependent on the voltage and the current. It is probably no surprise, therefore, that power is equal to the product of voltage and current. To state the formula:

Power (Electric)
The amount of energy converted by a component or circuit in a unit of time, normally seconds. It is measured in watts or joules per second.

Power = voltage × current
$P = V \times I$

Quantity	Symbol	Unit	Symbol
Power	P	Watts	W
Voltage	V	Volts	V
Current	I	Amperes	A

This formula states that the amount of power delivered to a device is dependent on the electrical pressure or voltage applied across the device and the electric current flowing through the device.

■ **INTEGRATED MATH APPLICATION:**

What will be the electric current (I) in a circuit if a 60 watt lightbulb is connected across a 120 volt battery?

■ *Solution:*

Let us begin by inserting the known values in their appropriate places in the power formula.

Power (P) = voltage (V) × current (I)
60 W = 120 V × ? amperes

122   CHAPTER 5 / ALGEBRA, EQUATIONS, AND FORMULAS

CALCULATOR KEYS

Name: Reciprocal key

Function: Divides the value on the display into 1.

Example: $\dfrac{1}{3.2} = ?$

Press keys: 3 . 2 ⅟ₓ

Display shows: 0.3125

PI
The sixteenth letter of the Greek alphabet; the symbol π denoting the ratio of the circumference of a circle to its diameter; a transcendental number having a value to eight decimal places of 3.14159265.

The reciprocal of π is used in many different formulas. The Greek lowercase letter pi, which is pronounced "pie" and symbolized by π, is a constant value that describes how much bigger a circle's circumference is than its diameter.

The approximate value of this constant is 3.142, which means that the circumference of any circle is approximately 3 times bigger than its diameter.

Most calculators have a key for quickly obtaining the value of pi. It is called the pi (π) key and operates as follows.

CALCULATOR KEYS

Name: Pi key

Function: Enters the value of pi correct to eight decimal digits.

π = 3.1415927

Example: $2 \times \pi = ?$

Press keys: 2 × π =

Display shows: 6.2831853

36   CHAPTER 2 / DECIMAL NUMBERS

Comfortable mastery of the material is achieved through numerous in-chapter **examples, applications, self-test evaluation questions,** and end-of-chapter **multiple-choice, communication skill,** and **practice problem questions.**

...th or magnitude of the resultant vectors in the following vector diagrams.

29. Calculate the value of the following trigonometric functions.

a. sin 0°
b. sin 30°
c. sin 45°
d. sin 60°
e. sin 90°
f. cos 0°
g. cos 30°
h. cos 45°
i. cos 60°
j. cos 90°
k. tan 0°
l. tan 30°
m. tan 45°
n. tan 60°
o. tan 90°

30. Calculate angle θ from the function given.

a. $\sin \theta = 0.707$, $\theta = ?$
b. $\sin \theta = 0.233$, $\theta = ?$
c. $\cos \theta = 0.707$, $\theta = ?$
d. $\cos \theta = 0.839$, $\theta = ?$
e. $\tan \theta = 1.25$, $\theta = ?$
f. $\tan \theta = 0.866$, $\theta = ?$

31. Name each of the triangle's sides, and calculate the unknown values.

32. If a cylinder has a radius of 3 cm and a height of 10 cm, calculate its:

a. Lateral surface area
b. Total surface area
c. Volume

33. If a sphere has a diameter of 2 inches, what would be its surface area and volume?

34. What is the lateral surface area of a rectangular prism that has a base perimeter of 40 cm and a height of 12 cm?

35. Calculate the lateral surface area of a pyramid that has a base perimeter of 24 inches and a slant height of 8 inches.

Web Site Questions

Go to the web site http://www.prenhall.com/cook, select the textbook *Mathematics for Electronics and Computers,* select this chapter, and then follow the instructions when answering the multiple-choice practice problems.

REVIEW QUESTIONS   167

ILLUSTRATED TOUR OF TEXTBOOK FEATURES   xiii

Contents

Fractions

There's No Sleeping When He's Around!

Carl Friedrich Gauss was born April 30, 1777, to poor, uneducated parents in Brunswick, Germany. He was a child of precocious abilities, particularly in mental computation. In elementary school he soon impressed his teachers, who said that mathematical ability came easier to Gauss than speech.

In secondary school he rapidly distinguished himself in ancient languages and mathematics. At 14, Gauss was presented to the court of the duke of Brunswick, where he displayed his computing skill. Until his death in 1806, the duke generously supported Gauss and his family, encouraging the boy with textbooks and a laboratory.

In the early years of the nineteenth century, Gauss's interest was in astronomy, and his accumulated work on celestial mechanics was published in 1809. In 1828, at a conference in Berlin, Gauss met physicist Wilhelm Weber, who would eventually become famous for his work on electricity. They worked together for many years and became close friends, investigating electromagnetism and the use of a magnetic needle for the measurement of current. In 1833 they constructed an electric telegraph system that could communicate across Göttingen from Gauss's observatory to Weber's physics laboratory. (This telegraph system of communication was later developed independently by U.S. inventor Samuel Morse.)

Gauss conceived almost all of his fundamental mathematical discoveries between the ages of 14 and 17. There are many stories of his genius in his early years, one of which involved a sarcastic teacher who liked giving his students long-winded problems and then resting, or on some occasions sleeping, in class. On his first day with Gauss, who was 8 years old, the teacher began, as usual, by telling the students to find the sum of all the numbers from 1 to 100. The teacher barely had a chance to sit down before Gauss raised his hand and said "5050." The dumbfounded teacher, who believed Gauss must have heard the problem before and memorized the answer, asked Gauss to explain how he had solved the problem. He replied: "The numbers 1, 2, 3, 4, 5, and so on to 100 can be paired as 1 and 100, 2 and 99, 3 and 98, and so on. Since each pair has a sum of 101, and there are 50 pairs, the total is 5050."

Outline and Objectives

Introduction

Fraction

A numerical representation indicating the quotient of two numbers; a piece, fragment, or portion.

Half

Either of two equal parts into which a thing is divisible.

Quarter

One of four equal parts into which something is divisible; a fourth part.

Fourth

See *quarter*.

Another word for **fraction** is "piece"; therefore, if you have a fraction of something, you have a piece of that something. Having only a fraction or piece of something means that the object has been broken up into pieces, and therefore a fraction is something less than the whole thing. For example, if you wanted half of a pizza, a whole pizza would be cut into two pieces and you would be given one of the halves—a **half** is therefore a fraction. If, in another instance, you wanted only a quarter of an apple pie, a whole apple pie would be cut into four pieces and you would be given one of the quarters—a **quarter** is therefore also a fraction. A quarter (or a **fourth**) is a smaller fraction than a half, and therefore we can say that *the more cuts you make, the smaller the fraction.*

Since this is your first chapter, let me try to put you in the right frame of mind before you begin. Most people have trouble with mathematics because they never mastered the basics. As you proceed through this chapter and the succeeding chapters, it is imperative that you study every section, example, self-test evaluation point, and end-of-chapter question. If you cannot understand a particular section or example, go back and review the material that led up to the problem and make sure that you fully understand all the basics before you continue. Since each chapter builds on previous chapters, you may find that you need to return to an earlier chapter to refresh your understanding before moving on with the current chapter. This process of moving forward and then backtracking to refresh your understanding is very necessary and helps to engrave the basics of mathematics in your mind. Try never to skip a section or chapter because you feel that you already have a good understanding of that material. If it is a basic topic that you have no problem with, read it anyway to refresh your understanding about the steps involved and the terminology because these may be used in a more complex operation in a later chapter.

1-1   REPRESENTING FRACTIONS

Half and *quarter* (or *fourth*) are names that describe a fraction; however, some other system, sign, or symbol is needed to represent or show the fraction. This system will have to show two things:

1. The number of pieces into which the entire object was cut or divided up
2. The number of pieces we have or are concerned with

Numerator

The part of a fraction that is above the line and signifies the number of parts of the denominator taken.

Denominator

The part of a fraction below the line signifying division that functions as the divisor of the numerator.

Vinculum or Fraction Bar

A straight horizontal line placed between two or more numbers of a mathematical expression.

■ **EXAMPLE:**

Write down the proper or written fraction for one-quarter or one-fourth, and label its elements.

■ *Solution:*

Fraction bar (Its official name is **vinculum**; however, we will refer to it as the **fraction bar.**) $\longrightarrow$ $\dfrac{1}{4}$

Called the **numerator,** this tells us how many of the original pieces we have.

Called the **denominator,** this tells us into how many pieces the original object was cut.

The preceding example fraction represents an amount or quantity. In this example it represents "one-quarter" or "one-fourth" and indicates that the object was cut or divided into four pieces and that we have one of the four pieces.

■ **EXAMPLE:**

First, draw a picture of a pizza cut into six pieces; then show one of the six pieces separately and label it with its fraction.

■ *Solution:*

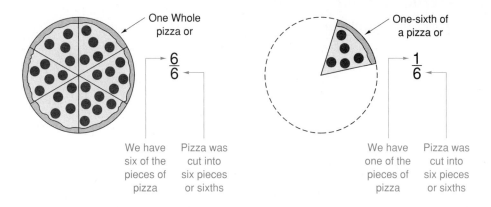

One Whole pizza or $\dfrac{6}{6}$

We have six of the pieces of pizza — Pizza was cut into six pieces or sixths

One-sixth of a pizza or $\dfrac{1}{6}$

We have one of the pieces of pizza — Pizza was cut into six pieces or sixths

■ **EXAMPLE:**

A pie is divided into eight pieces. If five of the pieces are eaten and three remain, how would we indicate or represent these fractions?

■ *Solution:*

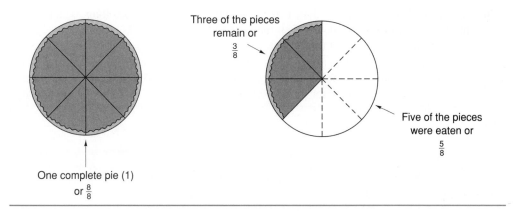

Three of the pieces remain or $\frac{3}{8}$

Five of the pieces were eaten or $\frac{5}{8}$

One complete pie (1) or $\frac{8}{8}$

In all technical fields you will have to interpret scales, such as the divisions on the straightedge shown next.

This ruler has one inch divided up into four quarter-inch increments. ($\frac{4}{4}$ = 1 in.)

One inch

Other rulers might have each inch broken up into halves (there would be two halves or $\frac{2}{2}$ in an inch), eighths (there would be eight eighths or $\frac{8}{8}$ in an inch), sixteenths (there would be sixteen sixteenths or $\frac{16}{16}$ in an inch), and so on.

■ INTEGRATED MATH APPLICATION:

Use the ruler to measure the length of the screws shown in the following figure.

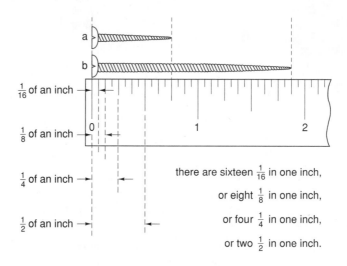

there are sixteen $\frac{1}{16}$ in one inch,

or eight $\frac{1}{8}$ in one inch,

or four $\frac{1}{4}$ in one inch,

or two $\frac{1}{2}$ in one inch.

■ *Solution:*

a. The smaller screw measures three-fourths of an inch ($\frac{3}{4}$ in.), or six-eighths of an inch ($\frac{6}{8}$ in.), or twelve-sixteenths ($\frac{12}{16}$ in.) of an inch.

b. The second screw is greater than 1 inch long but less than 2 inches long, which means that it is 1 inch plus a fraction of an inch long. Since its tip falls alongside an $\frac{1}{8}$ inch mark on the ruler, we will use the $\frac{1}{8}$ inch scale. The longer screw measures one and seven-eighths of an inch ($1\frac{7}{8}$ in.) long. We could also say that screw b measures $\frac{15}{8}$ of an inch, since the $\frac{8}{8}$ in the first inch, plus the $\frac{7}{8}$ in the second inch, equals a total of $\frac{15}{8}$. Using the sixteenths ruler marks, screw b measures $1\frac{14}{16}$ or $\frac{30}{16}$ of an inch long.

SELF-TEST EVALUATION POINT FOR SECTION 1-1

Now that you have completed this section, you should be able to:

■ **Objective 1.** *Describe what a fraction is, and name its elements.*

■ **Objective 2.** *Demonstrate how fractions are represented.*

Use the following questions to test your understanding of Section 1-1.

1. What fraction is shown shaded?
2. What would be the readings on the following scale?

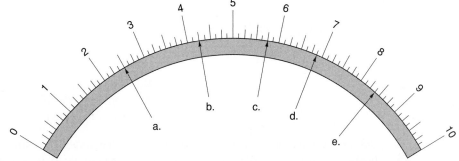

1-2 ADDING FRACTIONS

The *denominator* (number below the fraction bar) determines the fraction type. For example, $\frac{3}{9}$ tells us that the object was divided up into nine parts and that we have three of those nine parts. The number nine therefore tells us the type of fraction we are dealing with. When fractions are of the same type (have the same denominators), they are added as simply as adding 5 apples and 3 apples.

■ **EXAMPLE:**

Add the following fractions: $\dfrac{5}{9} + \dfrac{3}{9}$

■ *Solution:*

Since we are dealing with two fractions that are of the same type (ninths), the result is simply the sum or total of the two *numerators* (numbers above the fraction bar).

$$\text{Numerators} \rightarrow \quad \frac{5}{9} + \frac{3}{9} = \frac{8}{9}$$
$$\text{Denominators} \rightarrow$$

Notice that the numerators were added, whereas the denominators remained the same. This is because the type of fraction remained the same; only the number of pieces changed. To use an analogy, we could say:

$$\frac{5}{\text{apples}} + \frac{3}{\text{apples}} = \frac{8}{\text{apples}} \begin{array}{l} \leftarrow \text{Sum} \\ \leftarrow \text{Type} \end{array}$$

The next step is to address how we can add different types of fractions. For example, how would we add ½ plus ¼? In this instance we have two different types of fractions, and therefore they are as different as apples and oranges.

Using pictures, we know that ½ plus ¼ equal ¾.

The question is: What mathematical procedure do we follow when the answer is not this obvious? If you study the previous answer, you see that the answer was in fourths. Therefore, what we actually did was convert the one-half into two-fourths ($\frac{1}{2} = \frac{2}{4}$, one-half = two-fourths) and then add the other fourth to get three-fourths.

The written fractions would appear as follows:

Since one-half equals two-fourths, we can subsitute $\frac{2}{4}$ (two-fourths) for $\frac{1}{2}$ (one-half).

$$\frac{1}{2} + \frac{1}{4} = ?$$

$$\frac{2}{4} + \frac{1}{4} = \frac{3}{4}$$

Sum of numerators

Now the fractions are the same type; therefore the denominator answer indicates the same fraction type.

What we have noticed from this example is that to add fractions that are of different types, we need to find a *denominator that is common to both fractions*. There are two methods for finding a **common denominator** so that we can add different types of fractions.

Common Denominator

A common multiple of the denominators of a number of fractions.

1-2-1   *Method 1 for Finding a Common Denominator*

Probably the easiest way to find a common denominator is to multiply one denominator by another. For example, let us assume that we want to add $\frac{1}{2}$ plus $\frac{1}{3}$.

$$\frac{1}{2} + \frac{1}{3} = ?$$

Our common denominator can be obtained by multiplying one denominator by the other; therefore, $2 \times 3 = 6$.

Now that a common denominator has been found, how many sixths ($\frac{1}{6}$) are in one-half ($\frac{1}{2}$) and one-third ($\frac{1}{3}$)? This is determined in the following way.

$$\frac{1}{2} + \frac{1}{3} = \frac{?}{6}$$

(One-half plus one-third equals how many sixths?)

$$\frac{1}{2} + \frac{1}{3} = \frac{3}{6}$$

Step a: 2 into 6 = 3.
Step b: 3 × 1 = 3.
Step c: Place result in answers numerator.

$$\frac{1}{2} + \frac{1}{3} = \frac{3 + 2}{6}$$

Step a: 3 into 6 = 2.
Step b: 2 × 1 = 2.
Step c: Place result in answers numerator.

$$\frac{1}{2} + \frac{1}{3} = \frac{3 + 2}{6}$$

The last step is to add the two answers for the numerator (3 + 2 = 5).

$$\frac{1}{2} + \frac{1}{3} = \frac{5}{6}$$

■   **EXAMPLE:**

Add $\frac{1}{4}$ and $\frac{2}{3}$.

■ *Solution*

Common denominator = denominator × denominator

$$= \quad 4 \quad \times \quad 3$$
$$= \quad 12$$

$$\frac{1}{4} + \frac{2}{3} = \frac{3+8}{12} = \frac{11}{12}$$

Steps:
4 into 12 = 3,
3 × 1 = 3,
3 into 12 = 4,
4 × 2 = 8,
3 + 8 = 11.

1-2-2   *Method 2 for Finding a Common Denominator*

The problem with method 1 is that it often yields a common denominator that is larger than necessary. For example, what is the lowest-value common denominator for the following three-fraction addition?

$$\frac{3}{9} + \frac{1}{4} + \frac{1}{6} = \frac{?}{?}$$

Using method 1, would obtain the answer

$$9 \times 4 \times 6 = 216$$

Even though 216 will work as a common denominator, it is a very large number and therefore the fraction addition will be cumbersome.

The steps in the following method will always yield the **lowest common denominator.** Using the previous example, we first place all three denominators in an upside-down division box, as shown.

$$\underline{9 \qquad 4 \qquad 6}$$

The next step is to begin with 2 and see if it will divide exactly into any of the three denominators. Once 2 can no longer be used as a divisor, increase to 3, and then 4, and then 5, and so on. Let us follow through these steps in more detail.

2 does not go into 9 exactly.
2 goes into 4, giving 2,
and 2 goes into 6, giving 3.
Using 2 again as a divisor,
we can further reduce because 2
goes into 2, giving 1.
Because 2 will no longer reduce
any of the denominators, we increase
the divisor to 3. 3 goes into
9, giving 3, and into 3, giving 1.
Using 3 again as a divisor, we
can reduce because 3 goes into
3, giving 1.

Divisors to be
used as "factors."

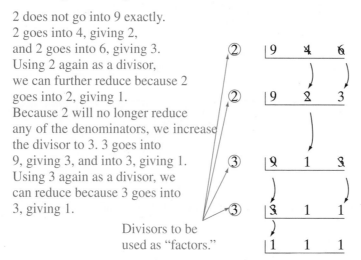

Looking at the previous process, you can see that the point is to keep dividing the denominators until they are all reduced to 1. Because no number is changed when it is multiplied or divided by 1 (for example, 22 × 1 = 22, 115 × 1 = 115, 17 ÷ 1 = 17, 63 ÷ 1 = 63), we can ignore the remaining 1s.

The divisors used in this process are called *factors.* A **factor** *is any number that is multiplied by another number and contributes to the result.* For example, consider 3 × 8. In this multiplication example, 3 is a factor and 8 is a factor that will contribute to the result of 24. Let us

Lowest Common Denominator

The lowest-value common denominator.

Factor

Any of the numbers or symbols in mathematics that when multiplied together form a product.

now take another example: What factors could contribute to a result of 12? In this case we must determine which small numbers, when multiplied together, will produce 12. The answer is

$$2 \times 6 = 12 \qquad \text{or} \qquad 3 \times 4 = 12$$

Are these, however, the smallest numbers that when multiplied together will produce 12? The answer is no because the 6 can be broken down into 2×3, and the 4 can be broken down into 2×2. Therefore, $2 \times 3 \times 2$ or $3 \times 2 \times 2$ shows that 12 has the three factors 2, 2, and 3.

Some numbers cannot be reduced or broken down into factors. For example, there are no smaller numbers that when multiplied together will produce 5. This number is called a **prime number,** and there are many, such as 3, 5, 7, 11, 13, and 17.

What we were doing, therefore, in the previous three-fraction addition example was extracting the factors from the denominators.

Prime Number

Any integer other than 0 or 1 that is not divisible without remainder.

These four factors, which were the divisors, can then be used to find the lowest common denominator.

$$
\begin{array}{c|ccc}
2 & 9 & 4 & 6 \\
2 & 9 & 2 & 3 \\
3 & 9 & 1 & 3 \\
3 & 3 & 1 & 1 \\
\end{array}
$$

(Example: $\dfrac{3}{9} + \dfrac{1}{4} + \dfrac{1}{6} = \dfrac{}{?}$)

$$\underbrace{2 \times 2 \times 3 \times 3}_{\text{Denominator factors}} = \underbrace{36}_{\substack{\text{Lowest common} \\ \text{denominator}}}$$

Factoring

Resolving into factors; a reducing process that extracts the common factors.

Now that we have completed the *factoring* step and determined the lowest common denominator, we can proceed with the example, which was to add ³⁄₉ + ¼ + ⅙:

$$\frac{3}{9} + \frac{1}{4} + \frac{1}{6} = \frac{}{36} \quad \longleftarrow \quad \text{Lowest common denominator}$$

Steps: 9 into 36 = 4, $4 \times 3 = 12$.

$$\left(\frac{3}{9}\right) + \frac{1}{4} + \frac{1}{6} = \frac{12+}{36}$$

$$\frac{3}{9} + \left(\frac{1}{4}\right) + \frac{1}{6} = \frac{12 + 9 +}{36}$$

4 into 36 = 9, $9 \times 1 = 9$.

$$\frac{3}{9} + \frac{1}{4} + \left(\frac{1}{6}\right) = \frac{12 + 9 + 6}{36}$$

6 into 36 = 6, $6 \times 1 = 6$.

$$= \frac{12 + 9 + 6}{36} = \frac{27}{36}$$

■ **EXAMPLE:**

Add $\dfrac{1}{4} + \dfrac{3}{5} + \dfrac{1}{8}$.

■ *Solution:*

The first step is to determine the factors in each denominator.

Divisors or factors 3 and 4 were ignored because 5 was a prime factor.

$$
\begin{array}{c|ccc}
2 & 4 & 5 & 8 \\
2 & 2 & 5 & 4 \\
2 & 1 & 5 & 2 \\
5 & 1 & 5 & 1 \\
 & 1 & 1 & 1 \\
\end{array}
$$

$2 \times 2 \times 2 \times 5 = 40$ (Lowest common denominator)

$$\frac{1}{4} + \frac{3}{5} + \frac{1}{8} = \frac{10 + 24 + 5}{40} = \frac{39}{40}$$

All the fractions discussed so far have a numerator that is smaller than the denominator. These are called **proper fractions.** For example:

Proper Fraction

A fraction in which the numerator is less or of lower degree than the denominator.

$$\text{Proper fractions: } \frac{1}{2} \quad \frac{3}{4} \quad \frac{5}{9} \quad \frac{7}{32} \quad \frac{15}{27}$$

$\leftarrow$ Smaller numerator
$\leftarrow$ Larger denominator

On some occasions, an addition of two or more proper fractions will produce a fraction with a numerator that is larger than the denominator. To explain this, let us use an example.

■ **EXAMPLE:**

Add $\dfrac{3}{4} + \dfrac{2}{4}$.

■ *Solution:*

$$\frac{3}{4} + \frac{2}{4} = \frac{3+2}{4} = \frac{5}{4} \leftarrow \text{Numerator is larger than denominator.}$$

To analyze this answer, ⁵⁄₄, we would say that the denominator 4 indicates that the original whole unit was divided into four pieces (quarters) and the numerator indicates that we have five of these pieces. Therefore, with four of the quarters (or fourths) we have one whole unit, leaving one extra fourth.

$$\frac{5}{4} = 1\frac{1}{4}$$

five-fourths = one whole unit and one-fourth

A fraction with a numerator that is larger than the denominator (for example, ⁵⁄₄) is called an **improper fraction** and needs to be converted to a **mixed number.** By definition, a mixed fraction or mixed number is a whole number and a fraction (for example, 1¼). Since the fraction bar actually indicates the arithmetic operation of division (for example, ⁵⁄₄ indicates 5 divided by 4), all we do is perform this division to convert an improper fraction to a mixed number. For example, 4 goes into 5 once with 1 remaining:

Improper Fraction

A fraction whose numerator is equal to or larger than the denominator.

Mixed Number

A number composed of an integer and a fraction.

$$\frac{5}{4} = 1\frac{1}{4}$$

four into five = one, with one-fourth remaining

■ **EXAMPLE:**

Add $\dfrac{2}{3} + \dfrac{4}{6}$.

■ *Solution:*

$$\frac{2}{3} + \frac{4}{6} = \frac{4+4}{6} = \frac{8}{6}$$

Lowest common denominator:

2	3	6
3	3	3

$\downarrow$ $2 \times 3 = 6$

Because ⁸⁄₆ is an improper fraction, the next step is to convert it to a mixed fraction.

$$\frac{8}{6} = 1\frac{2}{6}$$

six into eight = one, with two-sixths remaining

1-2-4   *Adding Mixed Numbers*

Now that we understand what a mixed number is, let us see how we would add two mixed fractions.

■ **EXAMPLE:**

Add $5\dfrac{1}{4} + 6\dfrac{3}{8}$.

■ *Solution:*

The easiest way to add mixed numbers is to deal with the whole numbers and fractions separately.

Step 1: Separate whole number and fractions.

Step 2: Add fractions.

Step 3: Add whole numbers.

Step 4: Combine whole-number result and fraction result.

$$5\frac{1}{4} + 6\frac{3}{8} = ?$$

$$\frac{1}{4} + \frac{3}{8} = \frac{2+3}{8} = \frac{5}{8}$$

$$5 + 6 = 11$$

$$\text{Answer} = 11\frac{5}{8}$$

■ **INTEGRATED MATH APPLICATION:**

If two objects measure $6\dfrac{3}{8}$ inches and $2\dfrac{1}{4}$ inches, what is the total length of both objects?

■ *Solution:*

$$6\frac{3}{8} + 2\frac{1}{4} = 8\frac{3+2}{8} = 8\frac{5}{8} \text{ inches}$$

$$6 + 2 \qquad \frac{3}{8} + \frac{1}{4} \qquad \frac{4}{8}$$

■ EXAMPLE:

Add $4\frac{15}{20} + 5\frac{4}{5}$.

■ Solution:

$$4\frac{15}{20} + 5\frac{4}{5} = 9\frac{15+16}{20} = 9\frac{31}{20}$$

Because the fraction $^{31}/_{20}$ is improper, we must convert it by dividing 20 into 31.

$$9\frac{31}{20} = 9 + \left(1\frac{11}{20}\right) \longleftarrow 31 \div 20 = 1\ ^{11}/_{20}$$

$$= 10\frac{11}{20} \longleftarrow \begin{array}{l}\text{9 whole objects plus}\\\text{1 whole object} = 10\end{array}$$

1-2-5   *Reducing Fractions*

To obtain half of a pizza politely, you could ask for two small quarters because two-quarters equals one-half.

$$\frac{2}{4} = \frac{1}{2}$$

Both fractions therefore indicate the same part of a whole object. In most instances when we have two fractions that describe the same portion, we would use the fraction that has the smallest denominator. For example, the denominator 2 in $^1/_2$ is smaller than the denominator 4 in $^2/_4$, and therefore we would use $^1/_2$ to represent the fraction. The next question is: How do we **reduce** a fraction so that it has the lowest possible denominator? The answer is that we use an algebraic rule that states: *If the same number is divided into both the numerator and denominator of a fraction, the value of the fraction remains the same, only reduced.*

For example, let us divide 2 into both the numerator and denominator of $^2/_4$:

$$\frac{2 \div 2}{4 \div 2} = \frac{1}{2}$$

To reduce a fraction we must therefore find the largest number that will divide evenly into both its numerator and denominator. This is best achieved by starting with 2, then trying 3, then 4, then 5, and so on. To demonstrate this, let us do the following example.

Reducing
The process of changing to an equivalent but more fundamental expression.

■ EXAMPLE:

Reduce the following fractions if possible.

a. $\frac{3}{6}$ **c.** $\frac{5}{45}$ **e.** $\frac{18}{90}$

b. $\frac{4}{16}$ **d.** $\frac{4}{9}$

Solution:

a. $\left(\dfrac{3}{6}\right) \dfrac{\div\ 3}{\div\ 3} = \dfrac{1}{2}$ (three-sixths = one-half)

└ Will both divide by 2? No. Will both divide by 3? Yes.

b. $\left(\dfrac{4}{16}\right) \dfrac{\div\ 4}{\div\ 4} = \dfrac{1}{4}$ (four-sixteenths = one-fourth)

└ Will both divide by 2? Yes. Will both divide by 3? No. Will both divide by 4? Yes. We will use 4 because dividing by a larger number will give us a smaller denominator in fewer steps. If we had used 2, we would have ended up with the same answer but would have had to reduce in two steps:

$$\left(\dfrac{4}{16}\right) \dfrac{\div\ 2}{\div\ 2} = \left(\dfrac{2}{8}\right) \dfrac{\div\ 2}{\div\ 2} = \dfrac{1}{4}$$

c. $\left(\dfrac{5}{45}\right) \dfrac{\div\ 5}{\div\ 5} = \dfrac{1}{9}$

└ Will both divide by 2? No. Will both divide by 3? No. Will both divide by 4? No. Will both divide by 5? Yes.

d. $\left(\dfrac{4}{9}\right)$

└ Will both divide by 2? No. Will both divide by 3? No. Will both divide by 4? No. This fraction cannot be reduced any further.

e. $\left(\dfrac{18}{90}\right) \dfrac{\div\ 18}{\div\ 18} = \dfrac{1}{5}$

└ Divide by 2? Yes. Divide by 3? Yes. Divide by 4? No. Divide by 5? No. Divide by 6? Yes. Divide by 7? No. Divide by 8? No. Divide by 9? Yes. . . . Divide by 18? Yes.

SELF-TEST EVALUATION POINT FOR SECTION 1-2

Now that you have completed this section, you should be able to:

■ **Objective 3.** *Describe how fractions are added.*

■ **Objective 4.** *Show two methods used for calculating the lowest common denominator.*

■ **Objective 5.** *Define a proper fraction, an improper fraction, and a mixed fraction.*

■ **Objective 6.** *Show how mixed fractions are added.*

■ **Objective 7.** *Describe how to reduce fractions.*

Use the following questions to test your understanding of Section 1-2.

Add the following fractions. Convert to proper fractions and reduce if necessary.

1. $\dfrac{4}{6} + \dfrac{1}{6} = ?$

2. $\dfrac{3}{64} + \dfrac{2}{64} + \dfrac{25}{64} = ?$

3. $\dfrac{3}{9} + \dfrac{6}{18} = ?$

4. $\dfrac{1}{3} + \dfrac{1}{4} + \dfrac{3}{15} = ?$

5. $\dfrac{3}{4} + \dfrac{4}{5} = ?$

6. $\dfrac{9}{12} + \dfrac{4}{24} + \dfrac{3}{4} = ?$

7. $\dfrac{7}{9} + \dfrac{5}{9} + \dfrac{4}{18} = ?$

(Convert to proper fractions and/or reduce if necessary.)

8. $\dfrac{15}{6} = ?$

9. $\dfrac{4}{24} = ?$

10. $\dfrac{25}{100} = ?$

1-3   SUBTRACTING FRACTIONS

To subtract one fraction from another, we follow the same steps as in addition except that we subtract the numerators. Let us examine a few examples, starting with a subtraction that involves two fractions with the same denominator.

■ EXAMPLE:

What is $\dfrac{9}{16} - \dfrac{4}{16}$?

■ *Solution:*

$$\frac{9}{16} - \frac{4}{16} = \frac{9-4}{16} = \frac{5}{16}$$

If the two fractions involved in the subtraction are of different types, we will need to determine the lowest common denominator so that we are subtracting one type of fraction from the same type of fraction.

■ EXAMPLE:

What is the result of $\dfrac{6}{10} - \dfrac{4}{20}$?

■ *Solution:*

$$\frac{6}{10} - \frac{4}{20} = \frac{12-4}{20} = \frac{8}{20}$$
$$= \frac{8 \div 4}{20 \div 4} = \frac{2}{5}$$

Steps:
Find lowest common denominator (20).
10 into 20 = 2,
2 × 6 = 12.
20 into 20 = 1,
1 × 4 = 4.
12 − 4 = 8.
Reduce result.

1-3-1   *Subtracting Mixed Fractions*

As with addition, the way to subtract one mixed fraction from another is to deal with the whole numbers and fractions separately.

■ EXAMPLE

Subtract $2\dfrac{1}{3}$ from $5\dfrac{4}{5}$.

■ *Solution:*

$$5\frac{4}{5} - 2\frac{1}{3} = 3\frac{12-5}{15}$$
$$= 3\frac{7}{15}$$

Steps:
Subtract whole numbers: 5 − 2 = 3.
Determine the lowest common denominator.

3 | 5   3
5 | 5   1
└→ 3 × 5 = 15
5 into 15 = 3, 3 × 4 = 12.
3 into 15 = 5, 5 × 1 = 5.
12 − 5 = 7.

Borrowing

The process of taking from the next higher order digit so that a subtraction can take place.

1-3-2   *Borrowing*

As is normal in subtraction, we sometimes have problems in which we need to subtract a large number from a small number. In these cases we will need to **borrow,** as shown in the following example.

■ **EXAMPLE:**

What is $6\dfrac{1}{3} - 2\dfrac{2}{3}$?

■ *Solution:*

Borrow 1 from 6 and convert it to ⅔.

$$6\frac{1}{3} - 2\frac{2}{3} = 4\frac{1-2}{3} = ?$$

$$6\frac{1}{3} = 5 + \left(\frac{3}{3}\right) + \frac{1}{3} = 5\frac{4}{3}$$

$$5\frac{4}{3} - 2\frac{2}{3} = 3\frac{4-2}{3} = 3\frac{2}{3}$$

Steps:
Subtract whole numbers; $6 - 2 = 4$.
Because fractions have the same denominator, subtract numerators.
You cannot subtract 2 from 1.
Borrow whole number from 6, and convert it to a fraction ($1 = ⅔$).
Add 3 pieces to 1 piece to get 4 pieces, or 4 thirds.
Perform subtraction.

■ **INTEGRATED MATH APPLICATION:**

The interest rate on a car loan has been reduced $2\dfrac{1}{4}$ percent from $6\dfrac{3}{16}$ percent. What is the new interest rate?

■ *Solution:*

$$6\frac{3}{16} - 2\frac{1}{4} = 4\frac{3-4}{16} \longleftarrow \text{You cannot subtract 4 from 3.}$$

$$5 + \frac{16}{16} + \frac{3}{16} - 2\frac{1}{4} =$$

$$5\frac{19}{16} - 2\frac{1}{4} = 3\frac{19-4}{16} = 3\frac{15}{16}$$

SELF-TEST EVALUATION POINT FOR SECTION 1-3

Now that you have completed this section, you should be able to:

■ *Objective 8.*   *Show how to subtract proper and mixed fractions.*

Use the following questions to test your understanding of Section 1-3.

1. $\dfrac{4}{32} - \dfrac{2}{32} = ?$

2. $\dfrac{4}{5} - \dfrac{3}{10} = ?$

3. $\dfrac{6}{12} - \dfrac{4}{16} = ?$

4. $15\dfrac{1}{3} - 2\dfrac{4}{12} = ?$

5. $5\dfrac{15}{36} - 2\dfrac{31}{36} = ?$

1-4   MULTIPLYING FRACTIONS

One of the best ways to understand something new is to use an example that you can imagine easily.

■ **EXAMPLE:**

What is half of a half?

■ *Solution:*

$$\frac{1}{2} \text{ of } \quad \big) \quad = \quad \triangleright$$

half of a half = a fourth

$$\frac{1}{2} \times \frac{1}{2} = \frac{1}{4}$$

It is not surprising that the product (multiplication result) of two fractions has a denominator that is smaller than the denominators of the two fractions, because multiplication is calculating the fractional part of a fraction. Let us look at another example.

■ **EXAMPLE:**

What is half of a fourth?

■ *Solution:*

$$\frac{1}{2} \text{ of } \quad \triangleright \quad = \quad \triangleright$$

half of one-fourth = one-eighth

$$\frac{1}{2} \times \frac{1}{4} \qquad = \frac{1}{8}$$

From the last two examples you can probably see that to obtain the product of two fractions, simply multiply the numerators and multiply the denominators, as shown.

$$\frac{1}{2} \times \frac{1}{2} = \frac{1 \times 1 = 1}{2 \times 2 = 4}$$ (one-half of one-half equals one-fourth)

$$\frac{1}{2} \times \frac{1}{4} = \frac{1 \times 1 = 1}{2 \times 4 = 8}$$ (one-half of one-fourth equals one-eighth)

■ **EXAMPLE:**

What is $\frac{3}{4}$ of $\frac{3}{5}$?

■ *Solution:*

$$\frac{3}{4} \times \frac{3}{5} = \frac{3 \times 3 = 9}{4 \times 5 = 20}$$

1-4-1   *Multiplying Mixed Fractions*

When adding or subtracting mixed fractions, we deal with the whole numbers and fractions separately. When multiplying mixed fractions, we must first convert the mixed fractions to improper fractions and then perform the multiplication. For example, what is half of one and one-third?

$$\frac{1}{2} \times 1\frac{1}{3} = ?$$

If we convert $1\frac{1}{3}$ to an improper fraction, we will be dealing only with thirds. Because there are three thirds in one whole ($\frac{3}{3} = 1$), these three thirds plus the additional one-third piece equals four-thirds.

$$1\frac{1}{3} = \frac{3}{3} + \frac{1}{3} = \frac{4}{3}$$

Now that the mixed fraction is an improper fraction, we can complete the original problem, which is: What is half of one and one-third, or four-thirds?

$$\frac{1}{2} \times \frac{4}{3} = \frac{4}{6} \begin{array}{l} \div 2 = 2 \\ \div 2 = 3 \end{array}$$

As expected, half of four-thirds equals two-thirds. As you can see, it was necessary to reduce the first result, $\frac{4}{6}$, to $\frac{2}{3}$. This often occurs due to the larger numbers in the improper fractions.

Before doing another example, let's develop a procedure for converting a mixed fraction to an improper fraction. As an example, we will convert the mixed fraction $2\frac{1}{2}$ to an improper fraction. Because the result will be all halves ($2\frac{1}{2} = \frac{?}{2}$), we simply have to determine how many halves are in 2 whole units and then add this value to the additional half. In summary:

$$2\frac{1}{2} = \frac{4}{2} + \frac{1}{2} = \frac{5}{2}$$

Two and $\begin{array}{c} \text{one-} \\ \text{half} \end{array}$ = $\begin{array}{c} \text{four-} \\ \text{halves} \end{array}$ + $\begin{array}{c} \text{one-} \\ \text{half} \end{array}$ = $\begin{array}{c} \text{five-} \\ \text{halves} \end{array}$

To change a mixed fraction to an improper fraction, therefore, you *multiply the whole number by the fraction's denominator and then add the fraction's numerator.*

$$2\frac{1}{2} = \frac{5}{2}$$

Steps:
Multiply whole number by fraction's denominator: $2 \times 2 = 4$.
Add result to fraction's numerator: $4 + 1 = 5$.

Now that we understand how to convert a mixed fraction to an improper fraction, we can do a multiplication example involving mixed fractions.

■   **EXAMPLE:**

What is $\dfrac{4}{9}$ of $5\dfrac{3}{4}$?

■   *Solution:*

Note that the term *of* implies multiplication. To determine the result, we must first convert the mixed fraction, $5\frac{3}{4}$, to an improper fraction.

$$5\frac{3}{4} = \frac{23}{4}$$

Steps:
$5 \times 4 = 20$.
$20 + 3 = 23$.

The next step is to perform the multiplication using the improper fraction.

$$\frac{4}{9} \times \frac{23}{4} = \frac{\boxed{92}}{\boxed{36}} \begin{array}{l} \div\ 4\ =\ \frac{23}{9} \\ \div\ 4\ = \end{array}$$

↖ Reduce

Once reduced, the answer may be an improper fraction and therefore will have to be converted back into a mixed fraction. With ²³⁄₉, the denominator 9 indicates that the original whole unit was divided into 9 pieces, and the numerator 23 indicates that we have 23 of these pieces. Dividing 9 into 23 will determine how many whole units we can obtain and how many ninths remain.

$$\frac{23}{9} = 2\frac{5}{9}$$

Steps:
9 into 23 = 2
remainder 5

$$\frac{23}{9} = 2\frac{5}{9}$$

■ **INTEGRATED MATH APPLICATION:**

The formula for converting a temperature in degrees Celcius (°C) to degrees Fahrenheit (°F) is:

$$°F = \left(\frac{9}{5} \times °C\right) + 32$$

If you were traveling in Europe and the temperature was 37 °C, what would that temperature be in degrees Fahrenheit?

$$°F = \left(\frac{9}{5} \times °C\right) + 32$$

$$= \left(\frac{9}{5} \times 37\ °C\right) + 32$$

$$= \left(\frac{9}{5} \times \frac{37}{1}\right) + 32$$

$$= \frac{333}{5} + 32$$

$$= 66\frac{3}{5} + 32$$

$$= 98\frac{3}{5}\ °F$$

SELF-TEST EVALUATION POINT FOR SECTION 1-4

Now that you have completed this section, you should be able to:

■ **Objective 9.** *Show how to multiply proper and mixed fractions.*

Use the following questions to test your understanding of Section 1-4.

1. $\frac{1}{3} \times \frac{2}{3} = ?$

2. $\frac{4}{8} \times \frac{2}{9} = ?$

3. $\frac{4}{17} \times \frac{3}{34} = ?$

4. $\frac{1}{16} \times 5\frac{3}{7} = ?$

5. $2\frac{1}{7} \times 7\frac{2}{3} = ?$

Once again let us understand what we are doing by using a simple example: How many quarters are in three-quarters? Stated mathematically, this problem appears as follows:

$$\frac{3}{4} \div \frac{1}{4} = ? \qquad \div \qquad = ?$$

The answer is straightforward and requires no mathematical processes or steps because we already know that there are three quarters in three-quarters.

$$\frac{3}{4} \div \frac{1}{4} = 3$$

Since most problems are never this simple, we need to follow a mathematical process to divide fractions. Because addition is the opposite of subtraction, and multiplication is the opposite of division, we can just invert the second fraction in a division problem and then multiply both fractions to obtain the result. Let us try this processs on the previous example:

$$\frac{3}{4} \div \frac{1}{4} = \frac{3 \times 4}{4 \times 1} = \frac{12 \div 4}{4 \div 4} = \frac{3}{1} = 3$$

Invert and multiply   Reduce   1 into 3 = 3

■   **EXAMPLE:**

How many twelfths are in two-thirds $\left(\dfrac{2}{3} \div \dfrac{1}{12}\right)$?

■   *Solution:*

$$\frac{2}{3} \div \left(\frac{1}{12}\right) = \frac{2 \times 12}{3 \times 1} = \frac{\left(24\right) \div 3}{\left(3\right) \div 3} = \frac{8}{1} = 8$$

Invert and multiply   Reduce

1-5-1   *Dividing Mixed Fractions*

As when we multiply mixed fractions, to divide mixed fractions we must first convert all mixed fractions to improper fractions. Once we have done this, we can divide as usual, which means: Invert the second fraction and multiply. To start, let us use a simple example and test the procedure to see how many halves are in 2½. As you have already determined, the result should indicate that there are five halves in two and one-half.

Steps
Convert mixed fraction to improper fraction:
$2 \times 2 = 4, 4 + 1 = 5.$

$$2\frac{1}{2} \div \frac{1}{2} = ?$$

$$\frac{5}{2} \div \left(\frac{1}{2}\right) = ?$$

Invert second fraction and multiply.

$$\frac{5 \times 2}{2 \times 1} = \frac{10}{2}$$

Reduce.

$$\frac{\left(10\right) \div 2}{\left(2\right) \div 2} = \frac{5}{1} = 5$$

EXAMPLE:

$6\frac{2}{3} \div 1\frac{1}{3} = ?$

Solution:

$$6\frac{2}{3} \div 1\frac{1}{3} = ?$$

$$\frac{20}{3} \div \frac{4}{3} = ?$$

$$\frac{20 \times 3}{3 \times 4} = \frac{60}{12}$$

$$\frac{60 \div 12}{12 \div 12} = \frac{5}{1} = 5$$

Steps:
Convert mixed fractions to improper fractions:
$6 \times 3 = 18; 18 + 2 = 20.$
$1 \times 3 = 3; 3 + 1 = 4.$

Invert second fraction and multiply.

Reduce.

Now let's do an example that does not work out so neatly.

EXAMPLE:

How many times will $3\frac{5}{9}$ go into $6\frac{1}{2}$?

Solution:

Note that the term *go into* implies division and is asking how many times the first number will go into the second number.

$$6\frac{1}{2} \div 3\frac{5}{9} = ?$$ *Steps:*

$$\frac{13}{2} \div \frac{32}{9} = ?$$ Convert mixed fractions to improper fractions.

$$\frac{13 \times 9}{2 \times 32} = \frac{117}{64}$$ Invert second fraction and multiply.

$$\frac{117}{64} = 1\frac{53}{64}$$ Convert improper fraction to mixed fraction.
$117 \div 64 = 1$ remainder 53.

INTEGRATED MATH APPLICATION:

If you are paying an annual percentage rate of $15\frac{1}{8}$ percent on a credit card, what is the monthly percentage rate?

Solution:

$$15\frac{1}{8} \div 12 = ?$$

$$\frac{120}{8} \div \frac{12}{1} = ?$$

$$\frac{120 \times 1}{8 \times 12} = \frac{120}{96} = 1\frac{24}{96} = 1\frac{1}{4}\%$$

Now that you have completed this section, you should be able to:

■ **Objective 10.** *Show how to divide proper and mixed fractions.*

Use the following questions to test your understanding of Section 1-5.

1. $\dfrac{2}{3} \div \dfrac{1}{12} = ?$

2. $3\dfrac{1}{4} \div 3\dfrac{3}{4} = ?$

3. $8\dfrac{16}{20} \div 4\dfrac{4}{20} = ?$

4. $4\dfrac{3}{8} \div 2\dfrac{5}{16} = ?$

1-6   CANCELING FRACTIONS

Canceling

The process of removing a common divisor from numerator and denominator.

Canceling is an operation that can be performed on fractions only when a multiplication symbol exists between the two. Its advantage is a reduction in the size of the numbers in the problem and therefore in time needed to solve, because fewer steps are needed. Let us demonstrate the process with a simple example.

$$\left(\frac{5}{12}\right) \times \frac{4}{20} = ?$$

$$\frac{5 \times 4}{12 \times 20} = \frac{20}{240} \div \frac{20}{20} = \frac{1}{12}$$

As you can see from this example, multiplying large numbers generates large results that generally need to be reduced. By canceling before we perform the multiplication, we start with smaller fractions and thus simplify the process.

$$\frac{5}{12} \times \frac{4}{20} \qquad = \qquad \frac{5}{3} \times \frac{1}{20} \qquad = \qquad \frac{1}{3} \times \frac{1}{4} = \frac{1}{12}$$

Four divides evenly into 12 and 4.
$12 \div 4 = 3$,
$4 \div 4 = 1$.

Five divides evenly into 5 and 20.
$5 \div 5 = 1$,
$20 \div 5 = 4$.

Multiplication is now easier with no need for reducing.

Up to this point we have "reduced" fractions only by finding a number that will divide evenly into the top and bottom part of the same fraction. In this example you can see that we could not reduce ⁵⁄12 because there was no number that would evenly divide into 5 and 12; however, we could reduce the numerator of one fraction with the denominator of another fraction. Reducing in this crisscross manner is called cancellation and can be done only when a multiplication sign is present between the two fractions.

■ **EXAMPLE:**

Calculate the following: $\dfrac{4}{9} \div \dfrac{8}{27}$

■ *Solution:*

$$\frac{4}{9} \div \frac{8}{27} = ?$$

Invert second fraction and multiply.

Now that a multiplication symbol exists between the two fractions, we can see if it is possible to crisscross cancel.

$$\frac{4}{9} \times \frac{27}{8} = ?$$

Cancellation: $4 \div 4 = 1, 8 \div 4 = 2$.

$$\frac{1}{9} \times \frac{27}{2} =$$

Cancellation: $9 \div 9 = 1, 27 \div 9 = 3$.

$$\frac{1}{1} \times \frac{3}{2} = \frac{3}{2} = 1\frac{1}{2}$$

Convert improper fraction to mixed fraction.

■ EXAMPLE:

Calculate the following: $3\frac{5}{7} \times 2\frac{2}{13}$

■ Solution:

$$3\frac{5}{7} \times 2\frac{2}{13} = ?$$

$$\frac{26}{7} \times \frac{28}{13} = \frac{2}{7} \times \frac{28}{1} = \frac{2}{1} \times \frac{4}{1}$$

$$\frac{2}{1} \times \frac{4}{1} = \frac{8}{1} = 8$$

Steps:
Convert mixed fractions to improper fractions:
$3 \times 7 = 21, 21 + 5 = 26$.
$2 \times 13 = 26, 26 + 2 = 28$.
Cancellation: $13 \div 13 = 1, 26 \div 13 = 2$.
Cancellation: $28 \div 7 = 4, 7 \div 7 = 1$.

SELF-TEST EVALUATION POINT FOR SECTION 1-6

Now that you have completed this section, you should be able to:

■ **Objective 11.** *Demonstrate how to use cancellation to simplify fraction multiplication or division.*

Use the following questions to test your understanding of Section 1-6.

1. $\frac{16}{27} \times \frac{27}{240} = ?$

2. $\frac{8}{17} \times \frac{3}{4} = ?$

3. $\frac{15}{32} \div \frac{30}{4} = ?$

4. $6\frac{2}{3} \times 2\frac{7}{10} = ?$

5. $6\frac{1}{2} \div \frac{9}{32} = ?$

REVIEW QUESTIONS

Multiple Choice Questions

1. A fraction is a piece; therefore if you have a fraction, you have something _____.
   a. Between 1 and 10   c. Less than 1
   b. Greater than 1   d. Between 1 and 100

2. The more cuts you make in the whole unit, the _____ the fraction.
   a. Larger
   b. Smaller

3. The number above the fraction bar is called the _____, while the number below the fraction bar is called the _____.
   a. Denominator, factor
   b. Numerator, denominator
   c. Denominator, numerator
   d. Numerator, factor

4. How many ninths are in a whole unit?

   a. 9
   b. Depends on the value of the numerator
   c. 16
   d. All the above

5. If you had $6\frac{3}{4}$ apples in one pocket and $2\frac{1}{2}$ apples in the other, what would you have?

   a. Big pockets
   b. $8\frac{1}{4}$ apples
   c. $9\frac{3}{4}$ apples
   d. $9\frac{1}{4}$ apples

6. What is the lowest common denominator of 3 and 15?

   a. 45   **c.** 5
   b. 15   **d.** 90

7. A(n) _____ fraction has a numerator that is smaller than the denominator.

   a. Mixed   **c.** Proper
   b. Improper   **d.** All the above

8. A(n) fraction contains both a whole number and a fraction.

   a. Mixed   **c.** Improper
   b. Proper   **d.** All the above

9. A(n) _____ fraction has a numerator that is larger than the denominator.

   a. Mixed   **c.** Improper
   b. Proper   **d.** All the above

10. Reduce the following fraction to its lowest terms: $^{32}/_{144}$.

   a. $\dfrac{4}{18}$ **b.** $\dfrac{2}{9}$ **c.** $\dfrac{1}{6}$ **d.** $\dfrac{8}{32}$

Communication Skill Questions

11. What is a fraction, and what do the two numbers above and below the fraction bar line indicate? (1-1)

12. Describe the best method for calculating the lowest-value common denominator. (1-2)

13. What is a prime factor? (1-2)

14. What is an improper fraction? (1-2)

15. What is a mixed fraction? (1-2)

16. Arbitrarily choose values, and describe the steps used to:

   a. Add fractions. (1-2)
   b. Subtract fractions. (1-3)
   c. Multiply fractions. (1-4)
   d. Divide fractions. (1-5)

17. How can fractions be reduced? (1-2)

18. Why do you need to sometimes borrow when subtracting fractions? (1-3)

19. Arbitrarily choose values, and describe the steps used to:

   a. Add mixed fractions. (1-2)
   b. Subtract mixed fractions. (1-3)
   c. Multiply mixed fractions. (1-4)
   d. Divide mixed fractions. (1-5)

20. How are fractions canceled? (1-6)

Practice Problems

21. Write the whole number and fraction for each of the following quantities.

a.

b.

22. Consider the value $5\frac{7}{9}$. State which of the numbers represents the:

   a. Whole number
   b. Denominator
   c. Numerator
   d. Number of original pieces
   e. Number of fractional pieces we have

23. Add the following fractions.

   a. $\dfrac{5}{8} + \dfrac{2}{8} = ?$

   b. $\dfrac{47}{76} + \dfrac{15}{76} + \dfrac{1}{76} = ?$

24. Calculate the lowest common denominator and then add the following fractions.

   a. $\dfrac{4}{9} + \dfrac{1}{3} = ?$

   b. $\dfrac{4}{7} + \dfrac{2}{21} = ?$

   c. $\dfrac{2}{14} + \dfrac{1}{8} + \dfrac{1}{4} = ?$

   d. $\dfrac{4}{10} + \dfrac{1}{5} + \dfrac{2}{15} = ?$

25. Convert the following improper fractions to mixed fractions.

a. $\dfrac{5}{2} = ?$

c. $\dfrac{25}{16} = ?$

b. $\dfrac{17}{4} = ?$

d. $\dfrac{37}{3} = ?$

26. Add the following mixed fractions.

a. $1\dfrac{3}{4} + 2\dfrac{1}{2} = ?$

b. $9\dfrac{4}{10} + 6\dfrac{7}{8} = ?$

27. Reduce the following fractions.

a. $\dfrac{4}{16} = ?$

c. $\dfrac{74}{148} = ?$

b. $\dfrac{16}{18} = ?$

d. $\dfrac{28}{45} = ?$

28. Perform the following subtraction of fractions.

a. $\dfrac{8}{14} - \dfrac{2}{7} = ?$

c. $4\dfrac{1}{16} - 2\dfrac{31}{32} = ?$

b. $3\dfrac{4}{9} - 1\dfrac{1}{3} = ?$

d. $5\dfrac{3}{5} - 2\dfrac{4}{15} = ?$

29. Multiply the following. Cancel and reduce if possible.

a. $\dfrac{1}{9} \times 4 = ?$

c. $\dfrac{1}{3} \times 4\dfrac{1}{2} = ?$

b. $\dfrac{3}{6} \times \dfrac{4}{5} = ?$

d. $2\dfrac{3}{4} \times 4\dfrac{4}{11} = ?$

30. Divide the following. Cancel and reduce if possible.

a. $\dfrac{1}{4} \div 8 = ?$

c. $3\dfrac{5}{7} \div 1\dfrac{6}{7} = ?$

b. $\dfrac{3}{5} \div \dfrac{5}{9} = ?$

d. $14\dfrac{8}{9} \div 2\dfrac{4}{5} = ?$

Web Site Questions

Go to the Web site http://www.prenhall.com/cook, select the textbook *Introductory Computer Mathematics, 2e,* select this chapter, and then follow the instructions when answering the multiple-choice practice problems.

Decimal Numbers

An Apple a Day

Electronics is the branch of science concerned with the use of electricity and electronic devices to manage information signals. As a science it has grown out of electrical engineering, which is concerned with the generation, control, distribution, storage, and applications of electrical power. Today, electronics is a separate science involving numerous scientific disciplines, including physics, chemistry, thermodynamics, quantum mechanics, and mathematics.

One of the early pioneers who laid the foundations of many branches of science was Isaac Newton. He was born in a small farmhouse near Woolsthorpe in Lincolnshire, England, on Christmas Day in 1642. He was an extremely small, premature baby, which worried the midwives who went off to get medicine and didn't expect to find him alive when they came back. Luckily for science, however, he did survive.

Newton's father was an illiterate farmer who died three months before he was born. His mother married the local vicar soon after Newton's birth and left him in the care of his grandmother. This parental absence while he was growing up had a traumatic effect on him and throughout his life affected his relationships with people.

At school in the nearby town of Grantham, Newton showed no interest in classical studies but rather in making working models and studying the world around him. When he was in his early teens, his stepfather died and Newton had to return to the farm to help his mother. Newton proved to be a hopeless farmer; in fact, on one occasion when he was tending sheep he became so engrossed with a stream that he followed it for miles and was missing for hours. Luckily, a schoolteacher recognized Newton's single-minded powers of concentration and convinced his mother to let him return to school, where he performed better and later went off to Cambridge University.

In 1665, in his graduation year, Newton left Cambridge to return home to escape an epidemic of bubonic plague that had spread throughout London. During this time, Newton reflected on his years of seclusion at his mother's cottage and called them the most significant time in his life. It was here on a warm summer's day that Newton saw the apple fall to the ground, leading him to develop his laws of motion and gravitation. It was here that he wondered about the nature of light and later built a prism and proved that white light contains all the colors in a rainbow.

Later, when Newton returned to Cambridge, he demonstrated many of his discoveries but was reluctant to publish the details and did so finally only at the insistence of others. Newton went on to build the first working reflecting astronomical telescope and wrote a paper on optics that was fiercely challenged by the physicist Robert Hooke. Hooke quarreled bitterly with Newton over the years, and there were also heated debates about whether Newton or the German mathematician Gottfried Leibniz invented calculus.

The truth is that many of Newton's discoveries roamed around with him in the English countryside, and even though many of these would, before long, have been put forward by others, it was Newton's genius and skill (and long walks) that tied together all the loose ends.

Outline and Objectives

Introduction

The *decimal system of counting and calculating* was developed by Hindu mathematicians in India around A.D. 400. The system involved using fingers and thumbs, so it was natural that it would have 10 numerals or *digits,* which is a word meaning "finger." The use of this 10-digit or 10-finger system became widespread and by A.D. 800 was used extensively by Arab cultures. The 10 digits were represented by combinations of the 10 arabic numerals: *0, 1, 2, 3, 4, 5, 6, 7, 8,* and *9.* This is still called the *arabic number system.* The system eventually found its way to England, where it was adopted by nearly all the European countries by A.D. 1200. The Europeans called it the *decimal number system* after the Latin word *decimus,* which means "ten."

In this chapter we will discuss the decimal system of counting and calculating. **Decimal** is a topic with which you probably already feel extremely comfortable because we make use of it every day in countless applications. For example, our money is based on the decimal number system (there are 100 cents in 1 dollar, or 10 dimes in 1 dollar), and the metric system of measurement is also based on the decimal system (there are 10 millimeters in 1 centimeter, and 100 centimeters in 1 meter). Basic as it seems, however, you should still review this chapter because it will discuss all the terminology and procedures regarding how the decimal system is used, and this will serve as our foundation for all subsequent chapters.

Decimal

Numbered or proceeding by tens; based on the number 10.

2-1 COUNTING IN DECIMAL

Counting was probably the first application for a number system such as decimal because people needed some way of keeping track of things. *Decimal is based on the number 10.* Let us first examine what is meant by the *base* of a number system.

2-1-1 *Base or Radix*

Base or Radix

A number equal to the number of units in a given number system. The decimal system uses a base of 10.

Subscript

A distinguishing letter or numeral written immediately below and to the right or left of another character.

The key element that distinguishes one number system from another is the **base** *or* **radix** *of the number system.* The base describes the number of digits that are used. The decimal number system has a base of 10, which means that it makes use of 10 digits (0 through 9) to represent the value of a quantity. A **subscript,** which is a smaller number written to the right or left of and below the main number, is sometimes included to indicate the number's base. For example, all the numbers following have the subscript "$_{10}$" to indicate that they are base 10, or decimal, numbers.

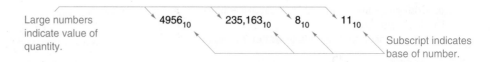

Large numbers indicate value of quantity. 4956_{10} $235,163_{10}$ 8_{10} 11_{10} Subscript indicates base of number.

If a subscript is not present, the number is assumed to be a base 10, or decimal, number. For example, the following are all decimal numbers, even though they do not include the subscript 10.

23   101   3,634,987   47

CHAPTER 2 / DECIMAL NUMBERS

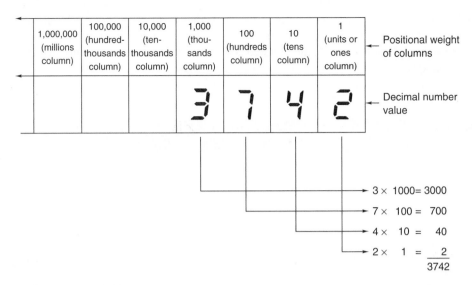

FIGURE 2-1   Positional Weight of Each Decimal Column.

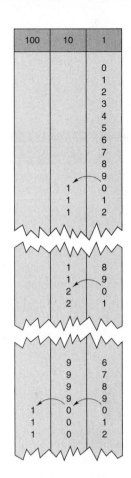

FIGURE 2-2   The Reset and Carry Action That Occurs When Counting in Decimal.

2-1-2   *Positional Weight*

The position of each digit of a decimal number determines the weight of that digit. For example, a 2 by itself indicates only a value of two, whereas a 2 to the left of three zeros (2000) indicates a value of two thousand. The position of a decimal digit therefore determines its weight, or value, with digits to the left carrying the greater weight. For example, the decimal number 3742 contains three thousands (3 × 1000), seven hundreds (7 × 100), four tens (4 × 10), and two ones (2 × 1). Figure 2-1 shows this decimal number with the positional weight of each decimal column.

You may have noticed that the weight of each decimal (base 10) column increases by a factor of 10 as you move to the left. For example, the units column has a weight of 1. The next column to the left of the units column (tens column) has a weight that is ten times (× 10) larger than the units column—10 × one = 10. The next column to the left of the tens column (hundreds column) has a weight ten times (× 10) larger than the tens column—10 × ten = 100. Therefore, each column in the decimal number system increases by ten times (× 10) as you move to the left.

2-1-3   *Reset and Carry*

Figure 2-2 shows the reset and carry action that occurs when we count in decimal. *This* **reset and carry action** *occurs whenever a column reaches the highest decimal digit of 9.* For example, when we begin to count from zero (0), the units column continues to advance by one or **increment** until it reaches 9. An increment beyond this point causes the units column to be reset to zero and a one to be carried into the tens column, producing 10 (ten). As this count continues, the units column will again repeat the cycle going from 0 to 9, which with the one in the tens column will be a count of 10 (ten) to 19 (nineteen). An increase by one beyond 19 will again cause the units column to reset to zero and carry a one into the tens column, resulting in a count of 20 (twenty). The units column will continue to cycle from 0 to 9 and reset and carry into the tens column until the tens column reaches the highest decimal digit of 9. As you can see in Figure 2-2, an advance by one beyond a count of 99 will cause the units column to reset and carry one into the tens column, and because the tens column is at its maximum of 9, it will also reset to zero and carry one into the hundreds column. This "reset to zero and carry one into the next left column action" will continue to occur in

---

Reset and Carry Action

An operation that occurs when counting, in which a digit resets after advancing beyond its maximum count and carries a one into the next higher order column.

Increment

The action or process of increasing a quantity or value by one.

a right-to-left motion whenever a column reaches a maximum digit of nine. In other words, the units column will reset and carry into the tens column, the tens column will reset and carry into the hundreds column, the hundreds column will reset and carry into the thousands column, and so on.

Now that you have completed this section, you should be able to:

■ **Objective 1.** *Describe the origination of the decimal number system.*

■ **Objective 2.** *Explain the basic process of counting in decimal, and the terms* base *or* radix, *positional weight,* and reset and carry.

Use the following questions to test your understanding of Section 2-1.

1. The decimal number system has a base or radix of _____.

2. Indicate the positional weight of each of the decimal digits in the following number: 35,729.

3. What is the reset and carry action?

4. Which columns will reset and carry when the number 499,997 is advanced or incremented by 5? What will be the final count?

2-2   THE DECIMAL POINT

Decimal Point

A period or centered dot between the parts of a decimal mixed number, separating the whole number from the fraction.

Decimal Fraction

A decimal value that is less than one.

The **decimal point (dp),** *symbolized by ".", is used to separate a whole decimal number from a* **decimal fraction.** Using a monetary example, we ask: What does the amount $12.75 actually mean? The number to the left of the decimal point, which is 12 in this example, indicates how many whole or complete dollars we have. On the other hand, the number to the right of the decimal point indicates the fraction, or the amount that is less than one dollar. Because there are 100 cents in $1, and this example indicates that we have 75 cents, this fraction is actually three-fourths of a dollar. In summary:

Decimal point

The numbers to the left of the decimal point indicate the amount of whole or complete units we have. In this monetary example, we have 12 complete or whole dollars.

The decimal point (dp) separates the whole numbers on the left from the fraction on the right. In this monetary example, the decimal point separates the complete dollars on the left from the fraction of a dollar or cents on the right.

The numbers to the right of the decimal point indicate the *decimal fraction,* or amount that is less than one. In this monetary example, we have 75 cents or $^{75}/_{100}$ (which reduces to $^{3}/_{4}$).

Most calculators have a key specifically for entering the decimal point. It operates as follows.

CALCULATOR KEYS

Name: Decimal-point key

Function: Used in conjunction with the digit keys 0 through 9 to enter a decimal point. The position of the decimal point separates the whole number on the left from the decimal fraction on the right.

Example: Enter the number 12.75.

Press keys: $\boxed{1}\boxed{2}\boxed{.}\boxed{7}\boxed{5}$

Display shows: 12.75

$$\frac{8}{9} = .888$$

$$\frac{6}{7} = .666$$

2-2-1 *Positional Weight*

Most of us know that $0.50 is half a dollar, or 50 cents; $0.25 is a fourth or a quarter of a dollar, or 25 cents; and $0.75 is three-fourths or three-quarters of a dollar, or 75 cents; however, when it comes to explaining why $½ = $0.50, $¼ = $0.25, and $¾ = $0.75, many people have difficulty. Why are these decimal fractions of 0.50, 0.25, and 0.75 equivalent to ½, ¼, and ¾? The answer is in the position that each digit occupies compared to the decimal point. To explain this further, Figure 2-3 repeats our previous example of $12.75, or twelve dollars and seventy-five cents. Once again, each place or position carries its own weight or value, and these positions are all relative to the decimal point. For instance, the digit in the first column to the left of the decimal point indicates how many units or ones are in the value, whereas the digit in the second column to the left of the decimal point indicates how many tens are in the value. Digits to the left of the decimal point all indicate the number of whole numbers in the value. On the other hand, digits to the right of the decimal point indicate the fraction or piece of a whole that is in the value. A value such as 3.5, therefore, indicates that we have three whole units and five-tenths ($^5/_{10}$). Remembering our proper fractions from Chapter 1,

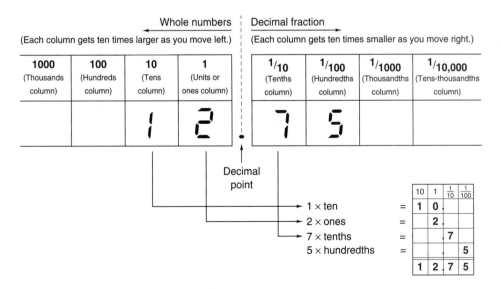

FIGURE 2-3 **The Positional Weight of Decimal Fractions.**

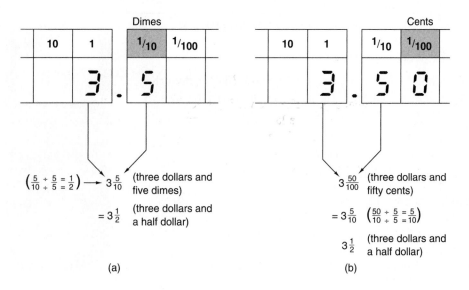

$$\left(\frac{5}{10} \div \frac{5}{5} = \frac{1}{2}\right) \longrightarrow 3\frac{5}{10} \quad \text{(three dollars and five dimes)}$$

$$= 3\frac{1}{2} \quad \text{(three dollars and a half dollar)}$$

$$3\frac{50}{100} \quad \text{(three dollars and fifty cents)}$$

$$= 3\frac{5}{10} \quad \left(\frac{50}{100} \div \frac{5}{5} = \frac{5}{10}\right)$$

$$3\frac{1}{2} \quad \text{(three dollars and a half dollar)}$$

(a)   (b)

FIGURE 2-4   **Interpreting the Decimal Fraction.**

we know that $\frac{5}{10}$ means that a whole unit was divided into ten parts and we have five of those parts. Because 5 is half of 10, we have half of a whole unit ($\frac{5}{10}$ can be reduced to $\frac{1}{2}$).

Is there a difference, therefore, between $3.5 and $3.50? The answer is no, which can be explained by looking at Figure 2-4. In Figure 2-4(a) we have interpreted the value as 3 dollars and $\frac{5}{10}$ of a dollar. Because $\frac{5}{10} = \frac{1}{2}$, the value in Figure 2-4(a) is 3 dollars and a half dollar. In Figure 2-4(b) we can interpret the decimal fraction as $\frac{5}{10}$ and $\frac{0}{100}$, or $\frac{50}{100}$. Because $\frac{50}{100} = \frac{5}{10} = \frac{1}{2}$, the value in Figure 2-4(b) is also 3 dollars and a half dollar. As far as money is concerned, we tend to refer to the fraction of a dollar in $\frac{1}{100}$ parts because a dollar is divided into 100 parts, and each of these parts is called a cent (there are 100 fractional parts or cents in a dollar).

Looking at this another way, we could label the $\frac{1}{10}$ column as dimes (10¢) and the $\frac{1}{100}$ column as cents (1¢). A value of 3.5 would then be interpreted as 3 dollars and 5 dimes, and a value of 3.50 would be interpreted as 3 dollars and 50 cents, as shown in Figure 2-4(a) and (b).

■   **EXAMPLE:**

State the individual digit weights of the following number: 7.25.

■   *Solution:*

The first step is to give each column its respective weight or value.

1	$\frac{1}{10}$	$\frac{1}{100}$
7	2	5

It is now easier to see that we have 7 whole units, $\frac{2}{10}$ (two-tenths), and $\frac{5}{100}$ (five-hundredths). Said another way, we have 7 whole units and $\frac{25}{100}$ (twenty-five hundredths). Because $\frac{25}{100}$ can be reduced to $\frac{1}{4}$,

$$\frac{25 \div 25 = 1}{100 \div 25 = 4}$$

we could also interpret this value as seven and one-fourth ($7\frac{1}{4}$).

2-2-2   *No Decimal Point and Extra Zeros*

What happens if a value such as

76

does not have a decimal point? The answer is: If you do not see a decimal point, always assume that it is to the right of the last digit in the number. In this example, therefore, the decimal point will follow the 6 in the number 76, as follows:

76.

Another point to remember: Always ignore extra zeros that are added to the left or right of a number. For example, to emphasize that a number such as .33 is a fraction, many people write this number as 0.33. The extra zero in this example is used to frame the decimal point so that it is not missed. You may have noticed that your calculator will add this zero if you press the keys [.] [3] [3] and will display 0.33. Sometimes, zeros are added to the left of decimal whole numbers as well as to the right of decimal fractions. For example:

1367.5
Extra zeros → 0293.3
are used to put digits 0006.67
in their correct column. 0049.24

These extra zeros are used to align the digits of each number in the appropriate column. Without them, digits may be placed in the wrong column, as follows:

1367.5
0293.3
6.67
49.24

In other instances, extra zeros are placed to the right of a decimal fraction to indicate how to interpret the fraction. For example, three dollars and fifty cents would be written as 3.50 instead of 3.5 even though the extra zero after the five has no bearing on the value. This is so the value will be interpreted as three dollars and fifty cents ($3 + ^{50}/_{100}$) instead of three dollars and five dimes ($3 + ^{5}/_{10}$).

In all instances, extra zeros that are outside the number (to the left of the decimal whole number or to the right of the decimal fraction) do not change the value of the number.

2-2-3   *Converting Written Fractions to Decimal Fractions*

To convert proper fractions (such as ½, ²⁄₆, ²⁄₃, ⁹⁄₁₆) to decimal fractions (such as 0.5, 0.3, 0.666, 0.562) we simply perform the operation that is indicated by the fraction bar. For example, when a fraction bar separates a numerator of 1 from a denominator of 2 (½), the arithmetic operation described by the fraction bar is a division of 2 into 1. Performing this operation, you will convert the proper fraction ½ to its decimal fraction equivalent.

$\dfrac{1}{2} = 1 \div 2 = 0.5$ *Calculator sequence:*

[1] ÷ [2] [=]

Answer: 0.5

■ **EXAMPLE:**

Convert the following proper fractions to decimal fractions.

a. $\dfrac{3}{4}$ .75   b. $\dfrac{1}{4}$ .25

c. $\dfrac{2}{3}$ **e.** $\dfrac{3}{9}$

d. $\dfrac{8}{16}$ **f.** $\dfrac{1}{8}$

■ *Solution:*

a. $\dfrac{3}{4} = 3 \div 4 = 0.75$

b. $\dfrac{1}{4} = 1 \div 4 = 0.25$

c. $\dfrac{2}{3} = 2 \div 3 = 0.666$

d. $\dfrac{8}{16} = 8 \div 16 = 0.5$ $\left(\dfrac{8 \div 8}{16 \div 8} = \dfrac{1}{2} = 0.5 \right)$

e. $\dfrac{3}{9} = 3 \div 9 = 0.333$ $\left(\dfrac{3 \div 3}{9 \div 3} = \dfrac{1}{3} = 0.3 \right)$

f. $\dfrac{1}{8} = 1 \div 8 = 0.125$

The next question is: How do we convert a mixed fraction such as $3\frac{1}{2}$ to its decimal equivalent? The answer is to do nothing with the whole number except place it in the whole-number decimal columns and then convert the proper fraction as before. Let us do the following example.

■ **EXAMPLE:**

Convert the following mixed fractions to their decimal equivalents.

a. $3\dfrac{1}{2}$ **c.** $27\dfrac{4}{9}$

b. $486\dfrac{1}{4}$ **d.** $31,348\dfrac{8}{9}$

■ *Solution:*

a. $3\dfrac{1}{2} = 3.5$ $(1 \div 2 = 0.5)$

b. $486\dfrac{1}{4} = 486.25$ $(1 \div 4 = 0.25)$

c. $27\dfrac{4}{9} = 27.444$ $(4 \div 9 = 0.444)$

d. $31,348\dfrac{8}{9} = 31,348.888$ $(8 \div 9 = 0.888)$

Finally, let us convert an improper fraction such as $\frac{4}{2}$ (four halves) to its decimal equivalent. This is the easiest operation to perform, as shown in the following example, because all you do is perform the division on your calculator to obtain a decimal whole number and fraction.

■ **EXAMPLE:**

Convert the following improper fractions to their decimal equivalents.

a. $\frac{4}{2}$ **b.** $\frac{7}{3}$ **c.** $6\frac{7}{2}$ **d.** $\frac{18}{16}$

■ *Solution:*

a. $\frac{4}{2} = 4 \div 2 = 2.0$

b. $\frac{7}{3} = 7 \div 3 = 2.333$

c. $6\frac{7}{2} = 6 + (7 \div 2) = 6 + 3.5 = 9.5$ $\left(\frac{7}{2} = 3\frac{1}{2} = 3.5\right)$

d. $\frac{18}{16} = 18 \div 16 = 1.125$ $\left(\frac{18 \div 2}{16 \div 2} = \frac{9}{8} = 1\frac{1}{8}\right)$

2-2-4   *Reciprocals*

Whenever you divide a number into 1, you get the **reciprocal** *of that number.* The word *reciprocal* means "inverse" or "opposite relation." This term can best be described by determining the reciprocal of two numbers, 2 and 2000. As just mentioned, to get the reciprocal of these numbers, we simply divide these numbers into 1, as shown:

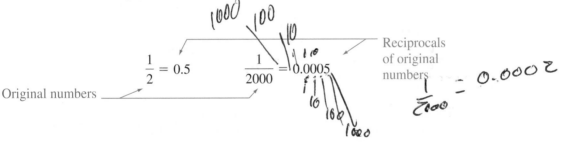

Original numbers

$\frac{1}{2} = 0.5$ $\frac{1}{2000} = 0.0005$ Reciprocals of original numbers

Reciprocal
Inversely related.

Comparing these two answers, you can see why the reciprocals, or result of dividing the numbers into 1, are inversely related to the original numbers. For instance, the reciprocal of a small number ($\downarrow$), such as 2, results in a large fraction ($\uparrow$), 0.5, whereas the reciprocal of a large number ($\uparrow$), such as 2000, results in a small fraction ($\downarrow$), 0.0005. This inverse relationship is even more evident when you consider that 2000 is one thousand times greater than 2, yet the reciprocal of 2000 is one thousand times smaller than the reciprocal of 2. To summarize:

	Original number						Reciprocal of number			
1000	100	10	1				$\frac{1}{10}$	$\frac{1}{100}$	$\frac{1}{1000}$	$\frac{1}{10,000}$
			2	0	$\frac{1}{2} =$		0	5		
2	0	0	0	0	$\frac{1}{2000} =$		0	0	0	0	5

2000 is three columns or 1000 times greater than 2.

0.0005 is three columns or 1000 times smaller than 0.5.

Most calculators have a key specifically for calculating the reciprocal of any number. It is called the reciprocal key and operates as follows.

CALCULATOR KEYS

Name: Reciprocal key

Function: Divides the value on the display into 1.

Example: $\dfrac{1}{3.2} = ?$

Press keys: $\boxed{3}\boxed{.}\boxed{2}\boxed{1/x}$

Display shows: 0.3125

The reciprocal of π is used in many different formulas. The Greek lowercase letter pi, which is pronounced "pie" and symbolized by π, is a constant value that describes how much bigger a circle's circumference is than its diameter.

The approximate value of this constant is 3.142, which means that the circumference of any circle is approximately 3 times bigger than its diameter.

Most calculators have a key for quickly obtaining the value of pi. It is called the pi (π) key and operates as follows.

CALCULATOR KEYS

Name: Pi key

Function: Enters the value of pi correct to eight decimal digits.

$\pi = 3.1415927$

Example: $2 \times \pi = ?$

Press keys: $\boxed{2}\boxed{\times}\boxed{\pi}\boxed{=}$

Display shows: 6.2831853

pi = 3.

The factor π is used frequently in many formulas involved with the analysis of circular motion. The reciprocal $\frac{1}{2}\pi$ (1 over $2 \times \pi$) is frequently used, so we should do a few problems involving the reciprocal and pi keys on the calculator.

■ **EXAMPLE:**

Calculate the reciprocals of the following numbers.

   a. 5   **b.** π **c.** 60   **d.** $2 \times \pi$

■ *Solution:*

   a. The reciprocal of $5 = \dfrac{1}{5} = 1 \div 5 = 0.2$.

      Calculator sequence: $\boxed{5}\boxed{1/x}$; display shows: 0.2

   b. The reciprocal of $\pi = \dfrac{1}{\pi} = \dfrac{1}{3.142} = 0.318$.

      Calculator sequence: $\boxed{\pi}\boxed{1/x}$; display shows: 0.3183098

   c. The reciprocal of $60 = \dfrac{1}{60} = 0.0166$.

      Calculator sequence: $\boxed{6}\boxed{0}\boxed{1/x}$; display shows: 0.0166667

   d. The reciprocal of $2 \times \pi = \dfrac{1}{2 \times \pi} = \dfrac{1}{6.28} = 0.159$.

      Calculator sequence: $\boxed{2}\boxed{\times}\boxed{\pi}\boxed{=}\boxed{1/x}$; display shows: 0.1591549

2-2-5   *Converting Decimal Fractions to Written Fractions*

First, let us discuss why we would want to convert a decimal fraction, such as 0.5, into a written fraction, such as $\frac{1}{2}$. The answer is so that we can convert from a scale of 10 (decimal) to a scale that is something other than 10. For example, imagine that you have a standard ruler with 1 inch divided into 16 parts (there are sixteen $\frac{1}{16}$ in 1 inch). Next, imagine that you have to measure a length of 0.125 of an inch. Since the scale of the ruler is dealing with 16 parts and your decimal value of 0.125 is based on 10 parts, the two are not compatible and a conversion is needed. The first step is to convert $\frac{1}{16}$ into its decimal equivalent:

$$\frac{1}{16} = 1 \div 16 = 0.0625$$

This answer tells us that one-sixteenth of an inch is equivalent to 0.0625 of an inch in decimal. This can be proved by multiplying 0.0625 by 16 to see if we have 16 of these parts in 1 inch.

$$16 \times 0.0625 = 1 \text{ inch}$$

Now that we know that the value 0.0625 inch is equal to $\frac{1}{16}$ inch, the next step is to find out how many 0.0625 parts (or $\frac{1}{16}$) are in our decimal measurement of 0.125.

$$\frac{0.125}{0.0625} = 0.125 \div 0.0625 = 2$$

The answer of 2 tells us that there are two 0.0625 parts (or sixteenths) in 0.125. Therefore, to measure off a length of 0.125 of an inch using a $\frac{1}{16}$ inch scale, simply measure a length of two-sixteenths ($\frac{2}{16}$).

$$0.125 \text{ in.} = \frac{2}{16} \text{ in.}$$

■ **EXAMPLE:**

Convert the following decimal fractions to proper fractions.

   a. $0.25 = \dfrac{?}{8}$ $\dfrac{1}{8}$

   b. $0.666 = \dfrac{?}{3}$

   c. $0.59375 = \dfrac{?}{32}$ $\dfrac{1}{32}$ - $1 -$

■ *Solution:*

   a. 0.25 equals how many eighths? $\frac{1}{8}$ has a decimal equivalent of 0.125 ($1 \div 8 = 0.125$). How many 0.125 parts are in 0.25?

$$\frac{0.25}{0.125} = 0.25 \div 0.125 = 2 \qquad \underline{0.25}$$

$$0.25 = \frac{2}{8} = \frac{1}{4}$$

   b. $\dfrac{1}{3} = 0.333$

$$\frac{0.666}{0.333} = 2$$

$$0.666 = \frac{2}{3}$$

   c. $\dfrac{1}{32} = 0.03125$

$$\frac{0.59375}{0.03125} = 19 \ .$$

$$0.59375 = \frac{19}{32}$$

CALCULATOR KEYS

Name: Conversion Functions

Display the answer as a fraction (▶ Frac).
Display the answer as a decimal (▶ Dec).

Function:

Value ▶ Frac (display as a fraction) displays an answer as its written fraction equivalent.
Value ▶ Dec (display as a decimal) displays an answer as its decimal equivalent.

Example: $\frac{1}{2} + \frac{1}{3}$ ▶ Frac

                       $\frac{5}{6}$

      Ans ▶ Dec

          .8333333333

Now that you have completed this section, you should be able to:

■ *Objective 3.* *Define the function of the decimal point.*

■ *Objective 4.* *Describe the positional weight of decimal fractions.*

■ *Objective 5.* *Demonstrate how to convert written fractions to decimal fractions, calculate reciprocals, and convert decimal fractions to written fractions.*

Use the following questions to test your understanding of Section 2-2.

1. What are the individual digit weights of the number 178.649?

2. Convert the following written fractions to decimal fractions.
   a. $1\frac{5}{2}$ b. $192\frac{3}{4}$ c. $67\frac{6}{9}$

3. Calculate the reciprocal of the following numbers.
   a. 2500   b. 0.25   c. 0.000025

4. Convert the following decimal fractions to written fractions.
   a. $7.25 = \frac{?}{16}$ b. $156.90625 = \frac{?}{32}$

2-3   CALCULATING IN DECIMAL

In this section we first review the basic principles of decimal addition, subtraction, multiplication, and division. As you will discover, there are really only *two basic operations: addition and subtraction.* If you can perform an addition, you can perform a multiplication because *multiplication is simply repeated addition.* For example, a problem such as 4 × 2 (4 multiplied by 2) is asking you to calculate the sum of four 2s (2 + 2 + 2 + 2). Similarly, *division is simply repeated subtraction.* For example, a problem such as 9 ÷ 3 (9 divided by 3) is asking you to calculate how many times 3 can be subtracted from 9 (9 − 3 = 6, 6 − 3 = 3, 3 − 3 = 0; therefore, there are three 3s in 9). In summary, therefore, the two basic mathematical operations (addition and subtraction) and their associated mathematical operations (multiplication and division) are as follows:

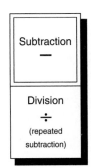

There is another relationship that needs to be discussed, and that is the inverse or opposite nature between addition and subtraction and between multiplication and division. For example, *subtraction is the opposite arithmetic operation of addition.* This can be proved by starting with a number such as 10, then adding 5, then subtracting 5.

$$\textcircled{10} + 5 = 15, \quad 15 - 5 = \textcircled{10}$$

Original number   Back to original number

Similarly, *multiplication is the opposite arithmetic operation of division.* This can be proved by starting with a number such as 4, then multiplying by 3, then dividing by 3.

$$\textcircled{4} \times 3 = 12, \quad 12 \div 3 = \textcircled{4}$$

Original number   Back to original number

In summary, therefore, addition is the opposite mathematical operation of subtraction, and multiplication is the opposite mathematical operation of division. This opposite relationship will be made use of in a subsequent chapter.

Addition +	Is the opposite mathematical operation	Subtraction −
Multiplication ×	Is the opposite mathematical operation	Division ÷

Every calculator has keys for performing these four basic arithmetic operations. The name, function, and procedure for using these keys are as follows.

CALCULATOR KEYS

Name: Decimal point key [.] (discussed previously)

Name: Equals key [=]

Function: Combines all previously entered numbers and operations. Used to obtain intermediate and final results.

Example: $3 \times 4 + 6 = ?$

Press keys: [3][×][4][=][+][6][=]

Display shows: 18.0

Name: Add key [+]

Function: Instructs calculator to add the next entered quantity to the displayed number.

Example: Add $31.6 + 2.9 = ?$

Press keys: [3][1][.][6][+][2][.][9][=]

Displays shows: 34.5

Name: Subtract key [−]

Function: Instructs calculator to subtract the next entered quantity from the displayed number.

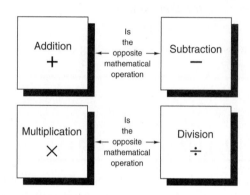

CALCULATOR KEYS *(continued)*

Example: $266.3 - 5.35 = ?$

Press keys: ⟦2⟧⟦6⟧⟦6⟧⟦.⟧⟦3⟧⟦−⟧⟦5⟧⟦.⟧⟦3⟧⟦5⟧⟦=⟧

Display shows: 260.95

Name: Multiply key ⟦×⟧

Function: Instructs calculator to multiply the displayed number by the next entered quantity.

Example: $3.85 \times 2.9 = ?$

Press keys: ⟦3⟧⟦.⟧⟦8⟧⟦5⟧⟦×⟧⟦2⟧⟦.⟧⟦9⟧⟦=⟧

Display shows: 11.165

Name: Divide key ⟦÷⟧

Function: Instructs calculator to divide the displayed number by the next entered quantity.

Example: $28.98 \div 2.3 = ?$

Press keys: ⟦2⟧⟦8⟧⟦.⟧⟦9⟧⟦8⟧⟦÷⟧⟦2⟧⟦.⟧⟦3⟧⟦=⟧

Display shows: 12.6

2-3-1   *Decimal Addition*

To review decimal **addition,** let us examine the steps involved in adding the following numbers.

$$66.332 + 285 + 0.002 + 182.788 = ?$$

First, list the numbers vertically, making sure that the decimal points are lined up as follows.

> **Addition**
> The operation of combining numbers so as to obtain an equivalent quantity.

Addition process:
  $(100)\ (10)\ (1)\ .\ (^1/_{10})\ (^1/_{100})\ (^1/_{1000})$

```
Carry →    2  1   1   1   1
           0  6 | 6.  3   3   2
           2  8 | 5.  0   0   0
           0  0 | 0.  0   0   2
          +1  8 | 2.  7   8   8
          _____
Sum  →     5  3   4.  1   2   2
```

The extra zeros are included to keep the digits in their proper columns.

Steps:
Starting at the right column ($^1/_{1000}$), add vertically.

$2 + 0 + 2 + 8 = 12,\ 2$ carry 1.
$1 + 3 + 0 + 0 + 8 = 12,\ 2$ carry 1.
$1 + 3 + 0 + 0 + 7 = 11,\ 1$ carry 1.
$1 + 6 + 5 + 0 + 2 = 14,\ 4$ carry 1.
$1 + 6 + 8 + 0 + 8 = 23,\ 3$ carry 2.
$2 + 0 + 2 + 0 + 1 = 5.$

Calculator sequence:

⟦6⟧⟦6⟧⟦.⟧⟦3⟧⟦3⟧⟦2⟧⟦+⟧⟦2⟧⟦8⟧⟦5⟧⟦+⟧⟦.⟧⟦0⟧⟦0⟧⟦2⟧⟦+⟧⟦1⟧⟦8⟧⟦2⟧⟦.⟧⟦7⟧⟦8⟧⟦8⟧⟦=⟧
Answer: 534.122

There are a few points to note regarding this previous addition. First, the number 285 did not have a decimal point indicated; therefore, it is assumed to be to the right of the last digit (285.0). Second, what is really happening when we **carry** or overflow into the next-left column? The answer is best explained by examining the addition of the units column in the

> **Carrying**
> The process of transferring from one column to another during an addition process.

previous example. In this column we had a result of $1 + 6 + 5 + 0 + 2 = 14$. As you know from our discussion on positional weight, 14 can be broken down into 1 ten (1×10) and 4 ones (4×1). This is why the 1 ten is carried over to the tens column, and the 4 ones or units are placed in the units sum column. Explained another way, whenever the sum of all the digits in any column exceeds the maximum single decimal digit of 9, a reset and carry action occurs. In the case of the units column with a sum of 14, we exceeded 9 and therefore the units sum was reset to 0 and then advanced to 4, and a 1 was carried into the tens column to make 14.

In summary, therefore, because each column to the left is 10 times larger, a reset and carry action will occur every time the sum of any column is incremented or advanced by one, from 9 to 10.

■ EXAMPLE:

How much is $976.73 + $998.28?

■ *Solution:*

Steps

Carry:	1111 1
Addend:	0976.73
Augend:	+ 0998.28
Sum:	$1975.01

$3 + 8 = 11$, 1 carry 1.
$1 + 7 + 2 = 10$, 0 carry 1.
$1 + 6 + 8 = 15$, 5 carry 1.
$1 + 7 + 9 = 17$, 7 carry 1.
$1 + 9 + 9 = 19$, 9 carry 1.
$1 + 0 + 0 = 1$, 1 carry 0.

Addend

The number or quantity to be added to an existing quantity called the augend.

Augend

The number to which a new quantity, called the addend, is to be added.

Calculator sequence: 9 7 6 . 7 3 +
9 9 8 . 2 8 =
Answer: 1975.01

■ INTEGRATED MATH APPLICATION:

Calculate the total cost of the following service station repair invoice:

Parts: $ 64.95
Labor: $226.35
Coolant: $ 4.25
Tax: $ 22.17

■ *Solution:*

```
    64.95
   226.35
     4.25
 + 22.17
  $317.72
```

■ INTEGRATED MATH APPLICATION:

How many hours did you work in the previous week if your time card shows the following:

Monday—6 hours
Tuesday—4.5 hours
Wednesday—7.5 hours
Thursday—5 hours
Friday—8 hours

■ *Solution:*

$6 + 4.5 + 7.5 + 5 + 8 = 31$ hours

2-3-2 *Decimal Subtraction*

To review decimal **subtraction,** let us examine the steps involved in the following example:

$$\$2005.23 - \$192.41 = ?$$

To begin, list the numbers vertically, making sure that the decimal points are lined up as follows:

Subtraction process:

	(1000)	(100)	(10)	(1)	Dimes ($^1/_{10}$)	Cents ($^1/_{100}$)	
Borrow:		1	9	4			
Minuend:	2	^{1}0	^{1}0	^{1}5.	^{1}2	3	
Subtrahend:	0	1	9	2.	4	1	
Difference:	1	8	1	2.	8	2	

Calculator sequence:

| 2 | 0 | 0 | 5 | . | 2 | 3 | − |

| 1 | 9 | 2 | . | 4 | 1 | = |

Answer: 1812.82

  or $1812.82

Steps:

Starting at the right column ($^1/_{100}$ or cent) subtract vertically:

$3 - 1 = 2$

$2 - 4$: impossible to subtract 4 from the smaller number 2. Therefore, we will *borrow* $1 from the $5 in the units column. This will leave $4 in the units column; because $1 = 10 dimes, we will be adding 10 dimes to the 2 dimes already present, giving a total of 12 dimes.

$12 - 4 = 8$

$4 - 2 = 2$

$0 - 9$: impossible to subtract 9 from nothing or zero.

Because we cannot borrow from the tens or hundreds columns, we will have to borrow from the thousands column.

Borrowing 1000 (10×100) gives 10 hundreds; borrowing 1 hundred from 10 hundreds gives 9 hundreds. Borrowing 100 (10×10) gives 10 tens.

$10 - 9 = 1$

$9 - 1 = 8$

$1 - 0 = 1$

Subtraction

The operation of deducting one number from another.

Minuend

A number from which the subtrahend is to be subtracted.

Subtrahend

A number that is to be subtracted from a minuend.

Difference

The amount by which things differ in quantity or measure.

As you can see from this example, *borrowing* is the exact opposite of carrying. With addition we sometimes end up with a sum of 10 or more, which results in an overflow or carry into the next-higher or next-left column, whereas with subtraction we sometimes end up with a larger number in the subtrahend than the minuend, and it becomes necessary to borrow a 1 from the next-higher place to the left in the minuend.

■ **EXAMPLE:**

Perform the following subtraction:

$$27,060.0143 - 963.86$$

■ *Solution:*

	(10,000)	(1000)	(100)	(10)	(1)	($^1/_{10}$)	($^1/_{100}$)	($^1/_{1000}$)	($^1/_{10,000}$)	
		6	9	^{1}5	9.	9				*Steps*
	2	7	0	6	0.	0	^{1}1	4	3	$3 - 0 = 3$
	−0	0	9	6	3.	8	6	0	0	$4 - 0 = 4$
Difference	2	6	0	9	6.	1	5	4	3	$1 - 6$: can't do; borrow.

Calculator sequence:

| 2 | 7 | 0 | 6 | 0 | . | 0 | 1 | 4 | 3 | − |

| 9 | 6 | 3 | . | 8 | 6 | = |

Answer: 26,096.1543

Borrowing 1 ten from 6 tens gives 5 tens and 10 ones. Borrowing 1 one from 10 ones gives 9 ones and 10 tenths.

Borrowing 1 tenth from 10 tenths gives 9 tenths and 10 hundredths.

10 hundredths + the original 1 hundredth = 11 hundredths:

$$11 - 6 = 5$$
$$9 - 8 = 1$$
$$9 - 3 = 6$$

5 − 6: can't do; borrow.

Borrowing 1 thousand from 7 thousands gives 6 thousands and 10 hundreds. Borrowing 1 hundred from 10 hundreds gives 9 hundreds and 10 tens. 10 tens + the original 5 tens = 15 tens:

$$15 - 6 = 9$$
$$9 - 9 = 0$$
$$6 - 0 = 6$$
$$2 - 0 = 2$$

■ INTEGRATED MATH APPLICATION:

If the regular price of a television is $856.99 and the sale price is $699.99, how much will you save?

■ *Solution:*

$$\begin{array}{r} \$859.99 \\ -\$699.99 \\ \hline \$160.00 \end{array}$$ — A savings of $160.00.

■ INTEGRATED MATH APPLICATION:

If your gross weekly salary is $695.00 and your deductions total $212.65, what is your take-home pay?

■ *Solution:*

$$\begin{array}{r} \$695.00 \\ -\$212.65 \\ \hline \$482.35 \end{array}$$ Your weekly take-home pay is $482.35.

2-3-3   *Decimal Multiplication*

Multiplication

The operation of adding an integer to itself a specified number of times.

Multiplicand

The number that is to be multiplied by another.

Multiplier

One that multiplies; a number by which another number is multiplied.

To review decimal **multiplication,** let us examine the steps involved in the following example:

$$27.66 \times 2.4 = ?$$

Unlike in the previous mathematical operations, we can ignore the alignment of the decimal point and simply make sure that the two numbers are flush on the right side.

Multiplicand:	27.66
Multiplier:	× 2.4

Two numbers are listed vertically so that they are flush on the right side.

We can now use the following multiplication steps:

Multiplication process:

Second carry:	11 1
First carry:	132 2
Multiplicand:	27.6 6
Multiplier:	× 2. 4
First partial product:	110 6 4
Second partial product:	553 2 0
Final product:	663 8 4

Steps:

To begin, multiply each digit in the multiplicand by the first digit of the multiplier (which is 4) to obtain the first partial product. Line up the rightmost digit of the first partial product under the multiplier digit being used (see arrow).

$4 \times 6 = 24$, 4 carry 2.

$4 \times 6 = 24$, 24 + previous carry of 2 = 26, 6 carry 2.

$4 \times 7 = 28$, 28 + carry of 2 = 30, 0 carry 3.

$4 \times 2 = 8$, 8 + carry of 3 = 11, 1 carry 1; bring 1 down.

The next step is to multiply each digit in the multiplicand by the second digit of the multiplier (which is 2) to obtain the second partial product. Line up the rightmost digit of the second partial product under the multiplier digit being used (see arrow).

$2 \times 6 = 12$, 2 carry 1.

$2 \times 6 = 12$, 12 + carry of 1 = 13, 3 carry 1.

$2 \times 7 = 14$, 14 + carry of 1 = 15, 5 carry 1.

$2 \times 2 = 4$, 4 + carry of 1 = 5.

The next step is to add the first and second partial products to obtain the final product.

$4 + 0 = 4$

$6 + 2 = 8$

$0 + 3 = 3$

$1 + 5 = 6$

$1 + 5 = 6$

Calculator sequence:

| 2 | 7 | . | 6 | 6 | × | 2 | . | 4 | = |

Answer: 66.384

27.6 6 The multiplicand has two digits to the right of the decimal point.

2.4 The multiplier has one digit to the right of the decimal point.

$2 + 1 = 3$ places.

Answer is 66.384

← The final step is to determine the position of the decimal point in the answer by counting the number of digits to the right of the decimal point in the multiplier and multiplicand. There are 3 in this example, so the final product will have three digits to the right of the decimal point.

66.3 8 4

■ **EXAMPLE:**

How much is 0.035 multiplied by 0.005?

■ *Solution:*

In this example we will take a few shortcuts. Since the decimal points do not make a difference at this stage, let us drop them completely. This will produce:

$$\begin{array}{r} 0035 \\ \times\ 0005 \end{array}$$

Second, we learned earlier that extra zeros appearing to the left or right of a number are not needed in the multiplication process, so let us remove these zeros:

$$\begin{array}{r} 35 \\ \times\ 5 \end{array}$$

The multiplication is now a lot easier because it is simply 35×5. Performing the multiplication, we obtain the product:

$$\begin{array}{r} \overset{2}{3}5 \\ \times\ \ 5 \\ \hline 175 \end{array}$$

The final step is to determine the position of the decimal point. The original multiplicand had three digits to the right of the decimal point (0.0 3 5), and the original multiplier also had three digits to the right of the decimal point ($0.\overset{\frown}{0}\,0\,\overset{\frown}{5}$), so the final product should have six digits to the right (3 + 3) of the decimal point.

$$\overset{\frown}{.0}\,\overset{\frown}{0}\,\overset{\frown}{0}\,\overset{\frown}{1}\,\overset{\frown}{7}\,\overset{\frown}{5}\ \text{or}\ 0.000175$$

■ **INTEGRATED MATH APPLICATION:**

If your car averages 26.3 miles per gallon, how far can you travel if your gas tank is filled with 12 gallons of gasoline?

■ *Solution:*

$$26.3\ \text{miles per gallon} \times 12\ \text{gallons} = 315.6\ \text{miles}$$

■ **INTEGRATED MATH APPLICATION:**

If an inch on a street map is equivalent to 0.5 mile, how many miles will you have to travel if the distance on the map measures 14.5 inches?

■ *Solution:*

$$14.5\ \text{inches} \times 0.5\ \text{mile} = 7.25\ \text{miles}$$

2-3-4   *Decimal Division*

Division

The operation of separating a number into equal parts.

Quotient

The number resulting from the division of one number by another.

Divisor

The number by which a dividend is divided.

Dividend

A number to be divided.

Remainder

The final undivided part after division that is less or of lower degree than the divisor.

To review decimal **division** (or *long division,* as it is sometimes called) let us step through the following division example: 1362 ÷ 12. The procedure is as follows.

Long-division process:

```
Quotient:              0113
Divisor: 12 √   1362
              − 12
Dividend         16
              − 12
                 42
              − 36
Remainder ──→     6
```

Steps:

The first step is to see how many times the divisor (12) can be subtracted from the first digit of the dividend (1). Because the dividend is smaller, the answer or quotient is zero (1 − 12 cannot be done). We place this zero above the dividend 1.

Next, we see how many times the divisor (12) can be subtracted from the first two digits of the dividend (13). 13 − 12 = 1, remainder 1. The answer or quotient of 1 is placed above the next digit of the dividend (3), and the remainder of the subtraction (1) is dropped below the 13 − 12 line.

We carry the next digit of the dividend (6) down to the remainder (1) to address the next part of the dividend, which is 16. Now we repeat the process by seeing how many times the divisor, 12, can be subtracted from these two digits, 16. The answer is a quotient of 1 (which is placed above the 6 dividend) with a remainder of 4 (which is dropped below the 16 − 12 line).

Calculator sequence:

[1][3][6][2][÷][1][2][=]

Answer: (113).(5)

The whole decimal number is telling you how many times 12 will go

The decimal fraction indicates how much of a fraction of the divisor is remaining.

Bringing down the next dividend digit of 2 and combining it with the previous remainder of 4, we now have the final part of the dividend, which is 42. Because 12 can be subtracted from 42 three times, a quotient of 3 is placed

into 1362.
Therefore,
1362 ÷ 12 = ⑾⑬
Answer to 1362 ÷ 12 is 113, remainder 6.

Therefore,
0.5 × 12 = 6

above the final digit of the dividend (2), and the remainder of the subtraction, 6, is dropped below the 42 − 36 line.
Therefore,
1362 ÷ 12 = 113, remainder 6. Said another way, 12 can be subtracted from 1362 113 times, with 6 remaining.

As another example, 13 ÷ 4 = ?

☐1☐3☐÷☐4☐=

Answer: ③.㉕

4 goes into 13 three times.
13 ÷ 4 = ③

Remainder is ¼ or 0.25 of 4; therefore, 0.25 × 4 = ①

Answer to 13 ÷ 4 is 3, remainder 1.

■ **EXAMPLE:**

Calculate the result of dividing 19,205.628 by 1.56.

■ *Solution:*

19,205.628 ÷ 1.56 = 12,311.3
Calculator sequence: ☐1☐9☐2☐0☐5☐.☐6☐2☐8☐÷☐1☐.☐5☐6☐=
Answer: 12,311.3

■ **INTEGRATED MATH APPLICATION:**

If your round-trip commute to work for five days totals 87.5 miles, how many miles are you driving per day?

■ *Solution:*

87.5 miles ÷ 5 days = 17.5 miles per day

■ **INTEGRATED MATH APPLICATION:**

If your gross pay for a week is $536.50 and you worked 37 hours, what is your hourly rate of pay?

■ *Solution:*

$536.50 ÷ 37 hours = $14.50 per hour

2-3-5   *Ratios*

A **ratio** is a comparison of one number to another number. For example, a ratio such as 7:4 ("seven to four") is comparing the number 7 to the number 4. This ratio could be written in one of the following three ways:

⁷⁄₄ ← Ratio is written as a fraction.
7:4 ← Ratio is written using the ratio sign (:).
1.75 to 1 ← Ratio is written as a decimal.

Ratio

The relationship in quantity, amount, or size between two or more things.

To express the ratio as a decimal, the number 7 was divided by the number 4, giving a result of 1.75 (7 ÷ 4 = 1.75). This result indicates that the number 7 is 1.75 times (or one and three-quarter times) larger than the number 4.

■ INTEGRATED MATH APPLICATION:

If one store has 360 items and another store has 100 of the same items, express the ratio of 360 to 100:

    a. Using the divide sign

    b. Using the ratio sign

    c. As a decimal

■ *Solution:*

    a. $\dfrac{360}{100} = \dfrac{360 \div 20}{100 \div 20} = \dfrac{18}{5}$ ← 360 to 100 and 18 to 5 are equivalent ratios.

    b. $360 : 100 = 18 : 5$

    c. 3.6 ← The first store has 3.6 times as many items as the other store ($18 \div 5 = 3.6$).

■ EXAMPLE:

Calculate the ratio of the following two like quantities. If necessary, reduce the ratio to its lowest terms.

    a. $0.8 \div 0.2 = ?$

    b. $0.0008 \div 0.0002 = ?$

■ *Solution:*

    a. How many 0.2s are in 0.8? The answer is the same as saying, How many 2s are in 8?

$$0.8 \div 0.2 = 4 \qquad \dfrac{8 \div 2}{2 \div 2} = \dfrac{4}{1} \qquad \text{(a ratio of 4 to 1)}$$

    b. How many 0.0002s are in 0.0008?

$$0.0008 \div 0.0002 = 4 : 1$$

The ratio in example (a) is the same as in (b). Another way to describe this is that because the size of each decimal fraction compared to the other decimal fraction (or the *ratio*) remained the same, the answer or quotient remained the same.

    8 is four times larger than 2, or 2 is four times smaller than 8.

    0.8 is four times larger than 0.2, or 0.2 is four times smaller than 0.8.

    0.0008 is four times larger than 0.0002, or 0.0002 is four times smaller than 0.0008. Therefore, all have the same 4 : 1 ratio.

■ INTEGRATED MATH APPLICATION:

In some instances, a ratio will be used to describe how a quantity is to be divided. For example, if 600 textbooks are to be divided between two schools in the ratio of 1 to 2 (1:2), how many textbooks will each school get? To calculate this, you should first add the terms of the ratio:

■ *Solution:*

    Step 1: The ratio is 1:2, so 1 + 2 = 3 (this means that the total is to be divided into 3 parts).

Next, you should multiply the total by each fractional part:

Step 2: The first school will receive ⅓ of the total 600 textbooks, which is 200 texts (⅓ × 600 = 200). The second school will receive ⅔ of the total 600 textbooks, which is 400 texts (⅔ × 600 = 400).

■ INTEGRATED MATH APPLICATION:

If three restaurants receive 216 hamburgers in the ratio of 6:14:7, how many hamburgers will each restaurant get?

The first step is to add the terms of the ratio:

■ Solution:

Step 1: The ratio is 6:14:7, so 6 + 14 + 7 = 27 (this means that the total is to be divided into 27 parts).

Next, you should multiply the total by each fractional part:

Step 2: The first restaurant will receive ⁶⁄₂₇ of the total 216 hamburgers, which is 48 (⁶⁄₂₇ × 216 = 48). The second restaurant will receive ¹⁴⁄₂₇ of the total 216 hamburgers, which is 112 (¹⁴⁄₂₇ × 216 = 112). The third restaurant will receive ⁷⁄₂₇ of the total 216 hamburgers, which is 56 (⁷⁄₂₇ × 216 = 56).

2-3-6   Rounding Off

In many cases we will **round off** *decimal numbers because we do not need the accuracy indicated by a large number of decimal digits.* For example, it is usually unnecessary to have so many decimal digits in a value such as 74.139896428. If we were to round off this value to the nearest hundredths place, we would include two digits after the decimal point, which is 74.13. This is not accurate, however, since the digit following 74.13 was a 9 and therefore one count away from causing a reset and carry action into the hundredths column. To take into account the digit that is to be dropped when rounding off, therefore, we follow this basic rule: *When the first digit to be dropped is 6 or more, increase the previous digit by 1. On the other hand, when the first digit to be dropped is 4 or less, do not change the previous digit.* Therefore, the value 74.139896428, rounded off to the nearest hundredths place, would equal:

┌─ Hundredths place

|74.13|9896428 = 74.14

First digit to be dropped is greater than 6, so previous digit should be increased by 1.

■ EXAMPLE:

Round off the value 74.139896428 to the nearest ten, whole number, tenth, hundredth, thousandth, and ten-thousandth.

■ Solution:

|7|4.139896428 rounded off to the nearest ten = 70

└─ First digit to be dropped is 4 or less; therefore, do not change previous digit.

|74.|139896428 rounded off to the nearest whole number = 74

└─ First digit to be dropped is 4 or less; therefore, do not change previous digit.

$\boxed{74.1}$ 39896428 rounded off to the nearest tenth = 74.1

⌐— First digit to be dropped is 4 or less; therefore, do not change previous digit.

$\boxed{74.13}$ 9896428 rounded off to the nearest hundredth = 74.14

⌐— First digit to be dropped is 6 or greater; therefore, increase previous digit by 1 (3 to 4).

$\boxed{74.139}$ 896428 rounded off to the nearest thousandth = 74.140

⌐— First digit to be dropped is 6 or greater; therefore, increase previous digit by 1. Since previous digit is 9, allow reset and carry action to occur. The steps are:

74.1398 ◂— 8 carries 1 into thousandths column.
74.130 ◂— 9 resets to 0, and carries 1 into hundredths column.
74.140 ◂— Hundredths digit is increased by 1.

$\boxed{74.1398}$ 96428 rounded off to the nearest ten-thousandth = 74.1399

⌐— First number to be dropped is 6 or greater; therefore, increase previous number by 1.

In the preceding example we discovered how to round off if a digit was over halfway up the decimal scale (greater than 6), or if a digit was below the halfway point on the decimal scale (less than 4). The next question, therefore, is: How do we round off when the digit to be dropped is a 5 and therefore midway in the decimal scale of 10? In this instance, we follow this rule: *When the 5 is followed by a digit that is more than zero, increase the previous digit by 1. On the other hand, when the 5 is followed by zero, do not change the previous digit.* Let us apply this to an example:

$\boxed{79.5}$ 3 rounded off to the nearest whole number = 80

Let us examine why this is correct. Because the digit following the 5 is more than zero, the decimal fraction (0.53) is in this case more than one-half or 0.5 (it is actually 5 tenths or $\frac{1}{2}$ plus 3 hundredths), and therefore the 9 should be changed to 0 and the 7 increased to 8.

■ **EXAMPLE:**

Round off the value 31,520.565 to the nearest ten thousand, thousand, hundred, whole number, tenth, and hundredth.

■ *Solution:*

$\boxed{3}$ 1,520.565 rounded off to the nearest ten thousand = 30,000

⌐ First digit to be dropped is 4 or less; therefore, do not change previous digit.

$\boxed{31,}$ 520.565 rounded off to the nearest thousand = 32,000

⌐ The 5 is followed by a digit that is more than zero; therefore, increase previous digit by 1.

$\boxed{31,5}$ 20.565 rounded off to the nearest hundred = 31,500

⌐ First digit to be dropped is 4 or less; therefore, do not change previous digit.

$\boxed{31,520}$ .565 rounded off to the nearest whole number = 31,521

⌐ The 5 is followed by a digit that is more than zero; therefore, increase previous digit by 1.

$\boxed{31,520.5}$ 65 rounded off to the nearest tenth = 31,520.6

⌐ Digit to be dropped is 6 or more; therefore, increase previous digit by 1.

$\boxed{31,520.56}$ 5 rounded off to the nearest hundredth = 31,520.56

‸ Because no digit follows 5, it is assumed to be a zero; therefore, do not change previous digit.

CALCULATOR KEYS

Name: Round function
round (*value*[,*#decimals*])

Function: Value is rounded to the number of decimals specified.

Example:
round (π, 4)

3.1416

2-3-7 *Significant Places*

The number of significant places describes how many digits are in the value and how many digits in the value are accurate after rounding off. For example, a number such as 347.63 is a five-significant-place (or significant-figure) value because it has five digits in five columns. If we were to round off this value to a whole number, we would get 348.00. This value would still be a five-significant-place number; however, it would now be accurate to only three significant places. Let us examine a few problems to practice using the terms *significant places* and *accurate to significant places*.

■ **EXAMPLE:**

On a calculator, the value of π will come up as 3.141592654.

a. Write the number π to six significant places.

b. Give π to five significant places, and also round off π to ten-thousandths.

c. 3.14159000 is the value of π to _____ significant places; however, it is accurate to only _____ significant places.

■ *Solution:*

a. The value π to six significant places or figures is $\boxed{3.14159}$ 27

3.14159

‸ Uses six digits or columns.

b. The value π to five significant places.
$\boxed{3.1415}$ 92 rounded off to ten-thousandths = 3.1416

c. 3.14159000 is the value of π to nine significant places (has nine digits); however, it is accurate to only six significant places (because the zeros to the right are just extra, and the value 3.14159 has only six digits).

2-3-8   *Percentages*

Percent

In the hundred; of each hundred.

The percent sign (%) means hundredths, which as a proper fraction is $^1\!/_{100}$ or as a decimal fraction is 0.01. For example, 50% means 50 hundredths ($^{50}\!/_{100} \div ^{50}\!/_{\div 50} = ^1\!/_2$), which is one-half. In decimal, 50% means 50 hundredths ($50 \times 0.01 = 0.5$), which is also one-half. The following shows how some of the more frequently used percentages can be expressed in decimal.

$$1\% = 1 \times 0.01 = 0.01$$ (For example, 1% is expressed mathematically in decimal as 0.01.)

$$5\% = 5 \times 0.01 = 0.05$$
$$10\% = 10 \times 0.01 = 0.10$$
$$25\% = 25 \times 0.01 = 0.25$$
$$50\% = 50 \times 0.01 = 0.50$$
$$75\% = 75 \times 0.01 = 0.75$$

Most calculators have a percent key that automatically makes this conversion to decimal hundredths. It operates as follows.

CALCULATOR KEYS

Name: Percent key

Function: Converts the displayed number from a percentage to a decimal fraction.

Example: 18.6% = ?

Press keys: $\boxed{1}\,\boxed{8}\,\boxed{.}\,\boxed{6}\,\boxed{\%}$

Display shows: 0.186

Now that we understand that the percent sign stands for hundredths, what does the following question actually mean: What is 50% of 12? We now know that 50% is 50 hundredths ($^{50}\!/_{100}$), which is one-half. This question is actually asking: What is half of 12? Expressed mathematically as both proper fractions and decimal fractions, it would appear as follows:

Proper Fractions	Decimal Fractions
50% of 12 = ?	50% of 12 = ?
$= \dfrac{50}{100} \times 12$	$= (50 \times 0.01) \times 12$
$= \dfrac{1}{2} \times 12$	$= 0.5 \times 12$
$= 6$	$= 6$

CHAPTER 2 / DECIMAL NUMBERS

Explaining this another way, we could say that percentages are fractions in hundredths. In the preceding example, therefore, the question is actually asking us to divide 12 into 100 parts and then to determine what value we would have if we had 50 of those 100 parts. Therefore, if we were to split 12 into 100 parts, each part would have a value of 0.12 ($12 \div 100 = 0.12$). Having 50 of these 100 parts would give us a total of $50 \times 0.12 = 6$.

■ **INTEGRATED MATH APPLICATION:**

If sales tax is 8% and the merchandise price is $256.00, what is the tax on the value, and what will be the total price?

■ *Solution:*

First, determine the amount of tax to be paid.

$$8\% \text{ of } \$256 =$$
$$0.08 \times 256 = \$20.48$$

Calculator sequence: $\boxed{8}\,\boxed{\%}\,\boxed{\times}\,\boxed{2}\,\boxed{5}\,\boxed{6}\,\boxed{=}$
Answer: 20.48
The total price paid will therefore equal

$$\$256.00 + \$20.48 = \$276.48$$

■ **INTEGRATED MATH APPLICATION:**

Which is the better buy—a $599.95 refrigerator at 25% off or a $499.95 refrigerator with a $50 manufacturer's rebate?

■ *Solution:*

$$25\% \text{ of } \$599.95 = \$150$$
$$\text{Refrigerator A: } \$599.95 - \$150.00 = \$449.95$$
$$\text{Refrigerator B: } \$499.95 - \$50.00 = \$449.95$$

The refrigerators are the same price.

2-3-9 *Averages*

A mean **average** *is a value that summarizes a set of unequal values.* This value is equal to the sum of all the values divided by the number of values. For example, if a team has scores of 5, 10, and 15, what is their average score? The answer is obtained by adding all the scores ($5 + 10 + 15 = 30$) and then dividing the result by the number of scores ($30 \div 3 = 10$).

Average

A single value that summarizes or represents the general significance of a set of unequal values.

■ **INTEGRATED MATH APPLICATION:**

The voltage at the wall outlet in your home was measured at different times in the day and equaled 108.6, 110.4, 115.5, and 123.6 volts (V). Find the average voltage.

Add all the voltages.

$$108.6 + 110.4 + 115.5 + 123.6 = 458.1$$

Divide the result by the number of readings.

$$\frac{458.1}{4} = 114.525 \text{ V}$$

The average voltage present at your home was therefore 114.525 volts.

Statistics

A branch of mathematics dealing with the collection, analysis, interpretation, and presentation of masses of numerical data.

Statistics are generally averages and can misrepresent the actual conditions. For example, if five people earn an average salary of $204,600 a year, one would believe that all five are very well paid. Studying the individual figures, however, we find out that their yearly earnings are $8000, $6000, $4000, $20,000, and $985,000 per year. Therefore,

$$8000 + 6000 + 4000 + 20,000 + 985,000 = 1,023,000$$
$$\frac{1,023,000}{5} = 204,600$$

If one value is very different from the others in a group, this value will heavily influence the result, creating an average that misrepresents the actual situation.

■ **INTEGRATED MATH APPLICATION:**

A football player's "average number of yards gained per carry" ranking is calculated by dividing the total number of yards gained by the total number of times the player carried the football. Which of the following players had the higher average number of yards gained per carry?

Player	Total Yards	Total Carries
Payton	16,726	3,838
Brown	12,312	2,359

■ *Solution:*

$$\text{Payton: } \frac{16,726}{3,838} = 4.4 \text{ yards per carry average}$$

$$\text{Brown: } \frac{12,312}{2,359} = 5.2 \text{ yards per carry average}$$

On average, Brown gained a higher number of yards per carry.

SELF-TEST EVALUATION POINT FOR SECTION 2-3

Now that you have completed this section, you should be able to:

■ *Objective 6.* *Describe the relationships among the four basic arithmetic operations: addition, subtraction, multiplication, and division.*

■ *Objective 7.* *Demonstrate how to perform the following arithmetic operations: decimal addition,*

decimal subtraction, decimal multiplication, and decimal division.

■ *Objective 8.* *Describe the following mathematical terms and their associated operations: ratios, rounding off, significant places, percentages, and averages.*

Use the following questions to test your understanding of Section 2-3.

1. The two basic arithmetic operations are _____ and _____.

2. Give the opposite of the following arithmetic operations.
   a. Addition
   b. Subtraction
   c. Multiplication
   d. Division

3. Perform the following arithmetic operations by hand, and then confirm your answers with a calculator.
   a. $26.443 + 197.1 + 2.1103 + 0.004 = ?$
   b. $19,637.224 - 866.43 = ?$
   c. $894.357 \times 8.6 = ?$
   d. $1.3397 \div 0.015 = ?$

4. Express a 176 meter to 8 meter ratio as a fraction, using the ratio sign and as a decimal.

5. Round off the following values to the nearest hundredth.
   a. 86.43760
   b. 12,263,415.00510
   c. 0.176600

6. Referring to the values in Question 5, describe:
   a. Their number of significant places
   b. To how many significant places they are accurate

7. Calculate the following.
   a. 15% of 0.5 = ?
   b. 22% of 1000 = ?
   c. 2.35% of 10 = ?
   d. 96% of 20 = ?

8. Calculate the average of the following values.
   a. 20, 30, 40, and 50
   b. 4000, 4010, 4008, and 3998

REVIEW QUESTIONS

Multiple Choice Questions

1. What is the base or radix of the decimal number system?
   a. 100
   b. 10
   c. 0 through 9
   d. 10, 100, 1000, 10,000

2. Because each column in the decimal number system increases by _____ times as you move left, two columns to the left will be _____ times greater.
   a. 100, 10
   b. 10, 1000
   c. 1000, 10
   d. 10, 100

3. The decimal point is used to:
   a. Separate the whole numbers from the fraction
   b. Separate the tens column from the units column
   c. Identify the decimal number system
   d. All the above

4. What is the positional weight of the second column to the right of the decimal point?
   a. Tenths
   b. Tens
   c. Units
   d. Hundredths

5. The Greek letter pi represents a constant value that describes how much bigger a circle's _____ is compared with its _____, and it is frequently used in many formulas involved with the analysis of circular motion.
   a. Circumference, radius
   b. Circumference, diameter
   c. Diameter, radius
   d. Diameter, circumference

6. _____ is the opposite arithmetic operation of _____, and _____ is the opposite arithmetic operation of _____.
   a. Multiplication, division, subtraction, addition
   b. Division, addition, multiplication, subtraction
   c. $+, \div, -, \times$
   d. $\times, -, \div, +$

7. Is π a ratio?
   a. Yes
   b. No

8. Round off the following number to the nearest hundredth: 74.8552.
   a. 74.85
   b. 74.86
   c. 74.84
   d. 74.855

9. The number 27.0003 is a _____ significant-place number and is accurate to _____ significant places.
   a. 6, 2
   b. 2, 6
   c. 6, 6
   d. 4, 6

10. Express 29.63% mathematically as a decimal.
    a. 29.63
    b. 2.963
    c. 0.02963
    d. 0.2963

Communication Skill Questions

11. Describe the base, positional weight, and reset and carry action of the decimal number system. (2–1)

12. How are whole numbers and fractions represented in decimal? (2–2)

13. Describe how written fractions are converted to decimal fractions and, conversely, how decimal fractions are converted to written fractions. (2–2)

14. How do you obtain the reciprocal of a number, and what is its relationship to the original number? (2–2)

15. Arbitrarily choose values, and describe the steps involved in: (2–2)
  a. Decimal addition
  b. Decimal subtraction
  c. Decimal multiplication
  d. Decimal division

16. What is a ratio, and how is it expressed? (2–3)

17. Briefly describe the rules for rounding off. (2–3)

18. What is a percentage? (2–3)

19. How is an average calculated? (2–3)

20. What is the approximate value of the constant pi (π), and what does it describe? (2–2)

Practice Problems

21. Indicate the positional weight of each of the following digits: 96,237.

22. Write each of the following values as a number.
  a. Two hundred and seventy-six
  b. Eight thousand and seven tens
  c. Twenty thousand, four tens, and nine units

23. Which columns will reset and carry when each of the following values is advanced or incremented by 1?
  a. 199
  b. 409
  c. 29,909
  d. 49,999

24. What are the individual digit weights of 2.7463?

25. Write each of the following values as a number.
  a. Two units and three-tenths
  b. Five-tenths and seven-thousandths
  c. Nine thousand, three tens, and four-hundredths

26. Indicate which zeros can be dropped in the following values (excluding the zero in the units place).
  a. 02.017
  b. 375.011020
  c. 0.0102000

27. Convert the following written fractions to their decimal equivalents.
  a. $\dfrac{16}{32}$
  b. $3\dfrac{8}{9}$
  c. $4\dfrac{9}{8}$
  d. $195\dfrac{7}{3}$

28. Calculate the reciprocal of the following values.
  a. π
  b. 0.707
  c. 0.3183

29. Convert the following decimal fractions to written fractions.
  a. $0.777 = \dfrac{?}{9}$
  b. $0.6149069 = \dfrac{?}{161}$
  c. $43.125 = \dfrac{?}{8}$

30. Perform the following arithmetic operations by hand, and then confirm your answers with a calculator.
  a. $764 + 37 + 2199 = ?$
  b. $22{,}763 + 4 + 56{,}003 = ?$
  c. $451 + 19 + 17 = ?$
  d. $5555 + 3333 + 9999 = ?$
  e. $83.57 + 4 + 0.663 = ?$
  f. $898 - 457 = ?$
  g. $7{,}660{,}232 - 23{,}000 = ?$
  h. $4.9 - 3.7 = ?$
  i. $22 - 14.7 = ?$
  j. $2.01 - 1.1 = ?$
  k. $4 \times 16 = ?$
  l. $488 \times 14.0 = ?$
  m. $0.3 \times 99 = ?$
  n. $22{,}000 \times 14{,}000 = ?$
  o. $0.5 \times 10{,}000 = ?$
  p. $488 \div 4 = ?$
  q. $64.8 \div 8.1 = ?$
  r. $99{,}000 \div 10 = ?$
  s. $123 \div 17.5 = ?$
  t. $0.662 \div 0.002 = ?$

31. Express the ratio of the following in their lowest terms. (Remember to compare like quantities.)
  a. The ratio of 20 feet to 5 feet
  b. The ratio of $2\frac{1}{2}$ minutes to 30 seconds

32. Round off as indicated.
  a. 48.36 to the nearest whole number
  b. 156.3625 to the nearest tenth
  c. 0.9254 to the nearest hundredth

33. Light travels in a vacuum at a speed of 186,282.3970 miles per second. This _____-significant-place value is accurate to _____ significant places.

34. Calculate the following percentages.

    a. 23% of 50 = ?
    b. 78% of 10 = ?
    c. 3% of 1.5 = ?
    d. 20% of 3300 = ?
    e. 5% of 10,000 = ?

35. Calculate the average value of the following groups of values (include units).

    a. 4, 5, 6, 7, and 8
    b. 15 seconds, 20 seconds, 18 seconds, and 16 seconds
    c. 110 volts, 115 volts, 121 volts, 117 volts, 128 volts, and 105 volts
    d. 33 ohms, 17 ohms, 1000 ohms, and 973 ohms
    e. 150 meters, 160 meters, and 155 meters

Web Site Questions

Go to the web site http://www.prenhall.com/cook, select the textbook *Introductory Computer Mathematics, 2e,* select this chapter, and then follow the instructions when answering the multiple-choice practice problems.

Positive and Negative Numbers

Blaise of Genius

Blaise Pascal, the son of a regional tax official, was born in France in 1623. His father realized early that his son was a genius when his son's understanding of mathematics at the age of 8 exceeded his own. By the age of 16, Pascal had published several mathematical papers on conic sections and hydraulics. While still a youth, he invented, among other things, the syringe, the hydraulic press, and the first public bus system in Paris.

At the age of 19 he began working on an adding machine in an attempt to reduce the computational drudgery of his father's job. Pascal's machine, which became known as the *Pascaline,* became the rage of European nobility and is recognized today as one of the first computers. The operator loaded a number into a machine by setting a series of wheels. Each wheel, marked with the digits zero through nine, stood for each of the decimal columns: 1s, 10s, 100s, and so on. A carry was achieved when a total greater than nine caused a complete revolution of a wheel, which advanced the wheel to the left by one digit.

Although widely praised, the Pascaline, whose adding principle of operation would remain the standard for the next 300 years, did not make Pascal rich. Like many geniuses, he was a troubled person whose ideas and problems would plague him until he could find solutions. At the age of 37, he dropped out of society to join a religious monastery that refrained from all scientific pursuits. He died just two years later at the age of 39.

In his lifetime he became famous as a mathematician, physicist, writer, and philosopher, and in honor of his achievements one of today's computer programming languages has been named after him—Pascal.

Outline and Objectives

Introduction

Until now we have been dealing only with **positive numbers.** In fact, all the values we have discussed so far have been positive. For example, even though 12 does not have a positive (+) sign in front of it, it is still a positive number and therefore could be written as +12. We do not normally include the positive sign in front of a positive value unless we have other values that are negative. In instances when a **negative number** is present, we include the positive sign (+) in front of positive numbers to distinguish them from negative numbers that have the negative sign (−) in front of them. For example, if a list contained both positive and negative values, we would have to include their signs, as follows:

$$-4 \qquad +12 \qquad -6 \qquad +156 \qquad -198,765 \qquad +1,000,000$$

On the other hand, if we were to list only the positive values, there would be no need to include the positive sign because all values are positive.

$$12 \qquad 156 \qquad 1,000,000 \qquad \text{or} \qquad +12 \qquad +156 \qquad +1,000,000$$

By definition, therefore, *a positive number is any value that is greater than zero, whereas a negative number is any value that is less than zero,* as shown in Figure 3-1.

The fractions discussed so far, which are pieces of a whole, are less than 1 and can be positive values or negative values. For example, some of the following fractions are positive (have a value between zero and positive one), and some are negative (have a value between zero and negative one).

$$-\frac{1}{2} \qquad +0.5 \qquad +\frac{1}{8} \qquad -0.0016$$

The next important question is: *Why do we need both positive and negative numbers? The answer is: to indicate values that are above and below a reference point.* For example, consider the thermometer shown in Figure 3-2, which measures temperature in degrees Celsius (symbolized °C). This scale is used to indicate the amount of heat. On this scale, a reference point of 0 °C (zero degrees Celsius) was assigned to the temperature at which water freezes or ice melts, and this is an easy point to remember. For example, let us assume that the ice cubes inside your freezer are at a temperature of −15 °C (negative 15 degrees Celsius), and you remove several ice cubes and place them in a glass on the kitchen counter. At the warmer room temperature the ice cubes will absorb heat, and their temperature value will increase, or become less negative, as follows: −15 °C, −12 °C, −10 °C, −6 °C, and so on.

As we cross this temperature scale's reference point of 0 °C, the ice cubes change from a solid (ice cubes) to a liquid (water). If we then pour the water into a kettle and heat it even further, its temperature value will increase or become more positive, as follows: +5 °C, +12 °C, +22 °C, and so on.

Negative and positive numbers therefore are used to designate some quantity or value relative to, or compared with, a zero reference point. Temperature is not the only application for these numbers. For example, positive values can be used to express money that you have, whereas negative values can be used to express money that you owe, or have paid out. Similarly, a zero reference point can be used to indicate the average rainfall or snowfall, with positive values indicating the amount of rain or snow above the average, and negative values indicating the amount of rain or snow below the average.

In Chapter 3 we will discuss how to add, subtract, multiply, and divide positive and negative numbers, or signed numbers (numbers with signs).

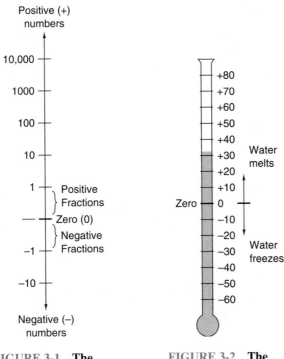

FIGURE 3-1   The Positive and Negative Number Line.

FIGURE 3-2   The Celsius Temperature Scale.

3-1   EXPRESSING POSITIVE AND NEGATIVE NUMBERS

Let us first discuss how we will mathematically express an addition, subtraction, multiplication, or division involving positive and negative numbers. As an example, let us assume that the rainfall over the last 2 years has been 3 and 9 graduations (marks on the scale) above the average reference point of zero. Because these two values of 3 and 9 are above the zero reference, they are positive numbers and can therefore be written as

$$3 \text{ and } 9 \quad \text{or} \quad +3 \text{ and } +9$$

To calculate the total amount of rainfall over the last 2 years, therefore, we must add these two values, as follows:

$$3 + 9 = 12 \quad \text{or} \quad +3 + (+9) = +12$$

If each graduation or mark on the scale was equivalent to 1 inch, over the past 2 years we would have had 12 inches of rain ($+12$) above the average of zero. Both preceding statements or expressions are identical, although the positive signs precede the values or numbers in the statement on the right. You can also see how the parentheses were used to separate the *plus sign* (indicating mathematical addition) from the *positive sign* (indicating that 9 is a positive number). If the parentheses had not been included, you might have found the statement confusing, because it would have appeared as follows:

$$+3 + +9 = +12$$

Plus sign ⌐   ⌐ Positive sign

The parentheses are included to isolate the sign of a number from a mathematical operation sign and therefore to prevent confusion.

As another example, let us imagine that last year the rainfall was 7 inches or graduations above the average, and the year before that there was a bit of a drought and the rainfall was 5 graduations or inches below the average. The total amount of rainfall for the last two years was therefore

$$+7 + (-5) = +2$$

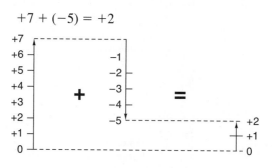

← Expression or meaningful combination of symbols

← Graphic representation of expression

Once again, notice how the parentheses were used to separate the addition symbol (plus sign) from the sign of the number (negative 5). This example is probably best understood by referring to the graphic representation below the expression, which shows how the −5 arrow cancels 5 graduations of the +7 arrow, resulting in a sum of 2 positive graduations (+2).

Before addressing the details of adding positive and negative numbers, let us discuss the *positive/negative* or *change-sign key* available on most calculators and used to change the sign of an inputted number.

CALCULATOR KEYS

Name: Change-sign, positive/negative, or negation key

Function: Generally used after a number has been entered to change the sign of the number. Numbers with no sign indicated are always positive. Normally you press the negation key $\boxed{(-)}$ before you enter the value.

Example: $\boxed{+/-}$ +7 + (−5) = ?

Press keys: $\boxed{7}\ \boxed{+}\ \boxed{5}\ \boxed{+/-}\ \boxed{=}$

Answer: 2 or +2

Example: $\boxed{(-)}$ −3 + (−15) = ?

Press keys: $\boxed{(-)}\boxed{3}\ \boxed{+}\ \boxed{(-)}\boxed{1}\boxed{5}\ \boxed{=}$

Answer: −18

Note: If the subtraction key $\boxed{-}$ is used in place of the negation key $\boxed{(-)}$, or vice versa, an error occurs.

SELF-TEST EVALUATION POINT FOR SECTION 3-1

Now that you have completed this section, you should be able to:

■ ***Objective 1.*** *Define the difference between a positive number and a negative number.*

■ ***Objective 2.*** *Describe some applications in which positive and negative numbers are used.*

■ ***Objective 3.*** *Explain how to express positive and negative numbers.*

Use the following questions to test your understanding of Section 3-1.

1. Use parentheses to separate the arithmetic operation sign from the sign of the number, and then determine the result.

    a. $+3 - -4 =$
    b. $12 \times +4 =$
    c. $-5 \div -7 =$
    d. $-0.63 \times +6.4 =$

2. Express the following statements mathematically.

    a. Positive six plus negative seven
    b. Negative zero point seven five multiplied by eighteen

3. Write the calculator key sequences to input the following problems.

    a. $-7.5 + (-4.6) =$
    b. $+5 \times (-2) =$
    c. $+316.629 \div (-1.44) =$

3-2   ADDING POSITIVE AND NEGATIVE NUMBERS

To begin, we will review how to add one positive number ($+$) to another positive number ($+$).

RULE 1: $(+) + (+) = (+)$

Description: A positive number ($+$) plus a positive number ($+$) equals a positive number ($+$).

Example:

$$5 + 15 = ?$$
$$+5 + (+15) = +20$$

Explanation:

A positive 5-unit arrow plus a positive 15-unit arrow equals a positive 20-unit arrow.

Calculator sequence: $\boxed{5}\boxed{+}\boxed{1}\boxed{5}\boxed{=}$
Answer: $20\ (+20)$

Other Examples:

$$+9 + 6 = ?$$
$$+9 + (+6) = +15$$
$$81 + +16 = ?$$
$$+81 + (+16) = +97$$
$$+(15) + +15 = ?$$
$$+15 + (+15) = +30$$

The preceding box helps to collect all the facts about this rule of addition. As you know, this basic rule of addition was covered in Chapter 2.

RULE 2: $(-) + (+)$ or $(+) + (-) =$ **(sign of larger number) and difference**

Description: A negative number ($-$) plus a positive number ($+$) or a positive number ($+$) plus a negative number ($-$) equals the sign of the larger number and the difference.

Two Examples:

$$-8 + (+15) = ? \qquad \text{or} \qquad +15 + (-8) = ?$$
$$-8 + (+15) = +7 \qquad\qquad\qquad +15 + (-8) = +7$$

Because 15
is the larger
number, sign
is positive.

Difference between
8 and 15 is 7.

Explanation:

A negative 8-unit arrow plus a
positive 15-unit arrow equals a
positive 7-unit arrow.

Calculator sequence: $\boxed{8}\ \boxed{+/-}\ \boxed{+}\ \boxed{1}\ \boxed{5}\ \boxed{=}$

Answer: $\qquad\qquad$ 7 (+7)

Other Examples:

$$6 + (-3) = ?$$
$$+6 + (-3) = +3 \qquad \text{(larger number is } +, \text{difference} = 3)$$
$$-112 + +59 = ?$$
$$-112 + (+59) = -53 \qquad \text{(larger number is } -, \text{difference} = 53)$$
$$-18 + +12 = ?$$
$$-18 + (+12) = -6 \qquad \text{(larger number is } -, \text{difference} = 6)$$

RULE 3: $(-) + (-) = (-)$

Description: A negative number $(-)$ plus a negative number $(-)$ equals a negative
number $(-)$.

Example:

$$-16 + (-4) = ?$$
$$-16 + (-4) = -20$$

Explanation:

A negative 16-unit arrow plus a
negative 4-unit arrow equals a
negative 20-unit arrow.

Calculator sequence: $\boxed{1}\boxed{6}\boxed{+/_-}\boxed{+}\boxed{4}\boxed{+/_-}\boxed{=}$

Answer: -20

Other Examples:

$$-15 + -25 = ?$$
$$-15 + (-25) = -40$$
$$(-1063) + (-1000) = ?$$
$$-1063 + (-1000) = -2063$$

SELF-TEST EVALUATION POINT FOR SECTION 3-2

Now that you have completed this section, you should be able to:

■ *Objective 4. Describe the three basic rules of positive and negative numbers as they relate to addition.*

Use the following questions to test your understanding of Section 3-2.

Add the following values.

1. $+15 + (+4) = ?$
2. $-3 + (+45) = ?$
3. $+114 + (-111) = ?$
4. $-357 + (-74) = ?$
5. $17 + 15 = ?$
6. $-8 + (-177) = ?$
7. $+4600 + (-3400) = ?$
8. $-6.25 + (+0.34) = ?$

3-3 SUBTRACTING POSITIVE AND NEGATIVE NUMBERS

In Chapter 2 we performed a basic subtraction in the following way:

$$19 - 12 = ?$$

$$\begin{array}{r} 19 \\ -12 \\ \hline 07 \end{array}$$

Steps:
$9 - 2 = 7.$
$1 - 1 = 0.$

Therefore, $19 - 12 = 7$.

If we wanted to, however, we could perform the same subtraction simply by changing the sign of the second number and then adding the two numbers. Let us apply this to the previous example of $19 - 12$ and see if it works.

$19 - 12 = ?$ ← Original example

$+19 - (+12) = ?$ ← Values written as signed numbers

Step 1: Change sign of second number.

Step 2: Add the two numbers.

$+19 + (-12) = +7$ ← Same result is achieved as performing the operation $19 - 12$.

Previous rule of addition
$(+) + (-) = $ (sign of larger number) and difference

In summary, therefore, *to subtract one number from another, we can change the sign of the second number and then add the two numbers together.* If you apply this "general rule of subtraction" to all the following positive and negative number subtraction examples, you will find that it simplifies the operation because all you have to do is change the sign of the second number and then apply the rules of addition previously discussed.

RULE 1: (+) − (+) or (−) − (−) = (sign of larger number) and difference

Description: A positive number (+) minus a positive number (+), or a negative number (−) minus a negative number (−), equals the sign of the larger number and the difference.

Two Examples:

$$+15 - (+8) = ? \qquad\qquad -10 - (-2) = ?$$
$$+15 + (-8) = ? \qquad\qquad -10 + (+2) = ?$$

Apply rule of subtraction:
Step 1: Change sign of second number.
Step 2: Add the two numbers.

$$+15 + (-8) = +7 \qquad\qquad -10 + (+2) = -8$$

Apply rule of addition:
(+) + (−) or (−) + (+) =
(sign of larger number)
and difference

$\Rightarrow 10+$

Explanation:

Calculator sequence: | 1 | 5 | − | 8 | = | | 1 | 0 | +/− | − | 2 | +/− | = |

Answer: 7(+7)   −8

Other Examples:

$$-15 - (+8) = ?$$
$$-15 + (-8) = -23 \qquad \text{(Change sign of second number and then add.)}$$

$$-2 - (-10) = ?$$
$$-2 + (+10) = +8 \qquad \text{(Change sign of second number and then add.)}$$

$$-654 - (-601) = ?$$
$$-654 + (+601) = -53 \qquad \text{(Change sign of second number and then add.)}$$

RULE 2: $(+) - (-) = (+)$

Description: A positive number $(+)$ minus a negative number $(-)$ will always equal a positive number $(+)$.

Example:

$$+6 - (-5) = ?$$ ⟵ Apply rule of subtraction:
— *Step 1:* Change sign of second number.
— *Step 2:* Add two numbers.
$$+6 + (+5) = +11$$ Apply rule of addition:
$$(+) + (+) = +.$$

Explanation: As you can see from the preceding example, subtracting a negative number is the same as adding a positive number, so the result will always be positive. To explain this further, the minus sign is immediately followed by a negative sign, causing two reversals in direction, and therefore the two numbers are combined.

This *double negative* (minus sign followed by a negative sign) can also be explained by using an English language analogy. If you were to use the expression "I don't have no money," you would be using two negatives (don't and no) or reversals and therefore be saying the opposite. If you don't have "no money," you must have some money.

Calculator sequence: $\boxed{6}\ \boxed{-}\ \boxed{5}\ \boxed{+/_-}\ \boxed{=}$
Answer: 11 (+11)

Other Examples:

$$+63 - (-14) = ?$$ ⟵ (Change sign of second number and then add.)
$$+63 + (+14) = +77$$
$$1 - -2 = ?$$
$$+1 - (-2) = ?$$ ⟵ (Change sign of second number and then add.)
$$+1 + (+2) = +3$$
$$(18) - -14 = ?$$
$$+(18) - (-14) = ?$$ ⟵ (Change sign of second number and then add.)
$$+18 + (+14) = +32$$

RULE 3: $(-) - (+) = (-)$

Description: A negative number $(-)$ minus a positive number $(+)$ equals a negative number $(-)$.

Example:

$$-7 - (+4) = ?$$ ⟵ Apply rule of subtraction:
— *Step 1:* Change sign of second number.
— *Step 2:* Add two numbers.
$$-7 + (-4) = -11$$ ⟵ Apply rule of addition:
$$(-) + (-) = -.$$

Explanation: A negative number minus a positive number $[(-) - (+) = (-)]$ is the same as a negative number plus a negative number $[(-) + (-) = (-)]$. The result will always be negative because the arithmetic operation is designed to calculate the sum of two negatives.

Calculator sequence: [7] [⁺/₋] [−] [4] [=]

Answer: −11

Other Examples:

$$-63 - (+15) = ?$$
$$-63 + (-15) = -78 \qquad \longleftarrow \text{(Change sign of second number and then add.)}$$
$$-(3) - (+15) = ?$$
$$-3 - (+15) = ?$$
$$-3 + (-15) = -18 \qquad \longleftarrow \text{(Change sign of second number and then add.)}$$

SELF-TEST EVALUATION POINT FOR SECTION 3-3

Now that you have completed this section, you should be able to:

■ ***Objective 4.*** *Describe the three basic rules of positive and negative numbers as they relate to subtraction.*

Use the following questions to test your understanding of Section 3-3.

Subtract the following values.

1. $+18 - (+7) = ?$
2. $-3.4 - (-5.7) = ?$
3. $19,665 - (-5031) = ?$
4. $-8 - (+5) = ?$
5. $467 - 223 = ?$
6. $-331 - (-2.6) = ?$
7. $8 - (+25) = ?$
8. $-0.64 - (-0.04) = ?$

3-4   MULTIPLYING POSITIVE AND NEGATIVE NUMBERS

Let us begin by reviewing how to multiply one positive number by another positive number. This basic rule of multiplication was covered in Chapter 2.

RULE 1: $(+) \times (+) = (+)$

Description: A positive number $(+)$ multiplied by a positive number $(+)$ will always yield a positive product or result $(+)$.

Example:

$$+6 \times (+3) = ? \qquad \leftarrow \text{(same as } 6 \times 3 = ?)$$
$$+6 \times (+3) = +18$$

Explanation: Multiplication is simply repeated addition, so adding a series of positive numbers will always produce a positive result.

Six positive 3s

Six positive 3-unit arrows equals a positive 18-unit arrow.

Calculator sequence: $\boxed{6}\;\boxed{\times}\;\boxed{3}\;\boxed{=}$
Answer: 18 (+18)

Other Examples:

$$24 \times 2 = ?$$
$$+24 \times (+2) = +48$$
$$+3 \times +7 = ?$$
$$+3 \times (+7) = +21$$
$$+26{,}397 \times +31 = ?$$
$$+26{,}397 \times (+31) = +818{,}307$$

RULE 2: $(+) \times (-)$ or $(-) \times (+) = (-)$

Description: A positive number (+) multiplied by a negative number (−), or a negative number (−) multiplied by a positive number (+), will always yield a negative result (−).

Example:

$$+6 \times (-4) = ? \qquad \text{or} \qquad -4 \times (+6) = ?$$
$$+6 \times (-4) = -24 \qquad\qquad\quad -4 \times (+6) = -24$$

Explanation: The result will always be negative because multiplication is simply repeated addition, and therefore adding a series of negative numbers will always result in a negative answer.

Calculator sequence: $\boxed{6}\;\boxed{\times}\;\boxed{4}\;\boxed{^+/_-}\;\boxed{=}$ or $\boxed{4}\;\boxed{^+/_-}\;\boxed{\times}\;\boxed{6}\;\boxed{=}$
Answer: −24   −24

Other Examples:

$$18 \times (-2) = ?$$
$$+18 \times (-2) = -36$$
$$-4 \times +3 = ?$$
$$-4 \times (+3) = -12$$
$$+21.3 \times -0.2 = ?$$
$$+21.3 \times (-0.2) = -4.26$$

RULE 3: $(-) \times (-) = (+)$

Description: A negative number (−) multiplied by a negative number (−) equals a positive number (+).

Example:

$$-3 \times (-5) = ?$$
$$-3 \times (-5) = +15$$

Explanation: The explanation for this is once again the double negative. After a series of negative numbers are added repeatedly, the negative answer is reversed to a positive. In summary, the reversals of direction in the operation coupled with the two negative numbers reverse the result to a positive value.

Calculator sequence:

Answer: 15 (+15)

Other Examples:

$$-30 \times (-2) = ?$$
$$-30 \times (-2) = +60$$
$$-(2) \times -3 = ?$$
$$-2 \times (-3) = +6$$
$$-76.44 \times -1.3 = ?$$
$$-76.44 \times (-1.3) = +99.37$$

Now that you have completed this section, you should be able to:

■ **Objective 4.** *Describe the three basic rules of positive and negative numbers as they relate to multiplication.*

Use the following questions to test your understanding of Section 3-4.

Multiply the following values.

1. $4 \times (+3) = ?$
2. $+17 \times (-2) = ?$
3. $-8 \times 16 = ?$
4. $-8 \times (-5) = ?$
5. $+12.6 \times 15 = ?$
6. $-3.3 \times +1.4 = ?$
7. $0.3 \times (-4) = ?$
8. $-4.6 \times -3.3 = ?$

3-5   DIVIDING POSITIVE AND NEGATIVE NUMBERS

Let us begin by reviewing how to divide one positive number by another positive number. This basic rule of division was covered in Chapter 2.

RULE 1: $(+) \div (+) = (+)$

Description: A positive number $(+)$ divided by a positive number $(+)$ will always result in a positive number $(+)$.

Example:

$$+12 \div (+3) = ? \quad \text{(same as } 12 \div 3\text{)}$$

or

$$\frac{+12}{+3} = ?$$

$$+12 \div (+3) = +4$$

Explanation: Because division is simply repeated subtraction, subtracting a series of positive numbers from a positive number will yield a positive answer.

Calculator sequence: $\boxed{1}\boxed{2}\boxed{\div}\boxed{3}\boxed{=}$

Answer: 4 (+4)

Other Examples:

$$+36 \div (+2) = ?$$
$$+36 \div (+2) = +18$$
$$\frac{+504}{(+8)} = ?$$
$$+504 \div (+8) = +63$$
$$+22.7 \div +3.6 = ?$$
$$+22.7 \div (+3.6) = +6.31$$

RULE 2: $(+) \div (-)$ or $(-) \div (+) = (-)$

Description: A positive number $(+)$ divided by a negative number $(-)$ or a negative number $(-)$ divided by a positive number $(+)$ will always yield a negative result $(-)$.

Example:

Dividend is larger than divisor:

$$+16 \div (-4) = ? \qquad \text{or} \qquad -16 \div (+4) = ?$$
$$+16 \div (-4) = -4 \qquad \qquad -16 \div (+4) = -4$$

Dividends (16) are larger than divisors (4).

Dividend is smaller than divisor:

$$+4 \div (-16) = -0.25 \qquad \text{or} \qquad -4 \div (+16) = -0.25$$

Even if dividends (4) are smaller than divisors (16), the answer is still negative.

Explanation: The result will always be negative because division is simply repeated subtraction, and therefore subtracting a series of negative numbers will always result in a negative answer.

Calculator sequence:

(Dividend is larger) → $\boxed{1}\,\boxed{6}\,\boxed{\div}\,\boxed{4}\,\boxed{+/-}\,\boxed{=}$ or $\boxed{1}\,\boxed{6}\,\boxed{+/-}\,\boxed{\div}\,\boxed{4}\,\boxed{=}$

Answer: -4 $\qquad\qquad\qquad\qquad\qquad -4$

(Divisor is larger) → $\boxed{4}\,\boxed{\div}\,\boxed{1}\,\boxed{6}\,\boxed{+/-}\,\boxed{=}$ or $\boxed{4}\,\boxed{+/-}\,\boxed{\div}\,\boxed{1}\,\boxed{6}\,\boxed{=}$

Answer: -0.25 $\qquad\qquad\qquad\quad -0.25$

Other Examples:

$$+48 \div (-4) = ?$$
$$+48 \div (-4) = -12$$
$$-4 \div +12 = ?$$
$$-4 \div (+12) = -0.333$$
$$\frac{+677.25}{-3.66} = ?$$
$$+677.25 \div (-3.66) = -185.04$$

RULE 3: $(-) \div (-) = (+)$

Description: A negative number $(-)$ divided by a negative number $(-)$ equals a positive number $(+)$.

Example:

$$-15 \div (-5) = ?$$
$$-15 \div (-5) = +3$$

Explanation: The explanation is once again the double negative. After a series of negative numbers are repeatedly subtracted, the negative answer is reversed to a positive. In summary, the reversals of direction in the operation coupled with the two negative numbers reverse the result to a positive value.

Calculator sequence: $\boxed{1}\,\boxed{5}\,\boxed{^+\!/\!_-}\,\boxed{\div}\,\boxed{5}\,\boxed{^+\!/\!_-}\,\boxed{=}$
Answer: 3 (+3)

Other Examples:

$$-4 \div (-2) = ?$$
$$-4 \div (-2) = +2$$
$$\frac{-96}{-3} = ?$$
$$-96 \div (-3) = +32$$
$$-2.2 \div -11.0 = ?$$
$$-2.2 \div (-11.0) = +0.2$$

SELF-TEST EVALUATION POINT FOR SECTION 3-5

Now that you have completed this section, you should be able to:

■ **Objective 4.**   *Describe the three basic rules of positive and negative numbers as they relate to division.*

Use the following questions to test your understanding of Section 3-5.

Divide the following values.

1. $\dfrac{+16.7}{+2.3} = ?$

2. $18 \div (+6) = ?$

3. $-6 \div (+2) = ?$

4. $+18 \div (-4) = ?$

5. $2 \div (-8) = ?$

6. $-8 \div +5 = ?$

7. $\dfrac{-15}{-5} = ?$

8. $0.664 \div (-0.2) = ?$

3-6   ORDER OF OPERATIONS

For easy access, Figure 3-3 summarizes all the rules for signed number addition, subtraction, multiplication, and division. Remember that it is more important to understand these rules rather than just to memorize them. Returning to the detailed in-chapter explanation for each of the end-of-chapter problems is not a waste of time but an investment, because you will be developing logic and reasoning skills that can be applied to all aspects of science and technology.

Addition	Subtraction
(+) + (+) = (+)	(+) − (+) or (−) − (−) = (sign of larger number) and difference
(−) + (+) or (+) + (−) = (sign of larger number) and difference	
(−) + (−) = (−)	(+) − (−) = (+)
	(−) − (+) = (−)

Multiplication	Division
(+) × (+) = (+)	(+) ÷ (+) = (+)
(+) × (−) or (−) × (+) = (−)	(+) ÷ (−) or (−) ÷ (+) = (−)
(−) × (−) = (+)	(−) ÷ (−) = (+)

FIGURE 3-3   Positive and Negative Number Rules of Addition, Subtraction, Multiplication, and Division.

When a string of more than two positive and negative numbers has to be combined by some arithmetic operation, the order in which the mathematical operations should be combined is as follows:

ORDER OF OPERATIONS	MEMORY AID
First: **P**arentheses	**Pl**ease
Second: **E**xponents	**E**xcuse
Third: **M**ultiplication	**M**y
Fourth: **D**ivision	**D**ear
Fifth: **A**ddition	**A**unt
Sixth: **S**ubtraction	**S**ally

This **order of operations** can more easily be remembered with the phrase "Please Excuse My Dear Aunt Sally."

■ **EXAMPLE:**

Calculate the following:
- **a.** $3 + 6 \cdot 4 = ?$
- **b.** $4 (4 + 3) = ?$
- **c.** $8^2 - 5 = ?$

■ *Solution:*

a. Problem has multiplication and addition operations. According to the order of operations (PEMDAS), multiplication should be performed first and then addition second.

$$3 + 6 \cdot 4 = ?$$
$$3 + 24 = \qquad (6 \cdot 4 = 24)$$
$$27 \qquad (3 + 24 = 27)$$

· = X Multiply

b. $4 (4 + 3) = ?$ Parentheses operation should be performed first, multiplication second.

$$4(4 + 3) = ?$$
$$4 (7) =$$
$$4 \cdot 7 =$$
$$28$$

c. $8^2 - 5 = ?$ Exponents first, subtraction second.

$$8^2 - 5 = ?$$
$$64 - 5 =$$
$$59$$

CALCULATOR KEYS

Name: Order of evaluation key

Function: Most calculators define the order in which functions in expressions are entered and evaluated. This order is as follows:

CALCULATOR KEYS (continued)

ORDER NUMBER	FUNCTION
1	Functions that precede the argument, such as $\sqrt{\ }$ (, **sin**(, or **log**(
2	Functions that are entered after the argument, such as ^{2}, $^{-1}$, ^{o}, ^{r}, and conversions
3	Powers and roots, such as **2^5** or $\mathbf{5^x\sqrt{32}}$
4	Permutations (**nPr**) and combinations (**nCr**)
5	Multiplication, implied multiplication, and division
6	Addition and subtraction
7	Relational functions, such as > or ≤
8	Logic operator **and**
9	Logic operators **or** and **xor**

Example: Calculate $3.76 \div (-7.9 + \sqrt{5}) + 2 \log 45$.

Press keys: ③ ． ⑦⑥ ÷ ((-) ⑦ ． ⑨ + [2nd] [√]
⑤) + ② × [LOG] (④⑤)
[ENTER]

```
3.76/(-7.9+√ 5)
+2log(45)
         2.642575252
```

SELF-TEST EVALUATION POINT FOR SECTION 3-6

Now that you have completed this section, you should be able to:

■ **Objective 5.** *Explain how to combine strings of positive and negative numbers.*

Use the following questions to test your understanding of Section 3-6.

1. $+6 + (+3) + (-7) + (-5) = ?$

2. $+9 - (+2) - (-13) - (-4) = ?$

3. $-6 \times (-4) \times (-5) = ?$

4. $+8 \div (+2) \div (-5) = ?$

or $\dfrac{\left(\dfrac{+8}{+2}\right)}{(-5)} = ?$

5. $-4 \div (-2) \times (+8) = ?$ or $\dfrac{-4}{-2} \times (+8) = ?$

6. $-9 + (+5) - (-7) = ?$

REVIEW QUESTIONS

Multiple Choice Questions

1. A _____ number is any value greater than zero, whereas a _____ number is any value less than zero.
  a. Positive, +   **c.** +, negative
  b. −, +   **d.** −, −

2. The fraction 0.3417 is a positive number.
  a. True   **b.** False

3. Negative and positive numbers are used to designate some value relative to a _____ reference point.
  a. 0   **c.** 10 (decimal)
  b. 0.5   **d.** π

4. Which of the following calculator sequences would be used to input the following expression? $+5 - (-15) =$
  a. ⑤ +/− − ① ⑤ =
  b. +/− ⑤ − − ① ⑤ =
  c. ⑤ − − ① ⑤ =
  d. ⑤ − ① ⑤ +/− =

5. $(+) + (+) = ?$
  a. $(+)$
  b. $(-)$
  c. (Sign of larger number) and difference
  d. None of the above

6. $(+) + (-) = ?$

    a. $(+)$
    b. $(-)$
    c. (Sign of larger number) and difference
    d. None of the above

7. $(-) + (-) = ?$

    a. $(+)$
    b. $(-)$
    c. (Sign of larger number) and difference
    d. None of the above

8. $(+) - (-)$ is equivalent to?

    a. $(-) + (-)$ **c.** $(+) + (+)$
    b. $(-) + (+)$ **d.** $(+) + (-)$

9. $(-) - (-)$ is equivalent to?

    a. $(-) + (+)$ **c.** $(-) + (-)$
    b. $(+) + (+)$ **d.** $(+) + (-)$

10. $(-) - (+)$ is equivalent to?

    a. $(-) + (+)$ **c.** $(+) + (+)$
    b. $(-) + (-)$ **d.** $(+) + (-)$

Communication Skill Questions

11. Define the following: (Introduction)

    a. Positive number   **c.** Positive fraction
    b. Negative number   **d.** Negative fraction

12. What method is used to show whether the number is positive or negative? (3-1)

13. Describe the three rules of adding positive and negative numbers. (3-2)

14. How can you achieve subtraction through addition? (3-3)

15. Describe the three rules of subtracting positive and negative numbers. (3-3)

16. What are the three rules of multiplying positive and negative numbers? (3-4)

17. List and describe the three rules of positive and negative number division. (3-5)

18. Arbitrarily choose values, and then describe the following: (3-2)

    a. $(+) + (-) = ?$
    b. $(-) + (-) = ?$
    c. $(+) + (+) = ?$

19. Arbitrarily choose values, and then describe the following: (3-3)

    a. $(+) - (+) = ?$
    b. $(-) - (+) = ?$
    c. $(+) - (-) = ?$

20. Arbitrarily choose values, and then describe the following: (3-4 and 3-5)

    a. $(+) \times (-) = ?$ **c.** $(+) \div (-) = ?$
    b. $(-) \times (-) = ?$ **d.** $(-) \div (-) = ?$

Practice Problems

21. Express the following statements mathematically.

    a. Positive eight minus negative five
    b. Negative zero point six times thirteen
    c. Negative twenty-two point three divided by negative seventeen
    d. Positive four divided by negative nine

22. Add the following values.

    a. $+8 + (+9) = ?$ **c.** $-6 + (+9) = ?$
    b. $+9 + (-6) = ?$ **d.** $-8 + (-9) = ?$

23. Subtract the following values.

    a. $+6 + (+8) = ?$ **c.** $-75 - (+62) = ?$
    b. $+9 - (-6) = ?$ **d.** $-39 + (-112) = ?$

24. Multiply the following values.

    a. $+18 \times (+4) = ?$ **c.** $-20 \times (+3) = ?$
    b. $+8 \times (-2) = ?$ **d.** $-16 \times (-10) = ?$

25. Divide the following values.

    a. $+19 \div (+3) = ?$ **c.** $-80 \div (+5) = ?$
    b. $+36 \div (-3) = ?$ **d.** $-44 \div (-2) = ?$

Calculate the answers to the following problems.

26. $+0.75 + (-0.25) - (+0.25) = ?$

27. $\left(\dfrac{+15}{+5}\right) \times (-3.5) = +15 \div (+5)$
    $\times (-3.5) = ?$

28. $\dfrac{+2 \times (-4) \times (-7)}{+3} = +2 \times (-4) \times (-7)$
    $\div (+3) = ?$

29. $-6 \div (-4) \div (-3) \times (-15)$
    $= \dfrac{\left(\dfrac{-6}{-4}\right)}{-3} \times (-15) = ?$

30. $[+15 - (-4) - (+5) - (-10)] \div 4 =$
    $\dfrac{+15 - (-4) - (+5) - (-10)}{4} = ?$

Web Site Questions

Go to the web site http://www.prenhall.com/cook, select the textbook *Introductory Computer Mathematics, 2e,* select this chapter, and then follow the instructions when answering the multiple-choice practice problems.

Exponents and the Metric System

Not a Morning Person

René Descartes was born in 1596 in Brittany, France, and at a very early age began to display an astonishing analytical genius. At the age of 8 he was sent to Jesuit College in La Flèche, then one of the most celebrated institutions in Europe, where he studied several subjects, none of them interesting him as much as mathematics. His genius impressed all his professors, who gave him permission, because of his delicate health, to study in bed until midday—a practice that he retained throughout his life. In 1612 he left La Flèche to go to the University of Poitiers, where he graduated in law in 1616, a profession he never practiced.

Wanting to see the world, he joined the army, which made use of his mathematical ability in military engineering. In 1619 he met Dutch philosopher Isaac Beeckman, who convinced Descartes to "turn his mind back to science and worthier occupations." After leaving the army, Descartes traveled to Neuberg, Germany, where he shut himself in a well-heated room for the winter. On the eve of St. Martin's Day (November 10, 1619), Descartes described a vivid dream that determined all his future endeavors. The dream clarified his purpose and showed him that physics and all sciences could be reduced to geometry, and therefore all were interconnected "as by a chain." From this point on, his genius was displayed in his invention of coordinate geometry and in contributions to theoretical physics, methodology, and metaphysics.

In his time he was heralded as an analytical genius, a reputation that has lasted to this day. His fame was so renowned that he was asked frequently to tutor royalty. When in Paris in 1649, Descartes was asked to tutor Queen Christina of Sweden. He did not want the opportunity to, in his words, "live in the land of bears among rock and ice, and lose my independence," but he was persuaded to do so by the French ambassador a month later. The queen chose five o'clock in the morning for her lessons, and on his travels one bitter morning Descartes caught a severe chill and died within two weeks.

Descartes's problem-solving ability was incredible. You may find his four-step method to solving a problem helpful:

1. Never accept anything as true unless it is clear and distinct enough to exclude all doubt from your mind.
2. Divide the problem into as many parts as necessary to reach a solution.
3. Start with the simplest things and proceed step by step toward the complex.
4. Review the solution so completely and generally that you are sure nothing was omitted.

Outline and Objectives

VIGNETTE: NOT A MORNING PERSON

INTRODUCTION

Objective 1: Define the term *exponent.*

4-1 RAISING A BASE NUMBER TO A HIGHER POWER

Objective 2: Describe what is meant by raising a number to a higher power.

4-1-1 Square of a Number

Objective 3: Explain how to find the square and root of a number.

4-1-2 Root of a Number

4-1-3 Powers and Roots in Combination

Objective 4: Calculate the result of problems with powers and roots in combination.

4-2 POWERS OF 10

Objective 5: Explain the powers-of-ten method and how to convert to powers of ten.

4-2-1 Converting to Powers of 10

4-2-2 Powers of 10 in Combination

Objective 6: Calculate the result of problems with powers of ten in combination.

4-2-3 Scientific and Engineering Notation

Objective 7: Describe the two following floating-point number systems:
   a. Scientific notation
   b. Engineering notation

4-3 THE METRIC SYSTEM

Objective 8: Define and explain the purpose of the metric system.

4-3-1 Metric Units and Prefixes

Objective 9: List the metric prefixes and describe the purpose of each.

4-3-2 Metric Units of Length

Objective 10: Describe the following metric units for:
   a. Length
   b. Area
   c. Weight
   d. Volume
   e. Temperature

Objective 11: Convert U.S. customary units to metric units, and metric units to U.S. customary units.

4-3-3 Metric Units of Area

4-3-4 Metric Units of Weight

4-3-5 Metric Units of Volume

4-3-6 Metric Unit of Temperature

4-3-7 Electrical Units, Prefixes, and Conversions

Objective 12: List and describe many of the more frequently used electrical units and prefixes, and describe how to interconvert prefixes.

MULTIPLE CHOICE QUESTIONS

COMMUNICATION SKILL QUESTIONS

PRACTICE PROBLEMS

WEB SITE QUESTIONS

Introduction

Like many terms used in mathematics, the word **exponent** sounds as if it will be complicated; however, once you find out that exponents are simply a sort of "math shorthand," the topic loses its intimidation. Many of the values used in science and technology contain numbers that have exponents. An exponent is a number in a smaller type size that appears to the right of and slightly higher than another number, for example:

$$2^2 \qquad 3^5 \qquad 10^6 \qquad 125^8$$

The number in the larger type size in these examples is called the *base number*.

All these numbers in the smaller type size are examples of an *exponent*.

However, what does a term like 2^3 mean? It means that the base 2 is to be used as a factor 3 times; therefore,

$$2^3 = 2 \times 2 \times 2 = 8$$

Similarly, 3^5 means that the base 3 is to be used as a factor 5 times; therefore,

$$3^5 = 3 \times 3 \times 3 \times 3 \times 3 = 243$$

As you can see, it is much easier to write 3^5 than to write $3 \times 3 \times 3 \times 3 \times 3$, and since both mean the same thing and equal the same amount (which is 243), exponents are a quick and easy math shorthand.

In this chapter we will examine the details relating to squares, roots, exponents, scientific and engineering notation, and prefixes.

4-1 RAISING A BASE NUMBER TO A HIGHER POWER

A base number's exponent indicates how many times the base number must be multiplied by itself. This is called *raising a number to a higher power*. For example, $5 \times 5 \times 5 \times 5$ can be written as 5^4, which indicates that the base number 5 is raised to the fourth power by the exponent 4. As another example, 3×3 can be written as 3^2, which indicates that the base number 3 is raised to the second power by the exponent 2. The second power is also called the **square** of the base number, and therefore 3^2 can be called "three squared" or "three to the second power."

■ **EXAMPLE:**

What does 10^3 mean?

■ *Solution:*

10^3 indicates that the base number 10 is raised to the third power by the exponent 3. Described another way, it means that the base 10 is to be used as a factor 3 times; therefore,

$$10^3 = 10 \times 10 \times 10 = 1000$$

Thus, instead of writing $10 \times 10 \times 10$, or 1000, you could simply write 10^3 (pronounced "ten to the three," "ten to the third power," or "ten cubed").

Because the square of a base number is used very frequently, let us begin by discussing raising a base number to the second power.

4-1-1   *Square of a Number*

The square of a base number means that the base number is to be multiplied by itself. For example, 4^2, which is pronounced "four squared" or "four to the second power," means 4×4. The squares of the first ten base numbers are used very frequently in numerical problems and are as follows:

$$
\begin{aligned}
0^2 &= 0 \times 0 = 0 \\
1^2 &= 1 \times 1 = 1 \\
2^2 &= 2 \times 2 = 4 \\
3^2 &= 3 \times 3 = 9 \\
4^2 &= 4 \times 4 = 16 \\
5^2 &= 5 \times 5 = 25 \\
6^2 &= 6 \times 6 = 36 \\
7^2 &= 7 \times 7 = 49 \\
8^2 &= 8 \times 8 = 64 \\
9^2 &= 9 \times 9 = 81 \\
10^2 &= 10 \times 10 = 100
\end{aligned}
$$

Many people get confused with the first three of the squares. Be careful to remember that $0^2 = 0$ because nothing $\times$ nothing equals nothing; $1^2 = 1$ because one times one is still one; and 2^2 means 2×2, not $2 + 2$, even though the answer works out both ways to be 4.

■ **EXAMPLE:**

Give the square of the following.
   a. 12^2 b. 7^2

■ *Solution:*

   a. $12^2 = 12 \times 12 = 144$
   b. $7^2 = 7 \times 7 = 49$

Finding the square of a base number is done so frequently in science and technology that most calculators have a special key just for that purpose. It is called the *square key* and operates as follows.

CALCULATOR KEYS

Name: Square key

Function: Calculates the square of the number in the display.

Example: $16^2 = ?$

Press keys: $\boxed{1}\;\boxed{6}\;\boxed{x^2}$

Display shows: 256

Most calculators also have a key for raising a base number of any value to any power. This is called the *y to the x power key,* and it operates as follows.

CALCULATOR KEYS

Name: *y* to the *x* power key

Function: Raises the displayed value *y* to the *x*th power.

Example: $5^4 = ?$

Press keys: $\boxed{5}\;\boxed{y^x}\;\boxed{4}\;\boxed{=}$

Display shows: 625

4-1-2   *Root of a Number*

What would we do if we had the result 64 and we didn't know the value of the number that was multiplied by itself to get 64? In other words, we wanted to *find the source or root number that was squared to give us the result.* This process, called **square root,** uses a special symbol called a *radical sign* and a smaller number called the *index.*

Square Root

A factor of a number that when squared gives the number.

NO # assume 2 → $\sqrt{64} = ?$

Smaller number is called the *index.* It indicates how many times a number was multiplied by itself to get the value shown inside the radical sign. A 2 index is called the *square root.*

↑ Radical sign ($\sqrt{}$) indicates that the value inside is the result of a multiplication of a number two or more times.

In this example the index is 2, which indicates that the number we are trying to find was multiplied by itself two times. Of course, in this example we already know that the answer is 8 because $8 \times 8 = 64$.

If squaring a number takes us forward ($8^2 = 8 \times 8 = 64$), taking the square root of a number must take us backward ($\sqrt[2]{64} = 8$). Almost nobody extracts the squares from square root problems by hand because calculators make this process more efficient. Most calculators have a special key just for determining the square root of a number. Called the *square root key,* it operates as follows.

CALCULATOR KEYS

Name: Square root key

Function: Calculates the square root of the number in the display.

Example: $\sqrt{81} = ?$

Press keys: $\boxed{8}\ \boxed{1}\ \boxed{\sqrt{x}}$

Display shows: 9

Some of the more frequently used *square root* values are as follows:

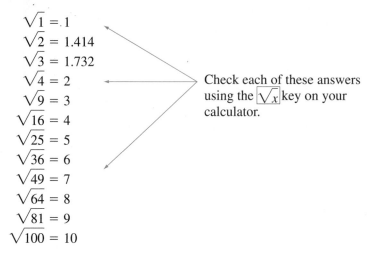

$$\sqrt{1} = .1$$
$$\sqrt{2} = 1.414$$
$$\sqrt{3} = 1.732$$
$$\sqrt{4} = 2$$
$$\sqrt{9} = 3$$
$$\sqrt{16} = 4$$
$$\sqrt{25} = 5$$
$$\sqrt{36} = 6$$
$$\sqrt{49} = 7$$
$$\sqrt{64} = 8$$
$$\sqrt{81} = 9$$
$$\sqrt{100} = 10$$

Check each of these answers using the $\boxed{\sqrt{x}}$ key on your calculator.

In some cases the index may be 3, meaning that some number was multiplied by itself three times to get the value within the radical sign; for example,

$$\sqrt[3]{125} = ?$$

Using the index 3 instead of the index 2 is called taking the **cube root.** The answer to this problem is 5 because 5 multiplied by itself three times equals 125.

$$\sqrt[3]{125} = 5$$

Cube Root

A factor of a number that when multiplied by itself three times gives the number.

Most calculators also have a key for calculating the root of any number with any index. It operates as follows.

CALCULATOR KEYS

Name: *x*th root of *y* key

Function: Calculate the *x*th root of the displayed value *y*.

Example: $\sqrt[3]{512} = ?$

Press keys: 5 1 2 $\sqrt[x]{y}$ 3 =

Display shows: 8

In most cases a radical sign will not have an index, in which case you can assume that the index is 2, or square root; for example,

$$\sqrt{16} = 4$$

Square root of sixteen = four

4-1-3   *Powers and Roots in Combination*

In some instances you may have several operations to perform before the result can be determined.

■ EXAMPLE:

What is the sum of $4^2 + 3^3$ (four squared plus three cubed)?

$$4^2 + 3^3 = ?$$

■ *Solution:*

The first step in a problem like this is to raise the base numbers to the power indicated by the exponent and then add the two values.

$4^2 + 3^3 =$
$16 + 27 = 43$

Steps:
$4^2 = 4 \times 4 = 16$
$3^3 = 3 \times 3 \times 3 = 27$
$16 + 27 = 43$
Calculator sequence: 4 x^2 + 3 y^x 3 =

Here are some other examples in which a mathematical operation needs to be performed on numbers with exponents to determine an answer or result.

EXAMPLE A:

$$(2 + 9)^2 = 11^2 = 121$$

Exponent 2 indicates that the sum within the parentheses is to be squared.

Parentheses are used to group term $2 + 9$.

Steps:
$2 + 9 = 11$
$11^2 = 11 \times 11 = 121$
Calculator sequence: $\boxed{2}\boxed{+}\boxed{9}\boxed{=}\boxed{x^2}$

[handwritten margin notes:] Same # √14 matching # cancel out Answer = 14

EXAMPLE B:

$$(8^2)^2 = 64^2 = 4096$$

Steps:
$8^2 = 8 \times 8 = 64$
$64^2 = 64 \times 64 = 4096$
Calculator sequence: $\boxed{8}\boxed{x^2}\boxed{x^2}$

EXAMPLE C:

$$(5^2 \times 3^3)^2 = (25 \times 27)^2$$
$$= 675^2 = 455,625$$

Steps:
$5^2 = 5 \times 5 = 25$
$3^3 = 3 \times 3 \times 3 = 27$
$25 \times 27 = 675, 675 \times 675 = 455,625$
Calculator sequence:
$\boxed{5}\boxed{x^2}\boxed{\times}\boxed{3}\boxed{y^x}\boxed{3}\boxed{=}\boxed{x^2}$

[handwritten margin notes:] to the Power y^x

EXAMPLE D:

$$(5^2 - 15)^3 = (25 - 15)^3$$
$$= 10^3 = 1000$$

Steps:
$5^2 = 5 \times 5 = 25$
$25 - 15 = 10$
$10^3 = 10 \times 10 \times 10 = 1000$
Calculator sequence:
$\boxed{5}\boxed{x^2}\boxed{-}\boxed{1}\boxed{5}\boxed{=}\boxed{y^x}\boxed{3}$

[handwritten margin notes:] Square Root $\sqrt[3]{1963}$ $9^{\frac{2}{x}}$ $y^x =$ to the Power

EXAMPLE E:

$$\left(\frac{4^2}{2}\right)^2 + 7^2 \text{ or } (4^2 \div 2)^2 + 7^2$$
$$= (16 \div 2)^2 + 7^2$$
$$= 8^2 + 7^2$$
$$= 64 + 49 = 113$$

Steps:
$4^2 = 4 \times 4 = 16$
$16 \div 2 = 8$
$8^2 = 8 \times 8 = 64$
$7^2 = 7 \times 7 = 49$
$64 + 49 = 113$
Calculator sequence: $\boxed{4}\boxed{x^2}\boxed{\div}\boxed{2}\boxed{=}$
$\boxed{x^2}\boxed{+}\boxed{7}\boxed{x^2}\boxed{=}$

Now let us examine a few examples involving powers and roots.

EXAMPLE A:

No index means square root.

$$\sqrt{3 + 7 + 6} = \sqrt{16} = 4$$

Radical sign extends over terms $3 + 7 + 6$.

Steps:
$3 + 7 + 6 = 16$
$\sqrt{16} = 4$

Calculator sequence: $\boxed{3}\boxed{+}\boxed{7}\boxed{+}\boxed{6}\boxed{=}\boxed{\sqrt{x}}$

EXAMPLE B:

$$\sqrt{14^2} = \sqrt{196} = 14$$

(Because square is the opposite of square root, the square root of fourteen squared $= 14$.)

Steps:
$14^2 = 14 \times 14 = 196$
$\sqrt{196} = 14$

Calculator sequence: $\boxed{1}\boxed{4}\boxed{x^2}\boxed{\sqrt{x}}$

EXAMPLE C:

$$\sqrt[2]{\frac{8}{9}} \text{ or } \sqrt[2]{8 \div 9}$$
$$= \sqrt{0.888} = 0.943$$

Steps:
$8 \div 9 = 0.889$
Calculator sequence: $\boxed{8}\boxed{\div}\boxed{9}\boxed{=}\boxed{\sqrt{x}}$

EXAMPLE D:

$$\sqrt[3]{5^2 \times 2^2} = \sqrt[3]{25 \times 4}$$
$$= \sqrt[3]{100} = 4.6416$$

Steps:
$5^2 = 25$
$2^2 = 4$
$25 \times 4 = 100$
$\sqrt[3]{100} = 4.6416$

Calculator sequence: $\boxed{5}\boxed{x^2}\boxed{\times}\boxed{2}\boxed{x^2}\boxed{=}\boxed{\sqrt[x]{y}}\boxed{3}\boxed{=}$

EXAMPLE E:

$$\sqrt{6^2 - 5} + \sqrt{121}$$
$$= \sqrt{36 - 5} + \sqrt{121}$$
$$= \sqrt{31} + \sqrt{121}$$
$$= 5.568 + 11 = 16.568$$

Steps:
$6^2 = 6 \times 6 = 36$
$36 - 5 = 31$
$\sqrt{31} = 5.568$
$\sqrt{121} = 11$
$5.568 + 11 = 16.568$

Calculator sequence: $\boxed{6}\boxed{x^2}\boxed{-}\boxed{5}\boxed{=}$

$\boxed{\sqrt{x}}\boxed{+}\boxed{1}\boxed{2}\boxed{1}\boxed{\sqrt{x}}\boxed{=}$

Now that you have completed this section, you should be able to:

- **Objective 1.** *Define the term* exponent.
- **Objective 2.** *Describe what is meant by raising a number to a higher power.*
- **Objective 3.** *Explain how to find the square and root of a number.*
- **Objective 4.** *Calculate the result of problems with powers and roots in combination.*

Use the following questions to test your understanding of Section 4-1.

1. Raise the following base numbers to the power indicated by the exponent, and give the answer.
   a. $16^4 = ?$
   b. $32^3 = ?$
   c. $112^2 = ?$
   d. $15^6 = ?$
   e. $2^3 = ?$
   f. $3^{12} = ?$

2. Give the following roots.
   a. $\sqrt[2]{144} = ?$
   b. $\sqrt[3]{3375} = ?$
   c. $\sqrt{20} = ?$
   d. $\sqrt[3]{9} = ?$

3. Calculate the following.
   a. $(9^2 + 14^2)^2 - \sqrt[3]{3 \times 7} = ?$
   b. $\sqrt{3^2 \div 2^2} + \dfrac{(151 \div 9^2)}{3.5^2} = ?$

(handwritten: $8/4 \; 196 = 76729$)

4-2   POWERS OF 10

(handwritten: $\sqrt[3]{125}$)

Many of the sciences deal with numbers that contain a large number of zeros, for example:

$$14{,}000$$
$$0.000032$$

(handwritten: Answer, what numbers X itself = 125)

By using exponents, we can eliminate the large number of zeros to obtain a shorthand version of the same number. This method is called *powers of 10.*

As an example, let us remove all the zeros from the number 14,000 until we are left with simply 14. However, this number (14) is not equal to the original number (14,000), and therefore simply removing the zeros is not an accurate shorthand. Another number needs to be written with the 14 to indicate what has been taken away—this is called a *multiplier.* The multiplier must indicate what you have to multiply 14 by to get back to 14,000; therefore,

$$14{,}000 = 14 \times 1000$$

As you know from our discussion on exponents, we can replace the 1000 with 10^3 because $1000 = 10 \times 10 \times 10$. Therefore, the powers-of-10 notation for 14,000 is 14×10^3. To convert 14×10^3 back to its original form (14,000), simply remember that each time a number is multiplied by 10, the decimal place is moved one position to the right. In this example 14 is multiplied by 10 three times (10^3, or $10 \times 10 \times 10$), and therefore the decimal point will have to be moved three positions to the right.

$$14 \times 10^3 = 14 \times 10 \times 10 \times 10 = 14{_0}0{_0}0{_0}0. = 14{,}000$$

As another example, what is the powers-of-10 notation for the number 0.000032? If we once again remove all the zeros to obtain the number 32, we will again have to include a multiplier with 32 to indicate what 32 has to be multiplied by to return it to its original form. In this case 32 will have to be multiplied by 1/1,000,000 (one millionth) to return it to 0.000032.

$$32 \times \frac{1}{1{,}000{,}000} = 0.000032$$

This can be verified because when you divide any number by 10, you move the decimal point one position to the left. Therefore, to divide any number by 1,000,000, you simply move the decimal point six positions to the left.

$$32 \times \frac{1}{1,000,000} = \frac{32}{1,000,000} = \frac{32}{10 \times 10 \times 10 \times 10 \times 10 \times 10} = 0.0000032$$

Once again an exponential expression can be used in place of the 1/1,000,000 multiplier, namely,

$$\frac{1}{1,000,000} = \frac{1}{10^6} = 0.000001 = 10^{-6}$$

Whenever you divide a number into 1, you get the *reciprocal* of that number. In this example, when you divide 1,000,000 into 1, you get 0.000001, which is equal to power-of-10 notation with a negative exponent of 10^{-6}. The multiplier 10^{-6} indicates that the decimal point must be moved back (to the left) by six places, and therefore

$$32 \times \frac{1}{1,000,000} = 32 \times 0.000001 = 32 \times 10^{-6} = 0.0000032$$

Now that you know exactly what a multiplier is, you have only to remember these simple rules:

1. A *negative exponent* tells you how many places *to the left* to move the decimal point.
2. A *positive exponent* tells you how many places *to the right* to move the decimal point.

Remember one important point: a negative exponent does not indicate a negative number; it simply indicates a fraction. For example, 4×10^{-3} meter means that 1 meter has been broken up into 1000 parts and we have 4 of those pieces, or 4/1000.

4-2-1   *Converting to Powers of 10*

To reinforce our understanding, let us try converting a few numbers to powers-of-10 notation.

EXAMPLE A:

$$230,000,000 = 23\overset{7\,6\,5\,4\,3\,2\,1}{0,000,000} = 23 \times 10^7$$

EXAMPLE B:

$$760,000 = 76\overset{4\,3\,2\,1}{0,000} = 76 \times 10^4$$

EXAMPLE C:

$$0.0019 = 0.0019 \overset{1\,2\,3\,4}{} = 19 \times 10^{-4}$$

EXAMPLE D:

$$\begin{array}{c} 1\ 2\ 3\ 4\ 5 \\ 0.00085 = 0.00085. = 85 \times 10^{-5} \end{array}$$

Most calculators have a key specifically for entering powers of 10. It is called the *exponent key* and operates as follows.

CALCULATOR KEYS

Name: Exponent entry key $\boxed{\text{EXP}}$ or $\boxed{\text{EE}}$

Function: Prepares calculator to accept next digits entered as a power-of-10 exponent. The sign of the exponent can be changed by using the change-sign key $\boxed{+/-}$.

Example: $76\ \boxed{\text{EXP}}^4$

Press keys: $\boxed{7}\ \boxed{6}\ \boxed{\text{EXP}}\ \boxed{4}$

Display shows: $\boxed{76.\ 04}$

Example: 85×10^{-5}

Press keys: $\boxed{8}\ \boxed{5}\ \boxed{\text{EXP}}\ \boxed{5}\ \boxed{+/-}$

Display shows: $\boxed{85.\ -05}$

4-2-2   *Powers of 10 in Combination*

A power-of-10 multiplier raises the base number to the power of 10 indicated by the exponent. To get you used to working with powers of 10, let us do a few examples and include the calculator sequences.

■ EXAMPLE A:

$(2 \times 10^2) + (3 \times 10^3)$
$= 200 + 3000 = 3200$

Steps:
$2 \times 10^2 = 2 \times 10 \times 10 = 200$
$3 \times 10^3 = 3 \times 10 \times 10 \times 10 = 3000$
$200 + 3000 = 3200$
Calculator sequence: $\boxed{2}\ \boxed{\text{EXP}}\ \boxed{2}\ \boxed{+}\ \boxed{3}\ \boxed{\text{EXP}}\ \boxed{3}\ \boxed{=}$
Answer: 3200 or
$\boxed{32.\ 02}$ (32×10^2 or 3.2×10^3)

■ EXAMPLE B:

$(3 \times 10^6) - (2 \times 10^4)$
$= 3,000,000 - 20,000$
$= 2,980,000$

Steps:
$3 \times 10^6 = 3 \times 10 \times 10 \times 10 \times 10 \times 10 \times 10$
$\quad = 3,000,000$
$2 \times 10^4 = 2 \times 10 \times 10 \times 10 \times 10$
$\quad = 20,000$
$3,000,000 - 20,000 = 2,980,000$
Calculator sequence: $\boxed{3}\ \boxed{\text{EXP}}\ \boxed{6}\ \boxed{-}\ \boxed{2}\ \boxed{\text{EXP}}\ \boxed{4}\ \boxed{=}$
Answer: 2,980,000 or
$\boxed{298.\ 04}$ (2.98×10^6)

EXAMPLE C:

$$\frac{1.6 \times 10^4}{4 \times 10^2} \text{ or}$$
$$(1.6 \times 10^4) \div (4 \times 10^2)$$
$$= 16{,}000 \div 400$$
$$= 40$$

Steps:
$1.6 \times 10^4 = 1.6 \times 10 \times 10$
$\qquad\qquad \times 10 \times 10 = 16{,}000$
$4 \times 10^2 = 4 \times 10 \times 10$
$\qquad\qquad = 400$
$16{,}000 \div 400 = 40$

Calculator sequence: $\boxed{1}\,\boxed{.}\,\boxed{6}\,\boxed{\text{EXP}}\,\boxed{4}\,\boxed{\div}$
$\boxed{4}\,\boxed{\text{EXP}}\,\boxed{2}\,\boxed{=}$

Answer: 40

EXAMPLE D:

$$(7.5 \times 10^{-6}) \times (1.86 \times 10^{-3})$$
$$= 0.0000075 \times 0.00186$$
$$= 0.00000001395 \text{ or}$$
$$1.395 \times 10^{-8}$$

Steps:
$7.5 \times 10^{-6} = 0.0000007.5$
$1.86 \times 10^{-3} = 0.001.86$
$0.0000075 \times 0.00186 =$
0.00000001395

Calculator sequence: $\boxed{7}\,\boxed{.}\,\boxed{5}\,\boxed{\text{EXP}}\,\boxed{6}$
$\boxed{+/-}\,\boxed{\times}\,\boxed{1}\,\boxed{.}\,\boxed{8}$
$\boxed{6}\,\boxed{\text{EXP}}\,\boxed{3}\,\boxed{+/-}\,\boxed{=}$

Answer: 1.395×10^{-8}

4-2-3   *Scientific and Engineering Notation*

As mentioned previously, powers of 10 are used in science and technology as a shorthand due to the large number of zeros in many values. There are basically two systems or notations used, involving values that have exponents that are a power of ten. They are called **scientific notation** and **engineering notation.**

A number in *scientific notation* is expressed as a base number between 1 and 10 multiplied by a power of 10. In the following examples, the values on the left have been converted to scientific notation.

Scientific Notation

A widely used floating-point system in which numbers are expressed as products consisting of a number between 1 and 10 multiplied by an appropriate power of 10.

Engineering Notation

A widely used floating-point system in which numbers are expressed as products consisting of a number that is greater than 1 multiplied by a power of 10 that is some multiple of 3.

EXAMPLE A:

$$32{,}000 = 3.2000_\circ = 3.2 \times 10^4$$
$\qquad\qquad\qquad\qquad\qquad$ Scientific notation

Decimal point is moved to a position that results in a base number between 1 and 10. If decimal point is moved left, exponent is positive. If decimal point is moved right, exponent is negative.

EXAMPLE B:

$$0.0019 = 0_\circ 001.9 = 1.9 \times 10^{-3}$$
$\qquad\qquad\qquad\qquad\qquad$ Scientific notation

EXAMPLE C:

$$114{,}300{,}000 = 1.143\,00\,00\,0_\circ = 1.143 \times 10^8$$
$\qquad\qquad\qquad\qquad\qquad\qquad$ Scientific notation

■ EXAMPLE D:

$$0.26 = 0_{\curvearrowleft}2.6 = 2.6 \times 10^{-1}$$

Scientific notation

As you can see from the preceding examples, the decimal point floats backward and forward, which explains why scientific notation is called a **floating-point number system.** Although scientific notation is used in science and technology, the engineering notation system, discussed next, is used more frequently.

In *engineering notation* a number is represented as a base number that is greater than 1 multiplied by a power of 10 that is some multiple of 3. In the following examples, the values on the left have been converted to engineering notation.

Floating-Point Number System

A system in which numbers are expressed as products consisting of a number and a power-of-10 multiplier.

■ EXAMPLE A:

$$32,000 = 32.\overset{\frown}{0\,0\,0}_{\circ} = 32 \times 10^3$$

Decimal point is moved to a position that results in a base number that is greater than 1, and a power-of-10 exponent that is some multiple of 3.

■ EXAMPLE B:

$$0.0019 = 0_{\circ}\overset{\frown}{0\,0\,1}.9 = 1.9 \times 10^{-3}$$

■ EXAMPLE C:

$$114,300,000 = 114.\overset{\frown}{3\,0\,0\,0\,0}_{\circ} = 114.3 \times 10^6$$

■ EXAMPLE D:

$$0.26 = 0_{\circ}\overset{\frown}{2\,6\,0}. = 260 \times 10^{-3}$$

CALCULATOR KEYS

Name: Normal, scientific, engineering modes

Function: Most calculators have different notation modes that affect the way an answer is displayed on the calculator's screen. Numeric answers can be displayed with up to 10 digits and a two-digit exponent. You can enter a number in any format.

Example: Normal notation mode is the usual way we express numbers, with digits to the left and right of the decimal, as in 12345.67.

Sci (scientific) notation mode expresses numbers in two parts. The significant digits display with one digit to the left of the decimal. The appropriate power of 10 displays to the right of E, as in 1.234567E4.

Eng (engineering) notation mode is similar to scientific notation; however, the number can have one, two, or three digits before the decimal, and the power-of-10 exponent is a multiple of three, as in 12.34567E3.

Now that you have completed this section, you should be able to:

- **Objective 5.** *Explain the powers-of-10 method and how to convert to powers of 10.*

- **Objective 6.** *Calculate the result of problems with powers of 10 in combination.*

- **Objective 7.** *Describe the two floating-point number systems: scientific notation and engineering notation.*

Use the following questions to test your understanding of Section 4-2.

1. Convert the following to powers of 10.
   - **a.** $100 = ?$
   - **b.** $1 = ?$
   - **c.** $10 = ?$
   - **d.** $1,000,000 = ?$
   - **e.** $\dfrac{1}{1,000} = ?$

   f. $\dfrac{1}{1,000,000} = ?$

2. Convert the following to common numbers without exponents.
   - **a.** $6.3 \times 10^3 = ?$
   - **b.** $114,000 \times 10^{-3} = ?$
   - **c.** $7,114,632 \times 10^{-6} = ?$
   - **d.** $6624 \times 10^6 = ?$

3. Perform the indicated operation on the following.
   - **a.** $\sqrt{3} \times 10^6 = ?$
   - **b.** $(2.6 \times 10^{-6}) - (9.7 \times 10^{-9}) = ?$
   - **c.** $\dfrac{(4.7 \times 10^3)^2}{3.6 \times 10^6} = ?$

4. Convert the following common numbers to engineering notation.
   - **a.** $47,000 = ?$
   - **b.** $0.00000025 = ?$
   - **c.** $250,000,000 = ?$
   - **d.** $0.0042 = ?$

4-3   THE METRIC SYSTEM

Metric System

A decimal system of weights and measures based on the meter and the kilogram.

*The **metric system** is a decimal system of weights and measures.* This system was developed to make working with values easier. Comparing two examples, you will see that working with the metric system, which uses multiples of 10 (decimal), is much easier than working with the *U.S. customary system of units.*

EXAMPLE: THE METRIC SYSTEM

The U.S. monetary system is like the metric system. It is a decimal system, which means that it is based on multiples of 10. For instance, there are 100 cents in 1 dollar ($100 \times 1¢ = \$1$), or 10 dimes in 1 dollar ($10 \times 10¢ = \$1$).

EXAMPLE: THE U.S. CUSTOMARY SYSTEM

The United States still has not adopted the metric system of measurements. For instance, there are 12 inches in 1 foot (12 in. = 1 ft), and 3 feet in 1 yard (3 ft = 1 yd). None of these values are multiples of 10. To add even more to the confusion of this system, inches are divided up into strange fractions such as fourths, sixteenths, thirty-seconds, and sixty-fourths.

From these two examples you can see that it is easier to count, represent, and perform mathematical operations on a system based on multiples of 10.

4-3-1   *Metric Units and Prefixes*

Unit

A determinate quantity adopted as a standard of measurement.

To help explain what we mean by units and prefixes, let us begin by examining how we measure length using the metric system. The standard **unit** of length in the metric system is the *meter* (abbreviated m). You have probably not seen the unit "meter" used a lot on its own. More frequently you have heard and seen the terms *centimeter, millimeter,* and *kilometer.* All

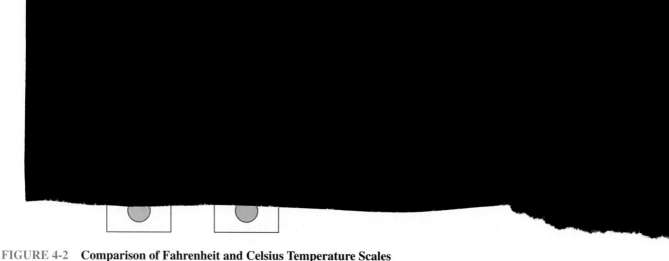

FIGURE 4-2   **Comparison of Fahrenheit and Celsius Temperature Scales**

■   INTEGRATED MATH APPLICATION:

If a thermometer reads 29 °C, what would this temperature be in degrees Fahrenheit?

■   *Solution:*

From Table 4-1, you can see that to convert a metric Celsius temperature to a U.S. customary temperature, you multiply the Celsius reading by ⁹⁄₅ and then add 32.

Put on

$$°F = \left(\frac{9}{5} \times °C\right) + 32$$
$$= (1.8 \times 29) + 32$$
$$= 52.2 + 32 = 84.2 \ °F$$

Steps:
$9 \div 5 = 1.82$
$1.8 \times 29 = 52.2$
$52.2 + 32 = 84.2 \ °F$

■   INTEGRATED MATH APPLICATION:

Convert 63 °F to Celsius.

dard

■   *Solution:*

Table 4-2 indicates that to convert °F to °C, you must subtract 32 from the Fahrenheit reading and then multiply the remainder by ⁵⁄₉.

$$°C = \frac{5}{9} \times (°F - 32)$$
$$= 0.55 \times (63 - 32)$$
$$= 0.55 \times 31 = 17.05 \ °C$$

Steps:
$5 \div 9 = 0.55$
$63 - 32 = 31 \ °C$

4-3-7   *Electrical Units, Prefixes, and Conversions*

Table 4-3 lists many of the more commonly used metric electrical quantities with their units and symbols. For example, electrical current (symbolized by I) is measured in amperes or amps (symbolized by A). As another example, electrical resistance (R) is measured in ohms (symbolized by the Greek capital letter omega, Ω).

   If electrical values are much larger than one or are a fraction of one, the unit of electrical quantity is preceded by a prefix. Most electrical and electronic applications use engineering notation, so these prefixes are power-of-10 exponents that are some multiple of three (10^3, 10^6, 10^9, 10^{12}, and so on). As an example, Table 4-4 shows how these prefixes precede ampere, which is the electrical current unit. To help our understanding of these prefixes, let us do a few conversion problems.

■ **INTEGRATED MATH APPLICATION:**

Convert the following.

a. 0.003 A = ___3___ mA (milliamperes)

b. 0.070 mA = _____ μA (microamperes)

c. 7333 mA = _____ A (amperes)

d. 1275 μA = _____ mA (milliamperes)

(handwritten margin notes: "move decimal 3 to right", "0.0700", "0.00000")

■ *Solution:*

a. 0.003 A = _____ mA. In this example, 0.003 A has to be converted so that it is represented in milliamperes (10^{-3}, or $^1/_{1000}$ of an ampere). The basic algebraic rule to remember is that both expressions on either side of the equals sign must be the same.

Left		Right	
Base	Exponent	Base	Exponent
0.003×10^0		=	_____ $\times 10^{-3}$

TABLE 4-4 **Electrical Current Prefixes**

NAME	ABBREVIATION	VALUE
Teraampere	TA	$10^{12} = 1,000,000,000,000$
Gigaampere	GA	$10^9 = 1,000,000,000$
Megaampere	MA	$10^6 = 1,000,000$
Kiloampere	kA	$10^3 = 1000$
Ampere	A	$10^0 = 1$
Milliampere	mA	$10^{-3} = \dfrac{1}{1000}$
Microampere	μA	$10^{-6} = \dfrac{1}{1,000,000}$
Nanoampere	nA	$10^{-9} = \dfrac{1}{1,000,000,000}$
Picoampere	pA	$10^{-12} = \dfrac{1}{1,000,000,000,000}$

The exponent on the right in this problem will be decreased 1000 times (10^0 to 10^{-3}), so for the statement to balance, the number on the right will have to be increased 1000 times; that is, the decimal point will have to be moved to the right three places (0.003 to 3). Therefore,

$$0.003 \times 10^0 = 3 \times 10^{-3}$$

$$\text{or} \qquad 0.003\,A = 3 \times 10^{-3}\,A \qquad \text{or} \qquad 3\,mA$$

b. 0.07 mA = _____ μA. In this example the exponent is going from milliamperes to microamperes (10^{-3} to 10^{-6}), or 1000 times smaller, so the number must be made 1000 times greater.

$$0.070. \quad \text{or} \quad 70.0$$

Therefore, 0.07 mA = 70 μA.

c. 7333 mA = _____ A. The exponent is going from milliamperes to amperes, increasing 1000 times, so the number must decrease 1000 times.

$$7333. \quad \text{or} \quad 7.333$$

Therefore, 7333 mA = 7.333 A.

d. 1275 μA = _____ mA. The exponent is changing from microamperes to milliamperes, an increase of 1000 times, so the number must decrease by the same factor.

$$1.2750 \quad \text{or} \quad 1.275$$

Therefore, 1275 μA = 1.275 mA.

CALCULATOR KEYS

Name: Convert key ▶

Function: Some calculators have a function that converts an expression from one unit to another.

expression_unit ▶ *_unit2* ⇒ *expression_unit2*

Examples:

How many feet are in 3 meters?

Press keys: 3_m ▶ _ft ENTER

Answer: 9.842_ft

How many kilometers are in 4 light-years?

Press keys: 4_ltyr ▶_km

Answer: 3.78421E13_km

How many ohms are in 4 kiloohms?

Press keys: 4_kΩ▶_Ω

Answer: 4000_Ω

To convert a temperature value, you must use tmpCnv( ) instead of the ▶ operator.

Example:

Press keys: tmpCnv(100_°C to _°F)

Answer: 212_°F

Now that you have completed this section, you should be able to:

■ **Objective 8.** *Define and explain the purpose of the metric system.*

■ **Objective 9.** *List the metric prefixes and describe the purpose of each.*

■ **Objective 10.** *Describe the metric units for length, area, weight, volume, and temperature.*

■ **Objective 11.** *Convert U.S. customary units to metric units, and metric units to U.S. customary units.*

■ **Objective 12.** *List and describe many of the more frequently used electrical units and prefixes, and describe how to interconvert prefixes.*

1. Give the power-of-10 value for the following prefixes.

    a. kilo   **d.** mega
    b. centi   **e.** micro
    c. milli

2. Convert the following metric units to U.S. customary units.

    a. 15 cm to inches
    b. 23 kg to pounds
    c. 37 L to gallons
    d. 23 °C to degrees Fahrenheit

3. Convert the following U.S. customary units to metric.

    a. 55 miles per hour (mph) to kilometers per hour (km/h)
    b. 16 gal to liters
    c. 3 yd^2 to square meters
    d. 92 °F to degrees Celsius

4. List the names of the metric units for the following quantities.

    a. Length   **f.** Current
    b. Weight   **g.** Voltage
    c. Temperature   **h.** Volume (liquid)
    d. Time   **i.** Energy
    e. Power   **j.** Resistance

5. Convert the following.

    a. 25,000 V = _____ kilovolts
    b. 0.014 W = _____ milliwatts
    c. 0.000016 μF = _____ nanofarads
    d. 3545 kHz = _____ megahertz

REVIEW QUESTIONS

Multiple Choice Questions

1. A base number's exponent indicates how many times the base number must be multiplied by:

    a. 2
    b. 3
    c. Itself
    d. Both (a) and (c)

2. What does 16^3 mean?

    a. 16 × 3
    b. 16 × 16 × 16
    c. 16 × (3 × 3 × 3)
    d. 3 × 3 × 3 × 3 × 3 ×3 × 3 × 3 × 3 × 3 × 3 × 3 × 3 × 3 × 3 × 3

3. If $x^2 = 16$, then $\sqrt[2]{16} = x$. What is the value of x?

    a. 4
    b. 2
    c. 16
    d. 8

4. Calculate $\sqrt[5]{32}$.

    a. 2
    b. 1.3
    c. 4
    d. 6.4

5. Raise the following number to the power of 10 indicated by the exponent, and give the answer: 3.6×10^2.

    a. 3600
    b. 36
    c. 0.036
    d. 360

6. Convert the following number to powers of 10: 0.00029.

    a. 0.29×10^{-3} *0.00029*
    b. 2.9×10^{-4} *0.00029*
    c. 29×10^{-5}
    d. All the above   *0.00029*

7. What is the power-of-10 value for the metric prefix *milli*?

    a. 10^{-3} **c.** 10^3
    b. 10^{-2} **d.** All the above

8. What is the metric unit for temperature?

    a. Fahrenheit
    b. Borg
    c. Celsius
    d. Rankine

9. Convert the following value of inductance to engineering notation: 0.016 henry (H).

    a. 1.6×10^2 H   **c.** 1.6 cH
    b. 16 mH   **d.** 16 kH

10. How many centimeters are in 8 inches?

    a. 2.03 m   **c.** 8 cm
    b. 3.2 cm   **d.** 20.32 cm

Communication Skill Questions

[handwritten: 10³ ← exponent]

11. What is an exponent? (Introduction)
12. Define the following terms: (4-1)
    a. Square of a number *[handwritten: Any # times itself]*
    b. Square root of a number
13. What is a power-of-10 exponent? (4-2) *[handwritten: any # multiplied 10 times]*
14. Describe the scientific and engineering notation systems. (4-2)
15. What is the metric system? (4-3) *[handwritten: a system of measurement]*
16. List the metric prefixes and their value. (4-3)
17. Explain how to convert the following: (4-3)
    a. Centimeters to inches *[handwritten: Inches x .5]*
    b. Ounces to grams
    c. Yards to meters
    d. Miles to kilometers

18. What is the metric unit of: (4-3)
    a. Area
    b. Length
    c. Weight
    d. Volume
    e. Temperature
19. Give the units for the following electrical quantities: (4-3)
    a. Current   c. Resistance
    b. Length   d. Power
20. Arbitrarily choose values, and then describe the following conversions: (4-3)
    a. Amps to milliamps
    b. Volts to kilovolts

Practice Problems

21. Determine the square of the following values.
    a. 9^2 d. 0^2
    b. 6^2 e. 1^2
    c. 2^2 f. 12^2
22. Determine the square root of the following values.
    a. $\sqrt[2]{81}$ d. $\sqrt[2]{36}$
    b. $\sqrt[2]{4}$ e. $\sqrt[2]{144}$
    c. $\sqrt[2]{0}$ f. $\sqrt[2]{1}$
23. Raise the following base numbers to the power indicated by the exponent, and give the answer.
    a. 9^3 c. 4^6
    b. 10^4 d. 2.5^3
24. Determine the roots of the following values.
    a. $\sqrt[3]{343} = ?$ c. $\sqrt[4]{760} = ?$
    b. $\sqrt{1296} = ?$ d. $\sqrt[6]{46,656} = 5 ?$
25. Convert the following values to powers of 10.
    a. $\dfrac{1}{100}$ c. $\dfrac{1}{1000}$
    b. 1,000,000,000   d. 1000
26. Convert the following to proper fractions and decimal fractions.
    a. 10^{-4} c. 10^{-6}
    b. 10^{-2} d. 10^{-3}
27. Convert the following values to powers of 10 in both scientific and engineering notation.
    a. 475   c. 0.07
    b. 8200   d. 0.00045

28. Convert the following to whole-number values with a metric prefix.
    a. 0.005 A
    b. 8000 m
    c. 15,000,000 Ω
    d. 0.000016 s
29. Convert the following metric units to U.S. customary units.
    a. 100 km   c. 67 kg
    b. 29 m^2 d. 2 L
30. Convert the following U.S. customary units to metric.
    a. 4 in.   c. 4 oz
    b. 2 mi^2 d. 10 gal
31. Convert the following temperatures from °F to °C.
    a. 32 °F   b. 72 °F
32. Convert the following temperatures from °C to °F.
    a. 6 °C   b. 32 °C
33. State the metric unit of:
    a. Length   d. Dry volume
    b. Area   e. Liquid volume
    c. Weight   f. Temperature
34. What is the metric unit of time?
35. Convert the following.
    a. 8000 ms = _____ microseconds
    b. 0.02 MV = _____ kilovolts
    c. 10 km = _____ meters
    d. 250 mm = _____ centimeters

Web Site Questions

Go to the web site http://www_____.com/cook, select the textbook *Introd___ ___mputer Mathematics, 2e,* select

this chapter, and then follow the instructions when answering the multiple-choice practice problems.

Algebra, Equations, and Formulas

Back to the Future

Each part of a computer system is designed to perform a specific task. You would think that all these units were first thought of by some recent pioneer in the twentieth century, but in fact, two of these elements were first described in 1833.

Born in England in 1791, Charles Babbage became very well known for both his mathematical genius and eccentric personality. For many years Babbage occupied the Cambridge chair of mathematics, once held by Isaac Newton, and although he never delivered a single lecture, he wrote several papers on numerous subjects, ranging from politics to manufacturing techniques. He also helped develop several practical devices, including the tachometer and the railroad cowcatcher.

Babbage's ultimate pursuit, however, was that of mathematical accuracy. He delighted in spotting errors in everything from log tables (used by astronomers, mathematicians, and navigators) to poetry. In fact, he once wrote to poet Alfred Lord Tennyson, pointing out an inaccuracy in his line "Every moment dies a man—every moment one is born." Babbage explained to Tennyson that since the world population was actually increasing and not, as he indicated, remaining constant, the line should be rewritten to read "Every moment dies a man—every moment one and one-sixteenth is born."

In 1822, Babbage described in a paper and built a model of what he called "a difference engine," which could be used to calculate mathematical tables. The Royal Society of Scientists described his machine as "highly deserving of public encouragement," and a year later the government awarded Babbage £1500 for his project. Babbage originally estimated that the project should take 3 years; however, the design had its complications, and after 10 years of frustrating labor, in which the government grants increased to £17,000, Babbage was still no closer to completion. Finally, the money stopped and Babbage reluctantly decided to let his brainchild go.

In 1833, Babbage developed an idea for a much more practical machine, which he named "the analytical engine." It was to be a more general machine that could be used to solve a variety of problems, depending on instructions supplied by the operator. It would include two units, called a "mill" and a "store," both of which would be made of cogs and wheels. The store, which was equivalent to a modern-day computer memory, could hold up to 100 forty-digit numbers. The mill, which was equivalent to a modern computer's arithmetic and logic unit (ALU), could perform both arithmetic and logic operations on variables or numbers retrieved from the store, and the result could be stored in the store and then acted upon again or printed out. The program of instructions directing these operations would be fed into the analytical engine in the form of punched cards.

The analytical engine was never built. All that remains are the volumes of descriptions and drawings, and a section of the mill and printer built by Babbage's son, who also had to concede defeat. It was, unfortunately for Charles Babbage, a lifetime of frustration to have conceived the basic building blocks of the modern computer a century before the technology existed to build it.

Outline and Objectives

Introduction

I contemplated calling this chapter "using letters in mathematics" because the word *algebra* seems to make many people back away. If you look up *algebra* in the dictionary, it states that it is "a branch of mathematics in which letters representing numbers are combined according to the rules of mathematics." In this chapter you will discover how easy algebra is to understand and then see how we can put it to some practical use, such as rearranging formulas. As you study this chapter you will find that algebra is quite useful, and because it is used in conjunction with formulas, an understanding is essential for anyone entering a technical field.

5-1   THE BASICS OF ALGEBRA

Algebra

A generalization of arithmetic in which letters representing numbers are combined according to the rules of arithmetic.

Formula

A general, fact, rule, or principle expressed usually in mathematical symbols.

As mentioned briefly in the introduction to this chapter, **algebra** by definition is a branch of mathematics in which letters representing numbers are combined according to the rules of mathematics. The purpose of using letters instead of values is to develop a general statement or **formula** that can be used for any values. For example, distance (*d*) equals velocity (*v*) multiplied by time (*t*), or

$$d = v \times t$$

where d = distance in miles
v = velocity or speed in miles per hour
t = time in hours

Using this formula, we can calculate how much distance was traveled if we know the speed or velocity at which we were traveling and the time for which we traveled at that speed. Therefore, if I were to travel at a speed of 20 miles per hour (20 mph) for 2 hours, how far, or how much distance, would I travel? Replacing the letters in the formula with values converts the problem from a formula to an **equation,** as shown:

$$d = v \times t \;\leftarrow\; \text{(Formula)}$$
$$d = 20 \text{ mph} \times 2 \text{ h} \;\leftarrow\; \text{(Equation)}$$
$$d = 40 \text{ mi} \;\leftarrow\; \text{(Answer)}$$

Equation

A formal statement of the equality or equivalence of mathematical or logical expressions.

Literal Number

A number expressed as a letter.

Letters such as *d, v, t* and *a, b, c* are called **literal numbers** (letter numbers), and are used, as we have just seen, in general statements showing the relationship between quantities. They are also used in equations to signify an *unknown quantity,* as will be discussed in the following section.

5-1-1   *The Equality on Both Sides of the Equals Sign*

All equations or formulas can basically be divided into two sections that exist on either side of an equals sign, as shown:

Left half   Right half

Fraction bar ⟶ $\dfrac{8 \times x}{2}$ = $\dfrac{16}{1}$ ⟵ Fraction bar

Equals sign

CHAPTER 5 / ALGEBRA, EQUATIONS, AND FORMULAS

Everything in the left half of the equation is equal to everything in the right half of the equation. This means that 8 times x (which is an unknown value) divided by 2 equals 16 divided by 1. Generally, it is not necessary to put 1 under the 16 in the right section because any number divided by 1 equals the same number ($^{16}/_1 = 16 \div 1 = 16$); however, it was included in this introduction to show that each section has both a top and a bottom. If the equation is written without the fraction bar or 1 in the denominator position, it appears as follows:

$$\frac{8 \times x}{2} = 16$$

Although the fraction bar is not visible, you should always assume that a number on its own on either side of the equals sign is above the fraction bar, as shown:

$$\frac{8 \times x}{2} = \frac{16}{}$$

Now that we understand the basics of an equation, let us see how we can manipulate it yet keep both sides equal to one another.

5-1-2   *Treating Both Sides Equally*

If you do exactly the same thing to both sides of an equation or formula, the two halves remain exactly equal, or in balance. This means that as long as you add, subtract, multiply, or divide both sides of the equation by the same number, the equality of the equation is preserved. For example, let us try adding, subtracting, multiplying, and dividing both sides of the following equation by 4 and see if both sides of the equation are still equal.

$$\boxed{2 \times 4 = 8} \quad \leftarrow \text{(Original equation)}$$

1. Add 4 to both sides of the equation:

$$2 \times 4 = 8 \quad \leftarrow \text{(Original equation)}$$
$$(2 \times 4)\,\boxed{+\,4} = 8\,\boxed{+\,4} \leftarrow \text{(Add 4 to both sides.)}$$
$$8 + 4 = 8 + 4$$
$$12 = 12$$

Both sides of the equation remain equal.

2. Subtract 4 from both sides of the equation:

$$2 \times 4 = 8 \quad \leftarrow \text{(Original equation)}$$
$$(2 \times 4)\,\boxed{-\,4} = 8\,\boxed{-\,4} \leftarrow \text{(Subtract 4 from both sides.)}$$
$$8 - 4 = 8 - 4$$
$$4 = 4$$

Both sides of the equation remain equal.

3. Multiply both sides of the equation by 4:

$$2 \times 4 = 8 \quad \leftarrow \text{(Original equation)}$$
$$(2 \times 4)\,\boxed{\times\,4} = 8\,\boxed{\times\,4} \leftarrow \text{(Multiply both sides by 4.)}$$
$$8 \times 4 = 8 \times 4$$
$$32 = 32$$

Both sides of the equation remain equal.

4. Divide both sides of the equation by 4:

$$2 \times 4 = 8 \qquad \leftarrow \text{(Original equation)}$$
$$(2 \times 4) \div 4 = 8 \div 4 \qquad \leftarrow \text{(Divide both sides by 4.)}$$
$$\frac{2 \times 4}{4} = \frac{8}{4}$$
$$\frac{8}{4} = \frac{8}{4}$$
$$2 = 2$$

Both sides of the equation remain equal.

As you can see from the four preceding procedures, *if you add, subtract, multiply, or divide both halves of an equation by the same number, the equality of the equation is preserved.* In the next section we will see how these operations can serve some practical purpose.

SELF-TEST EVALUATION POINT FOR SECTION 5-1

Now that you have completed this section, you should be able to:

■ **Objective 1.** *Define the terms* algebra, equation, formula, *and* literal numbers.

■ **Objective 2.** *Describe the rules regarding the equality on both sides of the equals sign.*

■ **Objective 3.** *Demonstrate how to remove a quantity by performing the same arithmetic operation on both sides of an equation or formula.*

1. Is the following equation true?

$$\frac{\frac{56}{14}}{2} = 2 \text{ or } \frac{56 \div 14}{2} = 2$$

2. If we were to multiply the left side of the equation in Question 1 by 3 and the right side of the equation by 2, would the two sides be equal to each other?

3. Fill in the missing values.

$$\frac{144}{12} \times \boxed{} = \frac{36}{6} \times 2 \times \boxed{}$$
$$60 = 60$$

4. Determine the result and state whether or not the equation is equal.

$$\frac{(8-4)+26}{5} = \frac{81-75}{2}$$

5. Is the following equation balanced?

$$3.2 \text{ k}\Omega = 3200 \text{ }\Omega$$

5-2 TRANSPOSITION

Transposition

The transfer of any term of an equation from one side over to the other side with a corresponding change of the sign.

It is important to know how to *transpose,* or rearrange, equations and formulas so that you can determine the unknown quantity. This process of rearranging, called **transposition,** is discussed in this section.

5-2-1 *Transposing Equations*

As an example, let us return to the original problem introduced at the beginning of this chapter and try to determine the value of the unknown quantity x.

$$\frac{8 \times x}{2} = 16$$

To transpose the equation we must follow two steps:

Step 1. Move the unknown quantity so that it is above the fraction bar on one side of the equals sign.

Step 2. Isolate the unknown quantity so that it stands by itself on one side of the equals sign.

Looking at the first step, let us see if our unknown quantity is above the fraction bar on either side of the equals sign.

The unknown quantity
x is above the fraction bar.

$$\frac{8 \times \textcircled{x}}{2} = 16$$

Since step 1 is done, we can move on to step 2. Looking at the equation, you can see that x does not stand by itself on one side of the equals sign. To satisfy this step, we must somehow move the 8 above the fraction bar, and the 2 below the fraction bar away from the left side of the equation, so that x is on its own.

Let us begin by removing the 2. To remove a letter or number from one side of a formula or equation, simply remember this rule: *To move a quantity, simply apply to both sides the arithmetic opposite of that quantity.* Multiplication is the opposite of division, so to remove a "divide by 2" ($\div$ 2), simply multiply both sides by 2 ($\times$ 2), as follows:

$$\frac{8 \times x}{2} = 16 \qquad \leftarrow \text{(Original equation)}$$

$$\frac{8 \times x}{2} \times 2 = 16 \times 2 \qquad \leftarrow \text{(Multiply both sides by 2.)}$$

$$\frac{8 \times x}{\cancel{2}} \times \frac{\cancel{2}}{1} = 16 \times 2 \qquad \leftarrow \text{(Because } \tfrac{2}{2} = 1, \text{ the two 2s on the left side of the equation cancel.)}$$

$$(8 \times x) \times 1 = 16 \times 2 \qquad \text{(Because anything multiplied by 1 equals the same number, the 1 on}$$

$$8 \times x = 16 \times 2 \qquad \text{the left side of the equation can be removed: } 8x \times 1 = 8x.)$$

Looking at the result so far, you can see that by multiplying both sides by 2, we effectively moved the 2 that was under the fraction bar on the left side to the right side of the equation above the fraction bar.

$$\frac{8 \times x}{2} = 16 \qquad \leftarrow \text{(Original equation)}$$

$$8 \times x = 16 \times 2 \qquad \leftarrow \text{(Result after both sides of the equation were multiplied by 2)}$$

However, we have still not completed step 2, which was to isolate x on one side of the equals sign. To achieve this we need to remove the 8 from the left side of the equation. Once again we will do the opposite: Because the opposite of multiply is divide, to remove a "multiply by 8" we must divide both sides by 8 ($\div$ 8), as follows:

$$\frac{8 \times x}{2} = 16 \qquad \leftarrow \text{(Original equation)}$$

$$8 \times x = 16 \times 2 \qquad \leftarrow \text{(Equation after both sides were multiplied by 2)}$$

$$\frac{8 \times x}{8} = \frac{16 \times 2}{8} \qquad \leftarrow \text{(Divide both sides by 8.)}$$

$$\frac{\cancel{8} \times x}{\cancel{8}} = \frac{16 \times 2}{8} \qquad \text{(Because } \tfrac{8}{8} = 1, \text{ the two 8s on the left side of the equation cancel.)}$$

$$1 \times x = \frac{16 \times 2}{8} \qquad \text{(Anything multiplied by 1 equals the same number, so the 1 on the left side of the equation can be removed: } 1 \times x = x.)$$

$$x = \frac{16 \times 2}{8}$$

[handwritten note: the Dot . means multiply]

Now that we have completed step 2, which was to isolate the unknown quantity so that it stands by itself on one side of the equation, we can calculate the value of the unknown x by performing the arithmetic operations indicated on the right side of the equation.

$$x = \frac{16 \times 2}{8} \quad (16 \times 2 = 32)$$

$$x = \frac{32}{8} \quad (32 \div 8 = 4)$$

$$x = 4$$

To double-check this answer, let us insert this value into the original equation to see if it works.

$$\frac{8 \times x}{2} = 16 \quad \text{(Replace } x \text{ with 4, or substitute 4 for } x.)$$

$$\frac{8 \times 4}{2} = 16 \quad (8 \times 4 = 32)$$

$$\frac{32}{2} = 16 \quad (32 \div 2 = 16)$$

$$16 = 16 \quad \text{(Answer checks out because } 16 = 16.)$$

1. Example with an Unknown Above the Fraction Bar

■ EXAMPLE:

Determine the value of the unknown a in the following equation.

$$(2 \times a) + 5 = 23$$

■ Solution:

Step 1. Is the unknown quantity above the fraction bar? Yes.

Step 2. Is the unknown quantity isolated on one side of the equals sign? No.

Use the following steps to isolate the unknown, a.

Steps:

$$(2 \times a) + 5 = 23$$
To remove the "$+ 5$," do the opposite: subtract 5 from both sides.

$$(2 \times a) + \cancel{5} - \cancel{5} = 23 - 5$$
$5 - 5 = 0$; any number plus 0 equals the same number:
$(2 \times a) + 0 = 2 \times a$

$$2 \times a = 23 - 5$$
To remove the "$2 \times$," do the opposite: divide both sides by 2.

$$\frac{\cancel{2} \times a}{\cancel{2}} = \frac{23 - 5}{2}$$
$2 \div 2 = 1$; any number multiplied by 1 equals the same number:
$1 \times a = a$

$$1 \times a = \frac{23 - 5}{2}$$

$$a = \frac{23 - 5}{2}$$
Perform the arithmetic operations indicated to determine the value a.

$$23 - 5 = 18 \qquad 18 \div 2 = 9$$

$$a = \frac{18}{2}$$

$$a = 9$$

Double-check your answer by inserting it into the original equation.

$$(2 \times a) + 5 = 23 \quad \text{(Replace } a \text{ with 9.)}$$
$$(2 \times 9) + 5 = 23$$
$$18 + 5 = 23$$
$$23 = 23 \quad \text{(Answer checks out because } 23 = 23.\text{)}$$

2. Example with an Unknown Below the Fraction Bar

EXAMPLE:

Determine the value of the unknown x in the following equation:

$$\frac{72}{x} = 12$$

Solution:

Step 1. Is the unknown quantity above the fraction bar? No.

Use the following steps to move x above the fraction bar.

$$\frac{72}{x} = 12$$

Steps:
To move the x do the opposite:
the opposite of "divide by x" is multiply by x.

$$\frac{72}{x} \left(\times x \right) = 12 \left(\times x \right)$$

Any number divided by itself $= 1$,
so $x \div x = 1$ and therefore the two x's on
the left side of the equation cancel.

$$72 \times 1 = 12 \times x$$
$$72 = 12 \times x$$

$72 \times 1 = 72$

Now that the unknown quantity x is above the fraction bar, we can proceed to step 2.

Step 2. Isolate the unknown quantity on one side of the equals sign.

$$72 = 12 \times x$$

$$\frac{72}{12} = \frac{\cancel{12} \times x}{\cancel{12}}$$

To remove the "$12 \times$," simply do the opposite: divide both sides by 12.
$12 \div 12 = 1$, $1 \times x = x$; therefore, the 12s cancel.

$$\frac{72}{12} = 1 \times x$$

$$\frac{72}{12} = x$$

$72 \div 12 = 6$, $x = 6$

$$6 = x \quad \text{or} \quad x = 6$$

To double-check your answer, replace x with 6 in the original equation.

$$\frac{72}{x} = 12 \quad \text{(Original equation)}$$

$$\frac{72}{6} = 12$$

$$12 = 12$$

3. Example with an Unknown on Both Sides of the Equals Sign

In this section we will see how to deal with equations containing an unknown on both sides of the equals sign. In all equations like this, the same letter or literal number will have the same value no matter where it appears.

Determine the value of the unknown y in the following equation:

$$(6 \times y) + 7 = y + 27$$

■ *Solution:*

Step 1. Is the unknown above the fraction bar? Yes.

Step 2. Is the unknown isolated on one side of the equals sign? No.

To fulfill step 2 we need to isolate the unknown on one side of the equals sign. Since y exists on both sides, we must remove one. Studying the left side of the equation you can see that it has six y's ($6 \times y$), whereas the right side of the equation has only one y (y). The first step could therefore be to subtract y ($-y$) from both sides.

$(6 \times y) + 7 = y + 27$	← (Original equation)
$(6 \times y) + 7\,\boxed{-y} = y + 27\,\boxed{-y}$	*Steps:*
$(5 \times y) + 7 = 0 + 27$	Subtract y from both sides:
$(5 \times y) + 7 = 27$	$(6 \times y) - y = 5 \times y,$
$(5 \times y) + 7\,\boxed{-7} = 27\,\boxed{-7}$	$y - y = 0,\ 0 + 27 = 27$
	To remove $+\,7$, subtract 7 from both sides.
	$7 - 7 = 0,\quad 27 - 7 = 20$
$5 \times y = 20$	The opposite of multiply is divide. To remove "$5 \times$," divide both sides by 5.
$\dfrac{\cancel{5} \times y}{\cancel{5}} = \dfrac{20}{5}$	$5 \div 5 = 1$
$1 \times y = \dfrac{20}{5}$	$1 \times y = y$
$y = \dfrac{20}{5}$	$20 \div 5 = 4$; therefore, $y = 4$.
$y = 4$	

To double-check your answer, replace y with 4 in the original equation.

$$(6 \times y) + 7 = y + 27 \quad \text{← (Original equation)}$$
$$(6 \times 4) + 7 = 4 + 27 \quad \text{← (Replace } y \text{ with 4.)}$$
$$24 + 7 = 4 + 27$$
$$31 = 31$$

4. Story Problems

Story problems are probably the best practice because they are connected to our everyday life and help us to see more clearly how we can apply our understanding of mathematics. I have found the following steps helpful with story problems:

a. Determine what is unknown and assign it a literal number such as x.

b. Determine how many other elements are involved and how they relate to the unknown.

c. Construct an equation using the unknown and the other elements with their associated arithmetic operations.

d. Calculate the unknown using transposition if necessary.

■ **INTEGRATED MATH APPLICATION:**

If you multiply the number of dollars in your wallet by 6 and then subtract $23, you would still end up with one fourth of $100. How many dollars do you have in your wallet?

■ *Solution:*

To start with step a, we must determine the unknown. The last sentence, "How many dollars do you have in your wallet?" states quite clearly that this is our unknown quantity. We will let x represent the number of dollars in our wallet.

Step b says to determine how many other elements are involved and how they relate. If we read again from the beginning, the example says:

"If you multiply the number of dollars in your wallet by 6 . . .	This part of the story indicates that our unknown x should be multiplied by 6; $x \times 6$.
. . . and then subtract \$23, . . .	This part states that after our unknown has been multiplied by 6, it should have \$23 subtracted; $(x \times 6) - 23$.
. . . you would still end up with . . .	This statement precedes a result and therefore an equals sign should follow our equation at this point; $(x \times 6) - 23 =$.
. . . one-fourth of \$100."	This is the result we end up with: ¼ of 100 or ¼ $\times$ 100.

Now, to complete step c, we must construct an equation using the unknown and the other elements with their associated arithmetic operations.

$$(x \times 6) - 23 = \frac{1}{4} \times 100$$

Now that we have an equation, we can move on to step d. Since x is not isolated on one side of the equals sign, we will have to transpose or rearrange the equation.

Step 1: Is the unknown above the fraction bar? Yes.

Step 2: Is the unknown isolated on one side of the equals sign? No.

	Steps:
$(x \times 6) - 23 = \dfrac{1}{4} \times 100$	Rather than multiply both sides by 4, let us reduce the fraction ¼ of 100. 100 is above the fraction bar, therefore it can be written
$(x \times 6) - 23 = \dfrac{1 \times 100}{4}$	$1 \times 100 = 100$
$(x \times 6) - 23 = \dfrac{100}{4}$	$100 \div 4 = 25$; the fraction is now eliminated.
$(x \times 6) - 23 = 25$ $(x \times 6) - 23 \enclose{circle}{+\ 23} = 25 \enclose{circle}{+\ 23}$	Add 23 to both sides to remove -23. $-23 + 23 = 0$, $(x \times 6) - 0 = x \times 6$, $25 + 23 = 48$
$x \times 6 = 48$	
$\dfrac{x \times 6}{\enclose{circle}{6}} = \dfrac{48}{\enclose{circle}{6}}$	Divide both sides by 6 to remove "multiply by 6". $6 \div 6 = 1, 1 \times x = x$
$x = \dfrac{48}{6}$	
$x = 8$	

You therefore started off with $8 in your pocket. To double-check this answer, replace *x* with 8 in the original equation.

$$(x \times 6) - 23 = \frac{1}{4} \times 100 \qquad \left(\frac{1}{4} \text{ of } 100 = 25\right)$$
$$(8 \times 6) - 23 = 25 \qquad (8 \times 6 = 48)$$
$$48 - 23 = 25 \qquad (48 - 23 = 25)$$
$$25 = 25$$

■ **INTEGRATED MATH APPLICATION:**

Store owner A has four times as many nails as store owner B, who has 12,000. Combined, A and B have 60,000 nails. How many nails does store owner A have?

■ *Solution:*

A is the unknown, and as stated in the question, A is four times larger than B.

$$A = 4 \times B$$

If B has 12,000 nails, we can substitute 12,000 for B.

$$A = 4 \times 12,000$$

To calculate how many nails store owner A has, all we have to do is a simple multiplication.

$$A = 4 \times 12,000$$
$$A = 48,000$$

To prove this answer is true, we can return to the part of the question that states "combined, both A and B have 60,000 nails."

$$A + B = 60,000$$

We know A has 48,000 nails and B has 12,000 nails, so we can replace A and B with their equivalent values as follows:

$$A + B = 60,000$$
$$48,000 + 12,000 = 60,000$$
$$60,000 = 60,000$$

5. Factoring and Removing Parentheses

A factor is any number multiplied by another number and contributing to the result. For example, consider 3×8. In this multiplication example, 3 is a factor and 8 is a factor, and both will contribute to a result of 24.

Now let us reverse the process. What factors contributed to a result of 12? In this case we must determine which small numbers, when multiplied together, will produce 12. The answer could be

$$2 \times 6 = 12 \qquad \text{or} \qquad 3 \times 4 = 12$$

Are these, however, the smallest numbers that, when multiplied together, will produce 12? The answer is no because the 6 (in 2×6) can be broken into 2×3, and the 4 (in 3×4) can be broken into 2×2. Therefore, 12 has the factors $2 \times 2 \times 3$.

Now consider the equation $12 + 9$. What smaller number is a factor of both these numbers? The answer is 3 because you can get results of 12 and 9 by multiplying some other number by 3. Therefore, both 12 and 9 contain the factor 3. If we now remove the common factor 3 from both the numbers 12 and 9, we end up with the following equation:

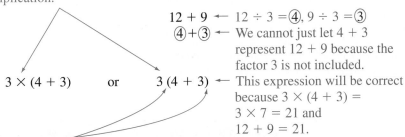

If no symbol appears, it is assumed that it is a multiplication.

$12 + 9$ ← $12 \div 3 = ④, 9 \div 3 = ③$

$④ + ③$ ← We cannot just let $4 + 3$ represent $12 + 9$ because the factor 3 is not included.

$3 \times (4 + 3)$ or $3(4 + 3)$ ← This expression will be correct because $3 \times (4 + 3) = 3 \times 7 = 21$ and $12 + 9 = 21$.

Parentheses are used to group the addition $4 + 3$. The factor 3 outside the parentheses applies to everything inside the parentheses.

Factoring is therefore a reducing process that extracts the common factor from two or more larger numbers. As another example, remove the factor from the following equation:

$27x - 9x$ *Steps:*

                    9 is a factor common to both 27 and 9.

                    $27x \div = 9 = 3x, 9x \div 9 = 1x$

$= 9(3x - 1x)$ Place 9 outside the parentheses and the results of the division inside the parentheses.

We can check if $9(3x - 1x) = 27x - 9x$ simply by performing the arithmetic operations indicated:

$27x - 9x = \boxed{18x}$

$9(3x - 1x)$

$= 9(3x - x)$ ← $3x - x = 2x$

     $1x = x$

$= 9 \times 2x$

$= \boxed{18x}$

Results are the same.

CALCULATOR KEYS

Name: Factor function

Function: Some calculators have a factor function that returns an expression factored with respect to all of its variables.

factor (expression)

Examples: Extract the common factor from $(64x - 8y)$.

Press keys: Factor $(64x - 8y)$
Answer: $8(8x - y)$

Press keys: Factor $(16a - 4)$
Answer: $4(4a - 1)$

To reverse the factoring process and convert $9(3x - 1x)$ back to the original equation $27x - 9x$, simply reverse the arithmetic steps. Because we began by dividing both numbers by the common factor, simply do the opposite and multiply both values within the parentheses by the factor 9, as follows:

$$9(3x - 1x)$$
$$= 27x - 9x$$

Steps:
$9 \times 3x = 27x,$
$9 \times 1x = 9x$

To make any equation with parentheses easier to solve, always begin by removing the parentheses before doing any other operation.

CALCULATOR KEYS

Name: Expand function

Function: Some calculators have a reverse factoring function that removes the parentheses.

Expand (expression)

Examples:

Press keys: Expand $(8(8x - y))$
Answer: $64x - 8y$

Press keys: Expand $(4(7a + 2b - 4c))$
Answer: $28a + 8b - 16c$

■ **EXAMPLE A:**

$$5(7x + 3x) \quad \leftarrow \quad 5 \times 7x = 35x, \; 5 \times 3x = 15x$$
$$= 35x + 15x$$

■ **EXAMPLE B:**

$$6(4a - 3) \qquad 6 \times 4a = 24a, \quad 6 \times 3 = 18$$
$$= 24a - 18$$

To summarize, we now know about two new processes called *factoring* and *removing parentheses*. To show how these two are the reverse of one another, let us do a few examples.

■ **EXAMPLE A:**

$$12 + 18$$
$$= 6(2 + 3)$$

Factoring:
Common factor is 6
$12 \div 6 = 2, \; 18 \div 6 = 3$

$$6(2 + 3)$$
$$= 12 + 18$$

Removing parentheses:
$6 \times 2 = 12, \; 6 \times 3 = 18$

CHAPTER 5 / ALGEBRA, EQUATIONS, AND FORMULAS

	Factoring:
$22y - 11$	Common factor is 11.
$= 11(2y - 1)$	$22y \div 11 = 2y, 11 \div 11 = 1$
$11(2y - 1)$	*Removing parentheses:*
$= \quad 22y - 11$	$11 \times 2y = 22y, 11 \times 1 = 11$

Now that we know how to perform both of these operations, let us look at a practical application of them.

■ INTEGRATED MATH APPLICATION:

A motorbike uses a certain number of gallons of gasoline a week, a car uses twice that amount, and a truck uses four times as much as the motorbike and car combined. If all three use 100 gallons of gasoline a week, what does each vehicle consume per week?

■ *Solution:*

Reading the problem carefully, you can see that each vehicle's consumption is related to the amount consumed by the motorbike. Therefore, if we can find the amount of gasoline consumed by the motorbike, we can calculate the amount consumed by the car and truck. We will represent the unknown quantity of gasoline that the bike consumes by x.

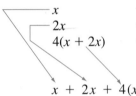

x ⟵ Bike consumption
$2x$ ⟵ Car consumption is twice that of the bike ($2 \times x$).
$4(x + 2x)$ ⟵ Truck consumption is four times that of the motorbike and car combined [$4 \times (x + 2x)$].
All three use 100 gallons per week.

$$x + 2x + 4(x + 2x) = 100 \text{ gal per week}$$

Now that we have an equation, the next step is to use transportation to determine the unknown value x.

	Steps:
$x + 2x + 4(x + 2x) = 100$	Remove parentheses:
	$4 \times x = 4x, 4 \times 2x = 8x$
$x + 2x + 4x + 8x = 100$	Combine all the unknowns:
	$x + 2x + 4x + 8x = 15x$
$15x = 100$	Divide both sides by 15 to
	remove the "multiply by 15."
$\dfrac{\cancel{15}x}{\cancel{15}} = \dfrac{100}{15}$	$15 \div 15 = 1, 1 \times x = x$
$x = \dfrac{100}{15}$	$100 \div 15 = 6.67$
$x = 6.67$	

Therefore,

$$\text{Motorbike consumption} = x = 6.67 \text{ gal per week}$$

The car consumes twice that amount:

$$\text{Car consumption} = 2 \times x = 2 \times 6.67 = 13.34 \text{ gal per week}$$

The truck consumes four times as much as both combined:

$$\begin{aligned}
\text{Truck consumption} = 4(x + 2x) &= 4(6.67 + 2 \times 6.67) \\
&= 4(6.67 + 13.34) \qquad 4 \times 6.67 = 26.68 \\
&\qquad\qquad\qquad\qquad\qquad 4 \times 13.34 = 53.36 \\
&= 26.68 + 53.36 \\
&= 80.0 \text{ gal per week}
\end{aligned}$$

To double-check our answer, let us add all the individual amounts to see if they equal the total:

$$\text{Motorbike} + \text{car} + \text{truck} = 100 \text{ gal per week}$$
$$6.67 + 13.34 + 80.0 = 100 \text{ gal per week}$$

Referring to the preceding example, you may have noticed that *you used factoring to develop the equation:*

$$\text{Truck consumption is 4 times that of bike and car} = 4(x + 2x)$$

and *you removed parentheses to transpose the equation:*

$$4(x + 2x) = 4x + 8x$$

6. Square and Square Root

How do we remove the square or square root on one side of an equation? The answer once again is: Perform the opposite function. For example, if we square a number and then find its square root, we end up with the number with which we started:

$$\textcircled{4}\text{squared} = 4^2 = 4 \times 4 = 16$$
$$\text{Square root of } 16 = \sqrt{16} = \textcircled{4}$$

■   **EXAMPLE:**

Determine the value of the unknown I in the following equation:

$$I^2 = 81$$

■   *Solution:*

To remove the square from the unknown quantity I, we simply take the square root of both sides.

$$\begin{aligned}
I^2 &= 81 \\
\sqrt{I^2} &= \sqrt{81} \quad \leftarrow \text{(Take square root of both sides.)} \\
I &= \sqrt{81} \quad \leftarrow \text{(Square and square root cancel.)} \\
I &= 9 \qquad \leftarrow (\sqrt{81} = 9, \text{ since } 9 \times 9 = 81)
\end{aligned}$$

■ **EXAMPLE:**

Determine the value of the unknown x in the following equation.

$$\sqrt{x} = 2 + 6$$

■ *Solution:*

$$\sqrt{x} = 2 + 6 \quad \leftarrow \text{(To remove the square root, square both sides.)}$$
$$\sqrt{x}^2 = (2 + 6)^2 \quad \leftarrow \text{(Square and square root cancel.)}$$
$$x = (2 + 6)^2$$
$$x = 8^2 \qquad (8 \times 8 = 64)$$
$$x = 64$$

5-2-2   *Transposing Formulas*

As mentioned previously, *a formula is a general statement using letters and sometimes numbers that enables us to calculate the value of an unknown quantity.* Some of the simplest relationships are formulas. For example, to calculate the distance traveled, you can use the following formula, which was discussed earlier.

$$d = v \times t$$

where d = distance in miles
         v = velocity in miles per hour
         t = time in hours

Using this formula we can calculate distance if we know the speed and time. If we need only to calculate distance, this formula is fine. But what if we want to know what speed to go (v = ?) to travel a certain distance in a certain amount of time, or what if we want to calculate how long it will take for a trip (t = ?) of a certain distance at a certain speed? The answer is: Transpose or rearrange the formula so that we can solve for any of the quantities in the formula.

■ **INTEGRATED MATH APPLICATION:**

As an example, calculate how long it will take to travel 250 miles when traveling at a speed of 50 miles per hour (mph). To solve this problem, let us place the values in their appropriate positions in the formula.

$$d = v \times t$$
$$250 \text{ mi} = 50 \text{ mph} \times t$$

The next step is to transpose just as we did before with equations to determine the value of the unknown quantity t.

Steps:

$$250 = 50 \times t \qquad \text{To isolate } t, \text{ divide both sides by 50.}$$
$$\frac{250}{50} = \frac{\cancel{50} \times t}{\cancel{50}} \qquad 50 \div 50 = 1, 1 \times t = t$$
$$\frac{250}{50} = t \qquad 250 \div 50 = 5$$
$$t = 5$$

It will therefore take 5 hours to travel 250 miles at 50 miles per hour ($250 \text{ mi} = 50 \text{ mph} \times 5 \text{ h}$).

All we have to do, therefore, is transpose the original formula so that we can obtain formulas for calculating any of the three quantities (distance, speed, or time) if two values are known. This time, however, we will transpose the formula instead of the inserted values.

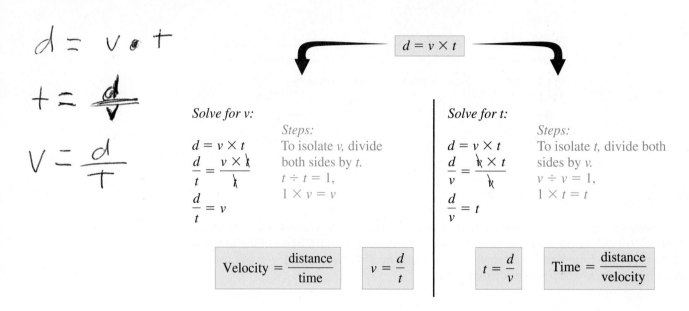

Solve for v:

Steps:
To isolate v, divide both sides by t.
$t \div t = 1$,
$1 \times v = v$

$d = v \times t$

$\dfrac{d}{t} = \dfrac{v \times t}{t}$

$\dfrac{d}{t} = v$

$\text{Velocity} = \dfrac{\text{distance}}{\text{time}}$ $v = \dfrac{d}{t}$

Solve for t:

Steps:
To isolate t, divide both sides by v.
$v \div v = 1$,
$1 \times t = t$

$d = v \times t$

$\dfrac{d}{v} = \dfrac{v \times t}{v}$

$\dfrac{d}{v} = t$

$t = \dfrac{d}{v}$ $\text{Time} = \dfrac{\text{distance}}{\text{velocity}}$

To test if all three formulas are correct, we can replace each letter with the values from the preceding example (also called "plugging in the values"), as follows:

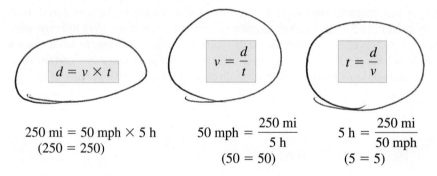

$d = v \times t$

250 mi = 50 mph × 5 h
(250 = 250)

$v = \dfrac{d}{t}$

$50 \text{ mph} = \dfrac{250 \text{ mi}}{5 \text{ h}}$
(50 = 50)

$t = \dfrac{d}{v}$

$5 \text{ h} = \dfrac{250 \text{ mi}}{50 \text{ mph}}$
(5 = 5)

1. Directly Proportional and Inversely Proportional

The word **proportional** is used to describe a relationship between two quantities. For example, reconsider the formula

$$\text{Time } (t) = \dfrac{\text{distance } (d)}{\text{velocity } (v)}$$

With this formula we can say that the time (t) it takes for a trip is directly proportional (symbolized by $\propto$) to the amount of distance (d) that needs to be traveled. In symbols, the relationship appears as follows:

$$t \propto d \qquad \text{(Time is directly proportional to distance.)}$$

The term **directly proportional** can therefore be used to describe the relationship between time and distance because an increase in distance will cause a corresponding increase in time ($d \uparrow$ causes $t \uparrow$), and similarly, a decrease in distance will cause a corresponding decrease in time ($d \downarrow$ causes $t \downarrow$).

Proportional

The relation of one part to another or to the whole with respect to magnitude, quantity, or degree.

Directly Proportional

The relation of one part to one or more other parts in which a change in one causes a similar change in the other.

CHAPTER 5 / ALGEBRA, EQUATIONS, AND FORMULAS

INTEGRATED MATH APPLICATION:

Using the previous example,

$$5 \text{ h } (t) = \frac{250 \text{ mi } (d)}{50 \text{ mph } (v)}$$

let us try doubling the distance and then halving the distance to see what effect it has on time. If time and distance are directly proportional to each other, they should change in proportion.

Original example $\rightarrow$ Time $= \dfrac{\text{distance}}{\text{velocity}} = \dfrac{250 \text{ mi}}{50 \text{ mph}} = 5 \text{ h}$

Doubling the distance from 250 mi to 500 mi should double the time it takes for the trip, from 5 h to 10 h.

$$t = \frac{d}{v} = \frac{\boxed{500 \text{ mi}}}{50 \text{ mph}} = \boxed{10 \text{ h}}$$

Doubling the distance did double the time ($d \uparrow, t \uparrow$).

Halving the distance from 250 mi to 125 mi should halve the time it takes for the trip, from 5 h to 2½ h.

$$t = \frac{d}{v} = \frac{\boxed{125 \text{ mi}}}{50 \text{ mph}} = \boxed{2.5 \text{ h}}$$

Halving the distance did halve the time ($d \downarrow, t \downarrow$).

As you can see from this exercise, time is directly proportional to distance ($t \propto d$). Therefore, *formula quantities are always directly proportional to one another when both quantities are above the fraction bars on opposite sides of the equals sign.*

On the other hand, **inversely proportional** means that the two quantities compared are opposite in effect. For example, consider again our time formula:

$$t = \frac{d}{v}$$

Observing the relationship between these quantities again, we can say that the time (t) it takes for a trip is *inversely proportional* (symbolized by $1/\propto$) to the speed or velocity (v) we travel. In symbols, the relationship appears as follows:

$$t \frac{1}{\propto} v \quad \text{(Time is inversely proportional to velocity.)}$$

The term *inversely proportional* can therefore be used to describe the relationship between time and velocity because an increase in velocity will cause a corresponding decrease in time ($v \uparrow$ causes $t \downarrow$), and similarly, a decrease in velocity will cause a corresponding increase in time ($v \downarrow$ causes $t \uparrow$).

> **Inversely Proportional**
> The relation of one part to one or more other parts in which a change in one causes an opposite change in the other.

INTEGRATED MATH APPLICATION:

Using the previous example,

$$5 \text{ h}(t) = \frac{250 \text{ mi } (d)}{50 \text{ mph } (v)}$$

let us try doubling the speed we travel, or velocity, and then halving the velocity to see what effect it has on time. If time and velocity are inversely proportional to each other, a velocity change should have the opposite effect on time.

$$\text{Original example j} \quad \text{Time} = \frac{\text{distance}}{\text{velocity}} = \frac{250 \text{ mi}}{50 \text{ mph}} = 5 \text{ h}$$

Doubling the speed from 50 mph to 100 mph should halve the time it takes to complete the trip, from 5 h to 2½ h.

$$\longrightarrow t = \frac{d}{v} = \frac{250 \text{ mi}}{\boxed{100 \text{ mph}}} = \boxed{2.5 \text{ h}}$$

Doubling the speed did halve the time ($v \downarrow, t \uparrow$).

Halving the speed from 50 mph to 25 mph should double the time it takes to complete the trip, from 5 h to 10 h.

$$\longrightarrow t = \frac{d}{v} = \frac{250 \text{ mi}}{\boxed{25 \text{ mph}}} = \boxed{10 \text{ h}}$$

Halving the speed did double the time ($v \downarrow, t \uparrow$).

As you can see from this exercise, time is inversely proportional to velocity ($t \, ^1/_\propto v$). Therefore, *formula quantities are always inversely proportional to one another when one of the quantities is above the fraction bar and the other is below the fraction bar on opposite sides of the equals sign.*

2. Ohm's Law Formula Example

In this section we will examine a law that was first discovered by Georg Ohm in 1826. It states that *the* **electric current** *in a circuit is directly proportional to the* **voltage** *source applied and inversely proportional to the* **resistance** *in the circuit.* Stated as a formula, it appears as follows:

Ohm's law formula: $\quad \boxed{\text{Current} = \dfrac{\text{voltage}}{\text{resistance}}}$

Quantity	Symbol	Unit	Symbol
Current	I	Amperes	A
Voltage	V	Volts	V
Resistance	R	Ohms	Ω

To explain this formula, let us use the fluid analogy shown in Figure 5-1(a). An object will move only when a force is applied, and as you can see in Figure 5-1(a), the greater the pressure applied to the piston, the greater the amount of water flow through the pipe. Water flow is therefore directly proportional to pressure applied:

$$\text{Water flow} \propto \text{pressure applied}$$

In a similar way, the electric current (I) shown in Figure 5-1(b) is directly dependent on the voltage (V) force applied by the battery. For instance, a smaller voltage ($V \downarrow$) will cause a corresponding smaller circuit current ($I \downarrow$), and similarly, a larger voltage ($V \uparrow$) will result in a larger circuit current ($I \uparrow$). It can therefore be said that electric current, which is the flow of electrons, is directly proportional to electric voltage, which is the force that moves electrons:

$$\text{Current } (I) \propto \text{voltage } (V)$$

Now, referring to the Ohm's law formula, you can see that current (I) and voltage (V) are directly proportional because these quantities are above the fraction bars on opposite sides of the equals sign.

$$\boxed{\text{Current } (I)} = \frac{\boxed{\text{voltage } (V)}}{\text{resistance } (R)}$$

The current (I) in an electric circuit is also inversely proportional to the opposition or resistance in the circuit. This can best be explained by the fluid analogy in Figure 5-2(a). If the valve is adjusted to offer a small opposition, a large amount of water will flow, whereas if the valve is adjusted to introduce a large amount of opposition, only a small amount of

CHAPTER 5 / ALGEBRA, EQUATIONS, AND FORMULAS

Small pressure

Large pressure

Small water flow

Large water flow

(a) Fluid system

Small voltage ($V\downarrow$)

Large voltage ($V\uparrow$)

V

V

Small current ($I\downarrow$)

Large current ($I\uparrow$)

(b) Electric circuit

FIGURE 5-1   Current Is Proportional to Voltage (a) Fluid System (b) Electric Circuit.

water will be permitted to flow. Water flow is therefore inversely proportional to the valve's opposition:

$$\text{Water flow} \frac{1}{\propto} \text{valve opposition}$$

In a similar way, the electric current (I) shown in Figure 5-2(b) is inversely dependent on the opposition or resistance (R) offered by the resistor in the circuit. For instance, a small-value resistor will offer only a small opposition or resistance ($R\downarrow$), which will permit a large circuit current ($I\uparrow$), whereas a large-value resistor will offer a large opposition or resistance ($R\uparrow$), which will allow only a small circuit current ($I\downarrow$). It can be said, therefore, that electric current, which is the flow of electrons, is inversely proportional to electric resistance, which is the restriction to current flow:

$$\text{Current } (I) \frac{1}{\propto} \text{resistance } (R)$$

Once again, referring to the Ohm's law formula, you can see that current (I) and resistance (R) are inversely proportional to each other because one quantity is above the fraction bar and the other is below the fraction bar on opposite sides of the equals sign:

$$\boxed{\text{Current } (I)} = \frac{\text{voltage } (V)}{\boxed{\text{resistance } (R)}}$$

Small opposition = large water flow   Large opposition = small water flow

(a) Fluid system

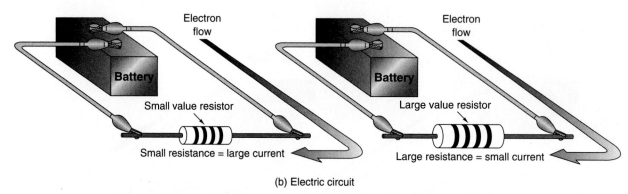

Small resistance = large current   Large resistance = small current

(b) Electric circuit

FIGURE 5-2   Current Is Inversely Proportional to Resistance (a) Fluid System (b) Electric Circuit.

■ INTEGRATED MATH APPLICATION:

Calculate what voltage would cause 2 amperes (A) of electric current in a circuit having an opposition or resistance of 3 ohms (Ω).

■ *Solution:*

To solve this problem, begin by placing the values in their appropriate position within the formula. In this example,

$$\text{Current } (I) = \frac{\text{voltage } (V)}{\text{resistance } (R)}$$

Therefore,

$$2\,A = \frac{V?}{3\,\Omega}$$

or

$$2 = \frac{V}{3}$$

As you can see, we will first need to transpose the current formula ($I = V/R$) so that we can find a formula for voltage (V), because V is the unknown in this example.

$$\text{Current } (I) = \frac{\text{voltage } (V)}{\text{resistance } (R)}$$

$$I = \frac{V}{R}$$

Solve for V	*Solve for R*

Solve for V

$$\frac{I}{-} = \frac{V}{R}$$

$$\frac{R \times I}{R} = \frac{V \times R}{R} \quad \text{(Multiply both sides by } R.\text{)}$$

$$\frac{R \times I}{\cancel{R}} = \frac{V \times \cancel{R}}{\cancel{R}} \quad (R \div R \text{ cancels.)}$$

$$R \times I = V$$

$$\boxed{V = I \times R}$$

Solve for R

$$\frac{I}{-} = \frac{V}{R}$$

$$\frac{R \times I}{R} = \frac{V \times R}{R} \quad \text{(Multiply both sides by } R \text{ to bring } R \text{ above the line.)}$$

$$\frac{R \times I}{\cancel{R}} = \frac{V \times \cancel{R}}{\cancel{R}} \quad (R \div R \text{ cancels.)}$$

$$\frac{R \times I}{I} = \frac{V}{I} \quad \text{(Divide both sides by } I \text{ to isolate } R.\text{)}$$

$$\frac{R \times \cancel{I}}{\cancel{I}} = \frac{V}{I} \quad (I \div I \text{ cancels.)}$$

$$\boxed{R = \frac{V}{I}}$$

Now that we have a formula for voltage ($V = I \times R$), we can complete this problem:

$$V = I \times R \quad \text{(Voltage} = \text{current} \times \text{resistance)}$$
$$= 2\,\text{A} \times 3\,\Omega$$
$$= 6\,\text{V}$$

This equation means that a voltage or electrical pressure of 6 volts would cause 2 amperes of electric current in a circuit having 3 ohms of resistance.

To test if these three formulas are correct, we can now plug into each formula the values of the preceding example, as follows:

$$I = \frac{V}{R} \qquad V = I \times R \qquad R = \frac{V}{I}$$

$$2\,A = \frac{6\,V}{3\,\Omega} \qquad 6\,V = 2\,A \times 3\,\Omega \qquad 3\,\Omega = \frac{6\,V}{2\,A}$$

As you can see from these formulas, the transposition has been successful.

■ **INTEGRATED MATH APPLICATION:**

Figure 5-3 shows a basic circuit for a home electrical system. If each household light and appliance has 120 volts connected across it, how much current will pass through lamp 1, lamp 2, the hair dryer, and the space heater?

■ *Solution:*

$$\text{Current through lamp 1 } (I_{L1}) = \frac{V_{L1}}{R_{L1}} = \frac{120\,\text{V}}{125\,\Omega} = 960\,\text{mA}$$

$$\text{Current through lamp 2 } (I_{L2}) = \frac{V_{L2}}{R_{L2}} = \frac{120\,\text{V}}{250\,\Omega} = 480\,\text{mA}$$

$$\text{Current through hair dryer } (I_{HD}) = \frac{V_{HD}}{R_{HD}} = \frac{120\,\text{V}}{40\,\Omega} = 3\,\text{A}$$

$$\text{Current through space heater } (I_{SH}) = \frac{V_{SH}}{R_{SH}} = \frac{120\,\text{V}}{12\,\Omega} = 10\,\text{A}$$

FIGURE 5-3 **Home Electrical System.**

3. Power Formula Example

Another formula frequently used in electricity and electronics is the power formula. We will also study this formula so that we can practice formula transposition again.

Electric power is the rate at which electric energy is converted into some other form of energy. Power, which is symbolized by *P*, is measured in *watts* (symbolized by **W**) in honor of James Watt. We hear watt used frequently in connection with the brightness of a lightbulb: for instance, a 60 watt bulb, a 100 watt bulb, and so on. This designation is a power rating and describes how much electric energy is converted every second. Because a lightbulb generates both heat and light energy, we would describe a 100 watt lightbulb as a device that converts 100 watts of electric energy into heat and light energy every second. Referring to Figure 5-1(b), you can see a lightbulb connected in an electric circuit. The amount of electric power supplied to that lightbulb is dependent on the voltage and the current. It is probably no surprise, therefore, that power is equal to the product of voltage and current. To state the formula:

> Power = voltage × current
> $P = V \times I$

Quantity	Symbol	Unit	Symbol
Power	*P*	Watts	W
Voltage	*V*	Volts	V
Current	*I*	Amperes	A

This formula states that the amount of power delivered to a device is dependent on the electrical pressure or voltage applied across the device and the electric current flowing through the device.

■ INTEGRATED MATH APPLICATION:

What will be the electric current (*I*) in a circuit if a 60 watt lightbulb is connected across a 120 volt battery?

■ *Solution:*

Let us begin by inserting the known values in their appropriate places in the power formula.

$$\text{Power } (P) = \text{voltage } (V) \times \text{current } (I)$$
$$60 \text{ W} = 120 \text{ V} \times ? \text{ amperes}$$

Power (Electric)

The amount of energy converted by a component or circuit in a unit of time, normally seconds. It is measured in watts or joules per second.

CHAPTER 5 / ALGEBRA, EQUATIONS, AND FORMULAS

As you can see, we first need to transpose the power formula ($P = V \times I$) so that we can find a formula for current (I), because I is the unknown in this example.

$$P = V \times I$$

Solve for V:

$$P = V \times I$$
$$\frac{P}{I} = \frac{V \times I}{I} \quad \text{(Divide both sides by } I.)$$
$$\frac{P}{I} = \frac{V \times \cancel{I}}{\cancel{I}} \quad (I \div I = 1, \text{ and } V \times 1 = V.)$$
$$\frac{P}{I} = V$$

$$\boxed{V = \frac{P}{I}}$$

This formula can be used to calculate V when P and I are known.

Solve for I:

$$P = V \times I$$
$$\frac{P}{V} = \frac{V \times I}{V} \quad \text{(Divide both sides by } V.)$$
$$\frac{P}{V} = \frac{\cancel{V} \times I}{\cancel{V}} \quad (V \div V = 1, \text{ and } I \times 1 = I.)$$
$$\frac{P}{V} = I$$

$$\boxed{I = \frac{P}{V}}$$

This formula can be used to calculate I when P and V are known.

Now that we have a formula for current ($I = P/V$), we can complete this problem:

$$I = \frac{P}{V} \quad \left(\text{Current} = \frac{\text{power}}{\text{voltage}}\right)$$
$$= \frac{60 \text{ W}}{120\text{V}}$$
$$= 0.5 \text{ A} \quad \text{or} \quad 500 \times 10^{-3} \text{ A} \leftarrow (0.500.0)$$
$$= 500 \text{ mA}$$

This result means that a current of 500 milliamperes will exist in an electric circuit if a 120 volt battery is connected across a 60 watt light bulb.

CALCULATOR KEYS

Name: Solve function key

Function: Some calculators have a function that solves an expression for a specified variable.

Solve (equation, variable)

Example: Solve for R, when $I = V/R$
Press keys: Solve($I = V/R,R$)
Answer: $R = V/I$

CALCULATOR KEYS *(continued)*

Solve for t, when $250 = 50 \times t$.
Press keys: Solve ($250 = 50 \times t,t$)
Answer: $t = 5$

Solve for y, when $(6 \times y) + 27 = y + 27$.
Press keys: Solve$((6 \times y) + 7 = y + 27, y)$
Answer: $y = 4$

SELF-TEST EVALUATION POINT FOR SECTION 5-2

Now that you have completed this section, you should be able to:

- **Objective 4.** *Explain the steps involved in transposing an equation to determine the unknown.*
- **Objective 5.** *Demonstrate how to develop an equation from a story problem.*
- **Objective 6.** *Describe the process of factoring and how to remove parentheses within a formula or expression.*
- **Objective 7.** *Demonstrate the process of transposing a formula.*
- **Objective 8.** *Describe how the terms* directly proportional *and* inversely proportional *are used to show the relationship between quantities in a formula.*
- **Objective 9.** *Explain the proportionality among the three quantities in the Ohm's law formula, and describe how to transpose the formula.*
- **Objective 10.** *Describe the electric power formula, and transpose the formula to develop alternative procedures.*

Use the following questions to test your understanding of Section 5-2.

1. Calculate the result of the following arithmetic operations involving literal numbers.
   a. $x + x =$ **d.** $2x - x =$
   b. $x \times x =$ **e.** $\dfrac{x}{x} =$
   c. $7a + 4a =$ **f.** $x - x =$

2. Transpose the following equations to determine the unknown quantity.
   a. $x + 14 = 30$
   b. $8 \times x = \dfrac{80 - 40}{10} \times 12$
   c. $y - 4 = 8$
   d. $(x \times 3) - 2 = \dfrac{26}{2}$
   e. $x^2 + 5 = 14$
   f. $2(3 + 4x) = (2x + 13)$

3. Transpose the following formulas.
   a. $x + y = z, y = ?$
   b. $Q = C \times V, C = ?$
   c. $x_L = 2 \times \pi \times f \times L, L = ?$
   d. $V = I \times R, R = ?$

4. Determine the unknown in the following equations by using transposition.
   a. $I^2 = 9$ **b.** $\sqrt{z} = 8$

5-3 SUBSTITUTION

Substitution

The replacement of one mathematical entity by another of equal value.

Substitution is a mathematical process used to develop alternative formulas by replacing or substituting one mathematical term with an equivalent mathematical term. To explain this process, let us try a couple of examples.

Circle

A closed plane curve, every point of which is equidistant from a fixed point within the curve.

5-3-1 *Circle Formula Example*

A **circle** is a perfectly round figure in which every point is at an equal distance from the center.

Center
Point

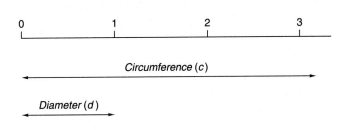

Some of the elements of a circle are as follows.

a. The **circumference** (C) of a circle is the distance around the perimeter of the circle.

b. The **radius** (r) of a circle is a straight line drawn from the center of the circle to any point of the circumference.

c. The **diameter** (d) of a circle is a straight line drawn through the center of the circle to opposite sides of the circle. The diameter of a circle is therefore equal to twice the circle's radius ($d = 2 \times r$).

To review, *pi* (symbolized by π) is the name given to the ratio, or comparison, of a circle's circumference to its diameter. This constant will always be equal to approximately 3.14 because the circumference of any circle will always be about 3 times larger than the same circle's diameter. Stated mathematically:

$$\pi = \frac{\text{circumference } (C)}{\text{diameter } (d)} = 3.14$$

Circumference
The perimeter of a circle.

Radius
A line segment extending from the center of a circle or sphere to the circumference.

Diameter
The length of a straight line passing through the center of an object.

■ INTEGRATED MATH APPLICATION:

Determine the circumference of a bicycle wheel that is 26 inches in diameter.

■ *Solution:*

To calculate the wheel's circumference, we will have to transpose the following formula to isolate C.

$$\pi = \frac{C}{d} \qquad \text{Multiply both sides by } d.$$

$$\pi \times d = \frac{C}{d} \times d \qquad d \div d = 1, \text{ and } C \times 1 = C.$$

$$\pi \times d = C$$

The circumference of the bicycle wheel is therefore:

$$C = \pi \times d$$
$$C = 3.14 \times 26 \text{ in.} = 81.64 \text{ in.}$$

In some examples, only the circle's radius will be known. In situations like this, we can substitute the expression $\pi = C/d$ to arrive at an equivalent expression comparing a circle's circumference to its radius.

$$\pi = \frac{C}{d} \qquad \text{Because } d = 2 \times r, \text{ we can substitute } 2 \times r \text{ for } d.$$

$$\pi = \frac{C}{d} \qquad \text{Therefore, this is equivalent to } \pi = \frac{C}{2 \times r}$$

How much decorative edging will you need to encircle a garden that measures 6 feet from the center to the edge?

■ *Solution:*

In this example *r* is known, so to calculate the garden's circumference, we will have to transpose the following formula to isolate *C*:

$$\pi = \frac{C}{2 \times r}$$ Multiply both sides by $2 \times r$.

$$\pi \times (2 \times r) = \frac{C}{2 \times r} \times (2 \times r)$$ $(2 \times r) \div (2 \times r) = 1$, and $C \times 1 = C$.

$$\pi \times (2 \times r) = C$$

The edging needed for the circumference of the garden will therefore be:

$$C = \pi \times (2 \times r)$$
$$C = 3.14 \times (2 \times 6 \text{ ft}) = 37.68 \text{ ft}$$

To continue our example of the circle, how could we obtain a formula for calculating the area within a circle? The answer is best explained with the diagram shown in Figure 5-4. In Figure 5-4(a) a circle has been divided into eight triangles, and then these triangles have been separated as shown in Figure 5-4(b). Figure 5-4(c) shows how each of these triangles can be thought of as having an A and a B section that, if separated and rearranged, will form a square. The area of a square is easy to calculate because it is simply the product of the square's width and height (area = $w \times h$). The area of the triangles shown in Figure 5-4(c), therefore, will be

$$\text{Area} = \frac{1}{2} \times (w \times h)$$

Referring to Figure 5-4(d), you can see that the width of all eight triangles within a circle is equal to the circle's circumference ($w = C$), and the height of the triangles is equal to the circle's radius ($h = r$). The area of the circle, therefore, can be calculated with the following formula:

$$\text{Area of a circle} = \frac{1}{2} \times (C \times r)$$

Using both substitution and transposition, let us now try to reduce this formula with $C = \pi \times (2 \times r)$.

$$\text{Area of a circle} = \frac{1}{2} \times (C \times r)$$ ← Because $C = \pi \times (2 \times r)$, we can replace C with $\pi \times (2 \times r)$.

$$= \frac{1}{2} \times [\pi \times (2 \times r) \times r]$$ ← We can now use transposition to reduce.

$$= \frac{1 \times \pi \times 2 \times r \times r}{2}$$ ← $2 \div 2 = 1$

$$= 1 \times \pi \times r \times r$$ ← $r \times r = r^2$

$$= 1 \times \pi \times r^2$$ ← $1 \times (\pi \times r^2) = \pi \times r^2$

$$\text{Area of a circle} = \pi \times r^2$$

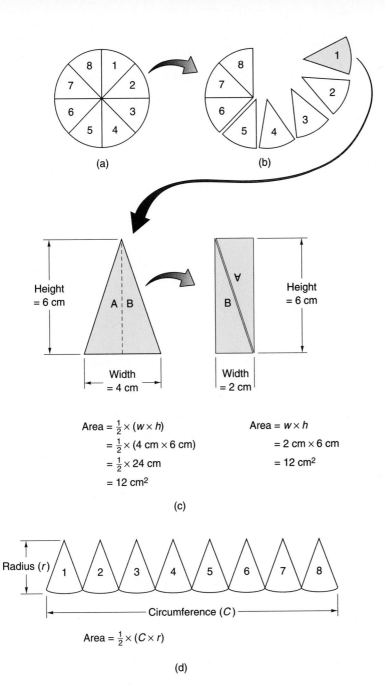

FIGURE 5-4 **Calculating the Area of a Circle.**

■ **EXAMPLE:**

Calculate the area of a circle that has a radius of 14 inches.

■ *Solution:*

$$\begin{aligned}
\text{Area of a circle} &= \pi \times r^2 \\
&= 3.14 \times (14 \text{ in.})^2 \\
&= 3.14 \times 196 \text{ in.}^2 \\
&= 615.4 \text{ in.}^2
\end{aligned}$$

■ **EXAMPLE:**

Calculate the radius of a circle that has an area of 624 square centimeters (cm^2).

■ *Solution:*

To determine the radius of a circle when its area is known, we have to transpose the formula area of a circle $= \pi \times r^2$ to isolate r.

$$\text{Area of a circle} = \pi \times r^2$$

$$A = \pi \times r^2 \qquad \leftarrow \text{Divide both sides by } \pi.$$

$$\frac{A}{\pi} = \frac{\pi \times r^2}{\pi} \qquad \leftarrow \pi \div \pi = 1,\ 1 \times r^2 = r^2$$

$$\frac{A}{\pi} = r^2 \qquad \leftarrow \text{Take the square root of both sides.}$$

$$\sqrt{\frac{A}{\pi}} = \sqrt{r^2} \qquad \leftarrow \sqrt{r^2} = r$$

$$\sqrt{\frac{A}{\pi}} = r$$

Now we can calculate the radius of a circle that has an area of 624 cm^2.

$$r = \sqrt{\frac{A}{\pi}} = \sqrt{\frac{624}{3.14}} = \sqrt{198.7} = 14.1 \text{ cm}$$

5-3-2 *Power Formula Example*

In Section 5-2 you were introduced to two electrical formulas: the Ohm's law formula, which states

$$\text{Current } (I) = \frac{\text{voltage } (V)}{\text{resistance } (R)}$$

and the power formula, which states

$$\text{Power } (P) = \text{voltage } (V) \times \text{current } (I)$$

Using transposition we were able to develop the following alternative formulas that allowed us to calculate any quantity if two quantities were known.

OHM'S LAW FORMULA			POWER FORMULA		
FORMULA	CALCULATES	WHEN THESE ARE KNOWN:	FORMULA	CALCULATES	WHEN THESE ARE KNOWN:
$I = \dfrac{V}{R}$	Current (I) in amperes (A)	Voltage (V) and resistance (R)	$P = V \times I$	Power (P) in watts (W)	Voltage (V) and current (I)
$V = I \times R$	Voltage (V) in volts (V)	Current (I) and resistance (R)	$V = \dfrac{P}{I}$	Voltage (V) in volts (V)	Power (P) and current (I)
$R = \dfrac{V}{I}$	Resistance (R) in ohms (Ω)	Voltage (V) and current (I)	$I = \dfrac{P}{V}$	Current (I) in amperes (A)	Power (P) and voltage (V)

Although it seems from this table that we have covered all possible alternatives, there are a few missing. For example, consider the power formula $P = V \times I$. When trying to calculate power, we may not always have voltage (V) and current (I). We may, for example, have values only for voltage (V) and resistance (R), or current (I) and resistance (R). In situations such as this the mathematical process called *substitution* can be used to develop alternative power formulas by replacing or substituting one mathematical term for an equivalent mathematical term. The following process shows how we can substitute terms in the $P = V \times I$ formula to arrive at alternative power formulas for wattage calculations.

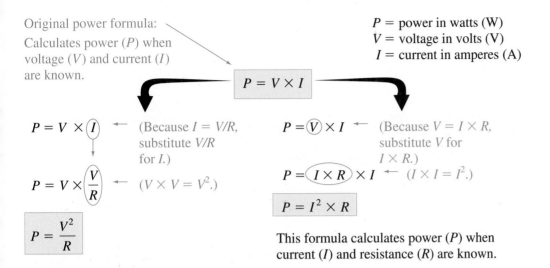

Original power formula:
Calculates power (P) when voltage (V) and current (I) are known.

$P =$ power in watts (W)
$V =$ voltage in volts (V)
$I =$ current in amperes (A)

$$P = V \times I$$

$P = V \times \widehat{I}$ ←   (Because $I = V/R$, substitute V/R for I.)

$P = V \times \left(\dfrac{V}{R}\right)$ ←   ($V \times V = V^2$.)

$$P = \dfrac{V^2}{R}$$

This formula calculates power (P) when voltage (V) and resistance (R) are known.

$P = \widehat{V} \times I$ ←   (Because $V = I \times R$, substitute V for $I \times R$.)

$P = \widehat{I \times R} \times I$ ←   ($I \times I = I^2$.)

$$P = I^2 \times R$$

This formula calculates power (P) when current (I) and resistance (R) are known.

Let us now look at two applications to see how we can use these formulas to calculate unknown quantities.

■ INTEGRATED MATH APPLICATION:

The household heater shown here has a resistance of 7 ohms and is connected to the 120 volt source at the wall outlet. Calculate the amount of power dissipated by the heater.

Electric heater

120 V

■ *Solution:*

In this example, both voltage (120 V) and resistance (7 Ω) are known, and the unknown to be calculated is power (P). We must therefore look at all the power formulas to select one that includes voltage (V) and resistance (R).

$$P = V \times I$$

$$\widehat{P = \dfrac{V^2}{R}}$$ ←   This power formula is the one to use because it includes V and R.

$$P = I^2 \times R$$

$$\text{Power } (P) = \frac{\text{voltage}^2 \, (V^2)}{\text{resistance } (R)}$$

Replace the literal numbers *V* and *R* with values.

$$= \frac{(120 \text{ V})^2}{7 \, \Omega}$$

$120^2 = 14{,}400$

$$= \frac{14{,}400}{7 \, \Omega}$$

$$= 2057 \text{ W}$$

2.057

or approximately 2 kW

■ INTEGRATED MATH APPLICATION:

The lightbulb shown here is rated at 6 volts, 60 milliamps.

  a. Using Ohm's law, calculate the resistance of the electric lamp or lightbulb.
  b. Using the power formula, calculate the lamp's wattage at its rated current and resistance.

■ *Solution:*

Part (a) asks us to use Ohm's law to calculate the resistance of the electric lamp.

Ohm's law

$$I = \frac{V}{R}$$

$$V = I \times R$$

$$R = \frac{V}{I}$$

$$\text{Resistance } (R) = \frac{\text{voltage } (V)}{\text{current } (I)}$$

$$= \frac{6 \text{ V}}{60 \text{ mA}}$$

$$= \frac{6 \text{ V}}{60 \times 10^{-3} \text{ A}}$$

$$= 100 \, \Omega$$

Light bulb

Part (b) directs us to use the power formula (*P*) to calculate the lamp's wattage at its rated current (*I*) and resistance (*R*). *I* and *R* are known, so we will use the following formula:

Power formula

$$P = V \times I$$

$$P = \frac{V^2}{R}$$

$$P = I^2 \times R$$

$$\text{Power } (P) = \text{current}^2 (I^2) \times \text{resistance}$$

$$= (60 \text{ mA})^2 \times 100 \, \Omega$$

$$= (60 \times 10^{-3})^2 \times 100 \, \Omega$$

$$= 0.0036 \times 100$$

$$= 0.36 \text{ W } (0.360)$$

$$\text{or } 360 \times 10^{-3} \text{ W}$$

$$\text{or } 360 \text{ mW}$$

As a review of the Ohm's law and power formula combinations, Figure 5-5 shows all the possible formulas for the four electrical properties.

Now that you have completed this section, you should be able to:

■ **Objective 12.** *Summarize the terms, rules, properties, and mathematical operations of algebra.*

Use the following questions to test your understanding of Section 5-4.

1. What is the order of operations?

2. Give an example of a:

   a. Monomial   **c.** Trinomial
   b. Binomial   **d.** Polynomial

3. Determine the following:

   a. $a \div 1 = ?$
   b. $a + a = ?$
   c. $a - a = ?$
   d. $a \times a = ?$
   e. $3x + b - 2x = ?$

4. Determine the unknown:

   a. $16 + a = 5 \times a, a = ?$
   b. $\dfrac{42}{b} = 3 \times 7$

REVIEW QUESTIONS

Multiple Choice Questions

1. If you perform exactly the same mathematical operation on both sides of an equation, the equality is preserved.

   a. True
   b. False

2. Are the two sides of the following equation, $(3 - 2) \times 16 = 32 \div 2$, equal?

   a. Yes
   b. No
   c. Only if the right half is doubled
   d. Only if the left half is doubled

3. $x + x =$ _____, $x \times x =$ _____, $x \div x =$ _____, and $x - x$ _____.

   a. $x^2, 2x, 1, 0$
   b. $2x, \sqrt{x}, 0, 1$
   c. $1x, 2x, 1, 0$
   d. $2x, x^2, 1, 0$

4. What is the value of the unknown in the following equation: $3a + 9 = 2a + 10$?

   a. 2   **c.** 1
   b. 10   **d.** 3

5. Calculate the value x for the following equation: $x + y^2 = y^2 + 9$.

   a. 3   **c.** 9
   b. 6   **d.** 12

6. What is the value of the unknown, s, in the following equation: $\sqrt{s^2} + 14 = 26$?

   a. 12   **c.** 4
   b. 26   **d.** 5

7. What is the correct transposition for the following formula: $a = \dfrac{b - c}{x}, x = ?$

   a. $x = \dfrac{b - a}{c}$ **c.** $x = \dfrac{b - c}{a}$
   b. $x = (b - a) \times c$ **d.** $x = x(b - c)$

8. If $V = I \times R$ and $P = V \times I$, develop a formula for P when I and R are known.

   a. $P = \dfrac{I}{R}$
   b. $P = I^2 \times R^2$
   c. $P = I^2 \times R$
   d. $P = \sqrt{\dfrac{I}{R}}$

9. If $E = M \times C^2, M = ?$ and $C = ?$

   a. $\sqrt{\dfrac{E}{C}}, \dfrac{E^2}{M}$
   b. $\dfrac{E}{C^2}, \sqrt{\dfrac{E}{M}}$
   c. $E \times C^2, \sqrt{C^2 \times E}$
   d. $\dfrac{C^2}{E}, \sqrt{\dfrac{M}{E}}$

10. If $P = \dfrac{V^2}{R}$, transpose to obtain a formula to solve for V.

   a. $V = \sqrt{P \times R}$
   b. $V = \dfrac{P^2}{R}$
   c. $V = P \times R^2$
   d. $V = \sqrt{P} \times R$

$$3 \times a + 9 = 2 \times a + 10$$

$$3x - 2$$

Communication Skill Questions

11. Define the following terms: (5-1)

    a. Algebra   **c.** Equation
    b. Formula   **d.** Literal number

12. Why is it important to treat both sides of an equation equally? (5-1)

13. Describe how the following three factors can be combined in a formula: distance (d), velocity (v), and time (t). (5-2)

14. What two steps must be followed to transpose an equation? (5-2)

15. Describe the factoring process. (5-2)

16. Define the terms *proportional and inversely proportional.* (5-2)

17. Explain how current (I), voltage (V), and resistance (R) can be combined in a formula. (5-2)

18. Transpose the formula from question 22 to solve for I, V, and R. (5-2)

19. If $P = V \times I$, describe how you would solve for V and I. (5-2)

20. Define substitution. (5-3)

21. If $\pi = C/d$, and area $= \frac{1}{2} \times (C \times r)$, describe how substitution and transposition can be employed to obtain a formula for area when only the radius of a circle is known. (5-3)

22. Explain how the parentheses can be removed from the following examples: (5-2)
    a. $4(3b - 1b)$ **b.** $2(3x \times 2x)$

23. If $a = b \times c$, explain how you would solve for b and c. (5-2)

24. Describe the relationship between square and square root. (5-2)

25. Why is formula manipulation important? (5-1)

Practice Problems

26. Calculate the result of the following arithmetic operations involving literal numbers.

    a. $x + x + x = ?$ **g.** $4y \times 3y^2 = ?$
    b. $5x + 2x = ?$ **h.** $a \times a = ?$
    c. $y - y = ?$ **i.** $2y \div y = ?$
    d. $2x - x = ?$ **j.** $6b \div 3b = ?$
    e. $y \times y = ?$ **k.** $\sqrt{a^2} = ?$
    f. $2x \times 4x = ?$ **l.** $(x^3)^2 = ?$

27. Transpose the following equations to determine the unknown value.

    a. $4x = 11$

    b. $6a + 4a = 70$

    c. $5b - 4b = \dfrac{7.5}{1.25}$

    d. $\dfrac{2z \times 3z}{4.5} = 2z$

28. Apply transposition and substitution to Newton's second law of motion.

    a. $F = ma$ [Force (F) = mass (m) $\times$ acceleration (a)] If $F = m \times a$, $m = ?$ and $a = ?$

    b. If acceleration (a) equals change in speed (Δv) divided by time (t), how would the $F = ma$ formula appear if acceleration (a) was substituted for $\dfrac{\Delta v}{t}$?

29. a. Using the formula power (P) = voltage (V) $\times$ current (I), calculate the value of electric current through a hairdryer that is rated as follows: power = 1500 watts and voltage = 120 volts.

$$\boxed{P = V \times I}$$

where P = power in watts
V = voltage in volts
I = current in amperes

    b. Using the Ohm's formula voltage (V) = current (I) $\times$ resistance (R), calculate the resistance or opposition to current flow offered by the hairdryer in Question 14a.

30. The following formula is used to calculate the total resistance (R_T) in a parallel circuit containing two parallel resistances (R_1 and R_2).

$$R_T = \cfrac{1}{\dfrac{1}{R_1} + \dfrac{1}{R_2}}$$

    a. Calculate R_T (total resistance) if the resistance of $R_1 = 6$ ohms and the resistance of $R_2 = 12$ ohms.

    b. Indicate the calculator sequence used for part (a).

    c. If R_2 and R_T were known and we needed to calculate the value of R_1, we would have to transpose the total parallel resistance formula to solve for R_1. The following steps show how to transpose the total resistance formula to calculate R_1 when R_T and R_2 are known. Describe what has occured in each of the steps.

$$R_T = \cfrac{1}{\dfrac{1}{R_1} + \dfrac{1}{R_2}} \quad \text{(Original formula)}$$

1. $R_T \times \left(\dfrac{1}{R_1} + \dfrac{1}{R_2}\right) = \cfrac{1}{\cancel{\dfrac{1}{R_1}} + \cancel{\dfrac{1}{R_2}}} \times \left(\cancel{\dfrac{1}{R_1}} + \cancel{\dfrac{1}{R_2}}\right)$

$R_T \times \left(\dfrac{1}{R_1} + \dfrac{1}{R_2}\right) = 1$

2. $\dfrac{\cancel{R_T} \times \left(\dfrac{1}{R_1} + \dfrac{1}{R_2}\right)}{\cancel{R_T}} = \dfrac{1}{R_T}$

$\dfrac{1}{R_1} + \dfrac{1}{R_2} = \dfrac{1}{R_T}$

or $\dfrac{1}{R_T} = \dfrac{1}{R_1} + \dfrac{1}{R_2}$

3. $\dfrac{1}{R_T} - \dfrac{1}{R_2} = \dfrac{1}{R_1} + \cancel{\dfrac{1}{R_2}} - \cancel{\dfrac{1}{R_2}}$

$\qquad \dfrac{1}{R_T} - \dfrac{1}{R_2} = \dfrac{1}{R_1}$

$\qquad \text{or} \quad \dfrac{1}{R_1} = \dfrac{1}{R_T} - \dfrac{1}{R_2}$

4. $\dfrac{1}{R_1} = \boxed{\dfrac{1}{R_T} - \dfrac{1}{R_2}}$

$\qquad \dfrac{1}{R_T} - \dfrac{1}{R_2} = \dfrac{}{\textit{lowest common denominator (LCD)}}$

$\qquad \dfrac{1}{R_T} - \dfrac{1}{R_2} = \dfrac{}{R_T \times R_2}$

$\qquad \dfrac{1}{R_T} - \dfrac{1}{R_2} = \dfrac{R_2 - R_T}{R_T \times R_2}$

$\qquad \dfrac{1}{R_T} - \dfrac{1}{R_2} = \boxed{\dfrac{R_2 - R_T}{R_T \times R_2}}$

$\qquad \dfrac{1}{R_1} = \dfrac{R_2 - R_T}{R_T \times R_2}$

(LCD) method 1: fractions

$R_T \times R_2 = R_T \times R_2$

Steps:

$(R_T \times R_2) \div R_T = R_2,$

$R_2 \times 1 = R_2$

$(R_T \times R_2) \div R_2 = R_T,$

$R_T \times 1 = R_T$

5. $\dfrac{1}{\cancel{R_1}} \times \cancel{R_1} = \dfrac{R_2 - R_T}{R_T \times R_2} \times R_1$

$\qquad 1 = \dfrac{R_2 - R_T}{R_T \times R_2} \times R_1$

6. $1 \times R_T \times R_2 = \dfrac{R_2 - R_T}{\cancel{R_T} \times \cancel{R_2}} \times R_1 \times \cancel{R_T} \times \cancel{R_2}$

$\qquad R_T \times R_2 = (R_2 - R_T) \times R_1$

7. $\dfrac{R_T \times R_2}{R_2 - R_T} = \dfrac{(\cancel{R_2} - \cancel{R_T}) \times R_1}{\cancel{R_2} - \cancel{R_T}}$

$\qquad \dfrac{R_T \times R_2}{R_2 - R_T} = R_1 \qquad \text{or} \qquad R_1 = \dfrac{R_T \times R_2}{R_2 - R_T}$

d. Using the values from part (a), double-check the new formula.

e. If we wanted to obtain a formula for calculating R_2 when both R_T and R_1 are known, which of the steps would we change, and in what way?

f. Give the formula for calculating R_2 when both R_T and R_1 are known.

g. Using the values from part (a), double-check this new formula.

Web Site Questions

Go to the web site http://www.prenhall.com/cook, select the textbook *Introductory Computer Mathematics, 2e,* select this chapter, and then follow the instructions when answering the multiple-choice practice problems.

Geometry and Trigonometry

Finding the Question to the Answer!

More than 350 years ago in 1637, Pierre de Fermat, a French mathematician and physicist, wrote an apparently simple theorem in the margins of a book. He also added that he had discovered marvelous proof of it but lacked enough space to include it in the margin. Later that night he died and took with him a mystery that mathematicians have been trying to solve ever since. Many of the brightest minds in mathematics have struggled to find the proof, and many have concluded that Fermat must have been mistaken despite his considerable mathematical ability.

Fermat's last theorem, as it has been appropriately named, has to do with equations of the form x to the nth power, plus y to the nth power = z to the nth power.

$$x^n + y^n = z^n$$

A power of 2 (square) is familiar as the Pythagorean theorem, which states that the sum of the squares of the lengths of two sides of a right-angle triangle is equal to the square of the length of the other side.

$$c^2 = a^2 + b^2$$

For example,

$$5^2 = 3^2 + 4^2$$
$$25 = 9 + 16$$

Fermat's last theorem seemed not to work because, for example,

$$? = 3^3 + 4^3$$
$$91 = 27 + 64$$

and there is not a whole number that can be cubed to equal 91.

Now, after thousands of claims of success that proved untrue, mathematicians say the daunting challenge, perhaps the most famous of unsolved mathematical problems, has been surmounted. The conqueror is Andrew Wiles, 40, a British mathematician who works at Princeton University. Wiles announced the result at the last of three lectures at Cambridge University in England. Within a few minutes of the conclusion of his final lecture, computer mail messages were winging around the world, sending mathematicians into a frenzy. It had apparently taken 7 years for Wiles to solve the problem. He continued a chain of ideas that was begun in 1954 by several other mathematicians.

After being criticized by many in the eighteenth and nineteenth centuries and ridiculed in the twentieth century by mathematicians with supercomputers, Fermat was elevated finally in 1993 to the status of "a genius before his time."

Outline and Objectives

Introduction

Geometry is one of the oldest branches of mathematics, and deals with the measurement, properties, and relationships of points, lines, angles, plane figures or surfaces, and solid figures. The practical attributes of geometry were used by the ancient Egyptians to build the pyramids, and today, it is widely used in almost every branch of engineering technology. Unlike algebra, which deals with unknowns that are normally represented by a letter, geometry deals with a physical point, line, surface, or figure.

The first section of this chapter deals with the basic elements of geometry, the second section covers plane figures, the third trigonometry, and the final section examines solid figures.

6-1 BASIC GEOMETRIC TERMS

Figure 6-1(a) shows some examples of line segments. In the first example you can see that a line segment exists between ending points A and B. An abbreviation is generally used to represent a line segment, in which a bar is placed over the two end-point letters, and so "line segment AB" could be written as $\overline{AB}$.

Figure 6-1(a) also shows an example of a vertical line that extends straight up and down, a horizontal line that extends from left to right, and two parallel lines that extend in the same direction while remaining at the same distance from one another.

In Figure 6-1(b) an angle is formed between two intersecting line segments or lines, which are also called **sides.** The common point at which they connect is called the **vertex.** The angle symbol $\angle$ is generally used as an abbreviation, and so "angle ABC" could be written as $\angle ABC$. The unit of measurement for the opening between the sides, or angle, is the degree (symbolized " ° ").

Figure 6-1(c) shows how a protractor can be used to measure angles. An image of a protractor has been printed in the back of this text, and can, if copied, be used for measuring and drawing angles. In this example, we are using the protractor to measure $\angle ABC$, from Figure 6-1(b). The 3-step procedure for measuring an angle using a protractor is as follows:

Sides

The lines that form an angle.

Vertex

A point where two lines meet.

MEASURING AN ANGLE WITH A PROTRACTOR

Step 1:   Place protractor's center mark on vertex.

Step 2:   Rotate protractor until zero-degree mark is on one line.

Step 3:   Read the measure in degrees where the line intersects the protractor's scale.

You may have noticed that the protractor has a reversing upper and lower scale, so that an angle can be read either clockwise or counter-clockwise. And so, if you were to measure the angle in Figure 6-1(b), you would use the lower scale of the protractor as shown in Figure 6-1(c), moving counterclockwise from 0°. On the other hand, if the angle was in a reverse direction as shown with the 90° angle in Figure 6-1(d), you would use the upper scale of the protractor moving clockwise from 0°.

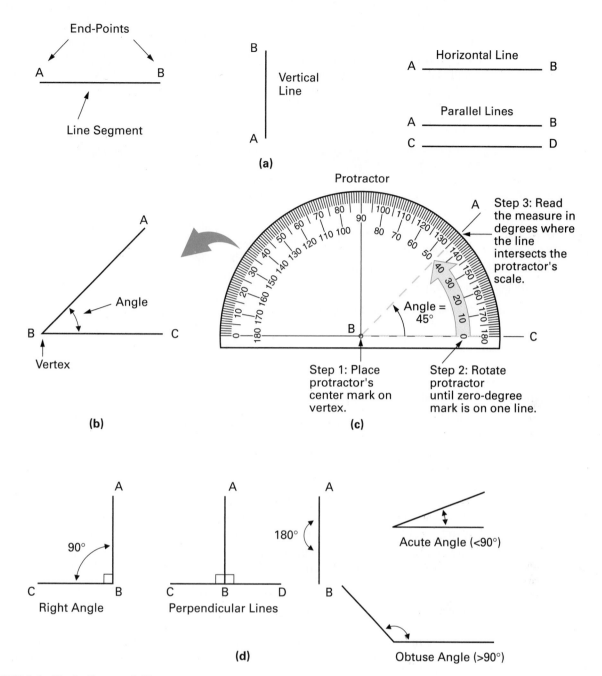

FIGURE 6-1   **Basic Geometric Terms.**

The protractor can also be used to draw an angle. The 3-step procedure for drawing an angle using a protractor is as follows:

DRAWING AN ANGLE WITH A PROTRACTOR

Step 1:   Draw a straight line on the paper representing one side of the angle.

Step 2:   Place the protractor's center mark on the end of the line that is to serve as the angle's vertex.

Step 3:   Draw a dot on the paper next to the protractor's scale equal to the measure of the desired angle, and then draw a line connecting this dot to the vertex.

Figure 6-1(d) shows a few more geometric basics. A right angle is one of 90°, and can be symbolized with a small square in the corner. Two lines are perpendicular to one another when the point at which they intersect forms right angles. The angle measure of a straight angle is 180°. The terms **acute angle** and **obtuse angle** describe angles that are relative to a right angle (90°). An acute angle is one that is less than a right angle (<90°), while an obtuse angle is one that is greater than a right angle (>90°).

SELF-TEST EVALUATION POINT FOR SECTION 6-1

Now that you have completed this section, you should be able to:

- ■ **Objective 1.** *Describe the purpose of geometry, and define many of the basic geometric terms.*

- ■ **Objective 2.** *Demonstrate how to use a protractor to measure and draw an angle.*

Use the following questions to test your understanding of section 6-1:

1. Using a protractor, measure:

a. ∠AXB	**d.** ∠AXE	**g.** ∠AXH	**j.** ∠IXG
b. ∠AXC	**e.** ∠AXF	**h.** ∠AXI	**k.** ∠IXF
c. ∠AXD	**f.** ∠AXG	**i.** ∠IXH	**l.** ∠IXE

2. Which of the angles in the previous question are right angles, obtuse angles, and acute angles?

3. Draw:

a. A vertical line	**e.** An angle of 35°
b. A horizontal line	**f.** An angle of 172°
c. Three parallel lines	**g.** An obtuse angle
d. A right angle	**h.** An acute angle

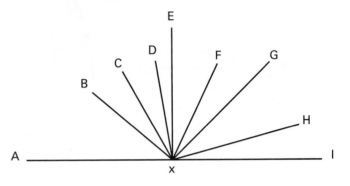

6-2   PLANE FIGURES

A plane figure is a perfectly flat surface, such as a tabletop, floor, wall, or windowpane. In this section we will examine the details of several flat plane surfaces, which, as you will discover, have only two dimensions—length and width.

6-2-1   *Quadrilaterals*

A **polygon** is a plane geometric figure that is made up of three or more line segments. Quadrilaterals, such as the square, rectangle, rhombus, parallelogram, and trapezoid, are all examples of four-sided polygons.

The information given in Figure 6-2 summarizes the facts and formulas for the square, which has four right angles and four equal length sides, and is a **parallelogram** (which means it has two sets of parallel sides). The box in Figure 6-2 lists the formulas for calculating a square's perimeter, area and side length.

Figures 6-3, 6-4, 6-5 and 6-6 on pp.141–143 summarize the facts and formulas for other quadrilaterals. In addition, these diagrams include, when necessary, a visual breakdown of the formulas so that it is easier to see how the formula was derived.

6-2-2   *Right Angle Triangles*

The **right-angle triangle,** or right triangle shown in Figure 6-7(a) (p. 145), has three sides and three corners. Its distinguishing feature, however, is that two of the sides of this triangle are at right angles (at 90°) to one another. The small square box within the triangle is placed in the corner to show that sides *A* and *B* are *square* or at *right angles* to one another.

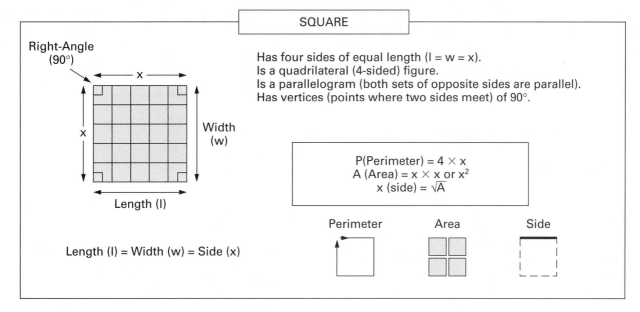

FIGURE 6-2   **The Square.**

If you study this right triangle you may notice two interesting facts about the relative lengths of sides *A, B,* and *C.* These observations are as follows.

1. Side *C* is always longer than side *A* or side *B.*
2. The total length of side *A* and side *B* is always longer than side *C.*

The right triangle in Figure 6-7(b) has been drawn to scale to demonstrate another interesting fact about this triangle. If side *A* were to equal 3 cm and side *B* were to equal 4 cm,

FIGURE 6-3   **The Rectangle.**

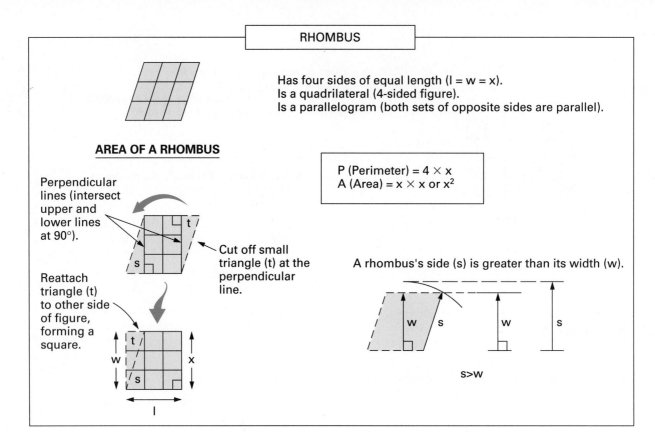

RHOMBUS

Has four sides of equal length (l = w = x).
Is a quadrilateral (4-sided figure).
Is a parallelogram (both sets of opposite sides are parallel).

AREA OF A RHOMBUS

Perpendicular lines (intersect upper and lower lines at 90°).

Cut off small triangle (t) at the perpendicular line.

Reattach triangle (t) to other side of figure, forming a square.

$$P \text{ (Perimeter)} = 4 \times x$$
$$A \text{ (Area)} = x \times x \text{ or } x^2$$

A rhombus's side (s) is greater than its width (w).

s>w

FIGURE 6-4 The Rhombus.

PARALLELOGRAM

Has opposite sides that are equal in length (w = w, l = l).
Is a quadrilateral (4-sided figure).
Is a parallelogram (both sets of opposite sides are parallel).

AREA OF A PARALLELOGRAM

Reattach triangle (t) to other side of figure, forming a rectangle.

Cut off small triangle (t) at the perpendicular line.

$$P \text{ (Perimeter)} = 2l \times 2w$$
$$A \text{ (Area)} = l \times w$$

A Parallelogram's side (s) is greater than its width (w).

s>w

FIGURE 6-5 The Parallelogram.

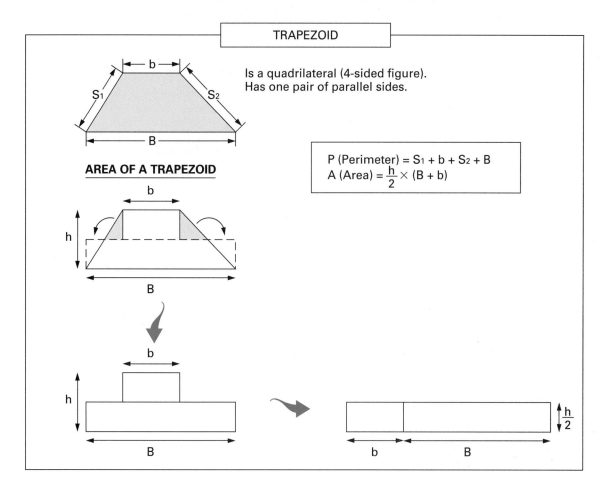

FIGURE 6-6   The Trapezoid.

side C would equal 5 cm. This demonstrates a basic relationship among the three sides and accounts for why this right triangle is sometimes referred to as a *3, 4, 5 triangle*. As long as the relative lengths remain the same, it makes no difference whether the sides are 3 cm, 4 cm, and 5 cm or 30 km, 40 km, and 50 km. Unfortunately, in most applications our lengths of *A, B,* and *C* will not work out as easily as 3, 4, 5.

Pythagoras's Theorem

When a relationship exists among three quantities, we can develop a formula to calculate an unknown when two of the quantities are known. It was Pythagoras who first developed this basic formula or equation (known as the **Pythagorean theorem**), which states that *the square of the length of the hypotenuse (side C) of a right triangle equals the sum of the squares of the lengths of the other two sides:*

$$C^2 = A^2 + B^2 \qquad \left(\begin{matrix} 5^2 = 3^2 + 4^2 \\ 25 = 9 + 16 \end{matrix} \right)$$

By using the rules of algebra, we can transpose this formula to derive formulas for sides *C, B,* and *A*.

Pythagorean Theorem
A theorem in geometry: The square of the length of the hypotenuse of a right triangle equals the sum of the squares of the lengths of the other two sides.

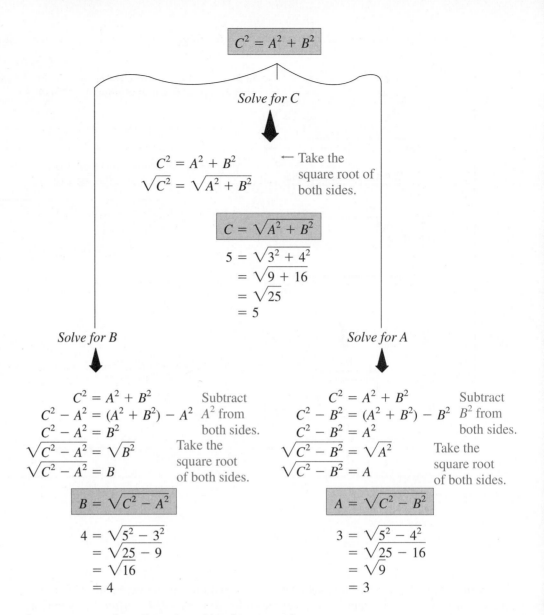

$$C^2 = A^2 + B^2$$

Solve for C

$$C^2 = A^2 + B^2$$
$$\sqrt{C^2} = \sqrt{A^2 + B^2}$$

← Take the square root of both sides.

$$C = \sqrt{A^2 + B^2}$$

$$5 = \sqrt{3^2 + 4^2}$$
$$= \sqrt{9 + 16}$$
$$= \sqrt{25}$$
$$= 5$$

Solve for B

$$C^2 = A^2 + B^2$$
$$C^2 - A^2 = (A^2 + B^2) - A^2$$
$$C^2 - A^2 = B^2$$
$$\sqrt{C^2 - A^2} = \sqrt{B^2}$$
$$\sqrt{C^2 - A^2} = B$$

Subtract A^2 from both sides.
Take the square root of both sides.

$$B = \sqrt{C^2 - A^2}$$

$$4 = \sqrt{5^2 - 3^2}$$
$$= \sqrt{25 - 9}$$
$$= \sqrt{16}$$
$$= 4$$

Solve for A

$$C^2 = A^2 + B^2$$
$$C^2 - B^2 = (A^2 + B^2) - B^2$$
$$C^2 - B^2 = A^2$$
$$\sqrt{C^2 - B^2} = \sqrt{A^2}$$
$$\sqrt{C^2 - B^2} = A$$

Subtract B^2 from both sides.
Take the square root of both sides.

$$A = \sqrt{C^2 - B^2}$$

$$3 = \sqrt{5^2 - 4^2}$$
$$= \sqrt{25 - 16}$$
$$= \sqrt{9}$$
$$= 3$$

Let us now test these formulas with a few examples.

■ **INTEGRATED MATH APPLICATION:**

A 4-foot ladder has been placed in a position 2 feet from a wall. How far up the wall will the ladder reach?

■ *Solution:*

In this example, A is unknown, and therefore

$$A = \sqrt{C^2 - B^2}$$
$$= \sqrt{4^2 - 2^2}$$
$$= \sqrt{16 - 4}$$
$$= \sqrt{12}$$
$$= 3.46 \quad \text{or} \quad \approx 3.5 \text{ ft}$$

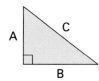

Is a triangle (3-sided figure).
Has one right angle (indicated by square).

(a) AREA OF A RIGHT TRIANGLE

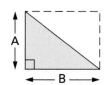

P (Perimeter) = A + B + C

$A\ (Area) = \dfrac{A \times B}{2}$

Sides (A, B, C)
Pythagorean theorem: $C^2 = A^2 + B^2$

$C = \sqrt{A^2 + B^2}$ $B = \sqrt{C^2 - A^2}$ $A = \sqrt{C^2 - B^2}$

(b) SIDES OF A RIGHT TRIANGLE

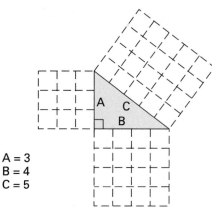

A = 3
B = 4
C = 5

Pythagorean theorem: $A^2 + B^2 = C^2$
Square of side A + Square of side B = Square of side C.
Example: 9 + 16 = 25

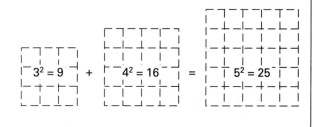

$3^2 = 9$ + $4^2 = 16$ = $5^2 = 25$

FIGURE 6-7   **The Right Triangle.**

■ **INTEGRATED MATH APPLICATION:**

A 16-foot flagpole is casting a 12-foot shadow. What is the distance from the end of the shadow to the top of the flagpole?

■ *Solution:*

In this example, *C* is unknown, and therefore

$$C = \sqrt{A^2 + B^2}$$
$$= \sqrt{16^2 + 12^2}$$
$$= \sqrt{256 + 144}$$
$$= \sqrt{400}$$
$$= 20\ ft$$

In Figure 6-2(c), a 200-foot piece of string is stretched from the top of a 125-foot cliff to a point on the beach. What is the distance from this point to the cliff?

■ *Solution:*

In this example, B is unknown, and therefore

$$B = \sqrt{C^2 - A^2}$$
$$= \sqrt{200^2 - 125^2}$$
$$= \sqrt{40,000 - 15,625}$$
$$= \sqrt{24,375}$$
$$= 156 \text{ ft}$$

6-2-3 *Vectors and Vector Diagrams*

Vector or Phasor

A quantity that has magnitude and direction and that is commonly represented by a directed line segment whose length represents the magnitude and whose orientation in space represents the direction.

A **vector** or **phasor** is an arrow used to represent the magnitude and direction of a quantity. Vectors are generally used to represent a physical quantity that has two properties. For example, Figure 6-8(a) shows a motorboat heading north at 12 miles per hour. Figure 6-8(b) shows how vector **A** could be used to represent the vessel's direction and speed. The size of the vector represents the speed of 12 mph by being 12 cm long, and because we have made the top

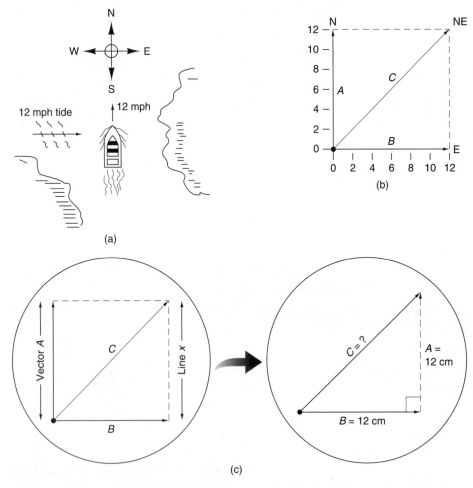

FIGURE 6-8   Vectors and Vector Diagrams.

of the page north, vector **A** should point straight up so that it represents the vessel's direction. Referring to Figure 6-8(a) you can see that there is another factor that also needs to be considered, a 12 mile per hour easterly tide. This tide is represented in our vector diagram in Figure 6-8(b) by vector **B,** which is 12 centimeters long and pointing east. Since the motorboat is pushing north at 12 miles per hour and the tide is pushing east at 12 miles per hour, the resultant course will be northeast, as indicated by vector **C** in Figure 6-8(b). Vector **C** was determined by **vector addition** (as seen by the dashed lines) of vectors **A** and **B**. The result, vector **C,** is called the **resultant vector.** This resultant vector indicates the motorboat's course and speed. The direction or course, we can see, is northeast because vector **C** points in a direction midway between north and east. The speed of the motorboat, however, is indicated by the length or magnitude of vector **C**. This length, and therefore the motorboat's speed, can be calculated by using the Pythagorean theorem. Probably your next question is: How is the **vector diagram** in Figure 6-8(b) similar to a right-angle triangle? The answer is to redraw Figure 6-8(b), as shown in Figure 6-8(c). Because the dashed line x is equal in length to vector **A,** vector **A** can be put in the position of line x to form a right-angle triangle with vector **B** and vector **C**. Now that we have a right-angle triangle with two known lengths and one unknown length, we can calculate the unknown vector's length and therefore the motorboat's speed.

$$
\begin{aligned}
C &= \sqrt{A^2 + B^2} \\
&= \sqrt{12^2 + 12^2} \\
&= \sqrt{144 + 144} \\
&= \sqrt{288} = 16.97
\end{aligned}
$$

The length of vector **C** is 16.97 cm, and because each 1 mile per hour of the motorboat was represented by 1 centimeter, the motorboat will travel at a speed of 16.97 miles per hour in a northeasterly direction.

■ INTEGRATED MATH APPLICATION:

An electrical circuit has two elements that oppose current flow; one form of opposition is called *resistance* (symbolized by R), and the other form of opposition is called *reactance* (symbolized by X). These two oppositions are out of sync, or not in phase, with each other, as seen by the vector diagram in Figure 6-9(a), and therefore cannot simply be added. Using the Pythagorean theorem, vectorially add the two forms of opposition (R and X) to determine the total opposition to current flow, which is called *impedance* (symbolized by Z).

■ Solution:

First, redraw the vector diagram to produce a right-angle triangle, as shown in Figure 6-9(b). Then calculate Z:

$$
\begin{aligned}
Z &= \sqrt{R^2 + X^2} \\
&= \sqrt{30^2 + 16^2} \\
&= \sqrt{900 + 256} \\
&= \sqrt{1156} \\
&= 34 \ \Omega
\end{aligned}
$$

FIGURE 6-9   An Electrical Application of the Pythagorean Theorem.

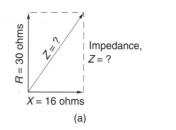

(a)   (b)

The total opposition or impedance (Z) to current flow is therefore 34 ohms.

Vector Addition

Determination of the sum of two out-of-phase vectors using the Pythagorean theorem.

Resultant Vector

A vector derived from or resulting from two or more other vectors.

Vector Diagram

A graphic drawing using vectors that shows arrangement and relations.

SCALENE TRIANGLE

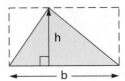

Is a triangle (3-sided figure).
Has no sides that are equal.

A different number of
check marks on the
sides of the triangle,
are used to indicate
sides of different sizes.

$$P \text{ (Perimeter)} = a + b + c$$
$$A \text{ (Area)} = \frac{b \times h}{2}$$

If the length of the sides are known:

$$A = \sqrt{s \times (s-a) \times (s-b) \times (s-c)}$$

In which s (half perimeter) $= \dfrac{a + b + c}{2}$

AREA OF A SCALENE TRIANGLE

Scalene triangle occupies half the
area of the rectangle below.

FIGURE 6-10 The Scalene Triangle.

6-2-4 *Other Triangles*

Figures 6-10, 6-11 and 6-12 summarize the facts and formulas for other triangle types.

6-2-5 *Circles*

Details relating to the circle have been discussed in previous chapters. These facts and formulas are summarized in Figure 6-13.

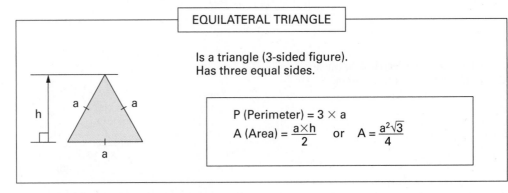

EQUILATERAL TRIANGLE

Is a triangle (3-sided figure).
Has three equal sides.

$$P \text{ (Perimeter)} = 3 \times a$$
$$A \text{ (Area)} = \frac{a \times h}{2} \quad \text{or} \quad A = \frac{a^2 \sqrt{3}}{4}$$

FIGURE 6-11 The Equilateral Triangle.

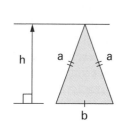

Is a triangle (3-sided figure).
Has two equal sides.

$$P \text{ (Perimeter)} = 2a + b$$
$$A \text{ (Area)} = \frac{b \times h}{2} \text{ or } A = \frac{b}{2}\sqrt{a^2 - \left(\frac{b}{2}\right)^2}$$

FIGURE 6-12   The Isosceles Triangle.

CIRCLE

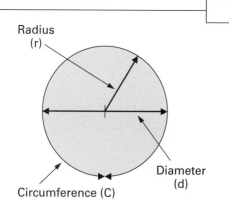

Radius (r)

Circumference (C)

Diameter (d)

Is a perfectly round figure.
Every point is at an equal distance from the center.

$$\text{Diameter (d)} = 2 \times r$$
$$\text{Circumference (c)} = \pi \times d \text{ or } c = \pi \times (2r)$$
$$\text{Area (A)} = \pi r^2$$

Every circle circumference is 3.14 times larger than its diameter. This constant is called pi (symbolized π).

$$\pi = \frac{\text{Circumference (c)}}{\text{Diameter (d)}} = 3.14$$

$$\pi = \frac{C}{d} \qquad C = \pi \times d \qquad d = \frac{c}{\pi}$$

Since d = 2r, then:

$$\pi = \frac{C}{2r} \qquad C = \pi \times 2r \qquad 2r = \frac{C}{\pi}$$

$$\text{or } r = \frac{C}{2\pi}$$

AREA OF A CIRCLE

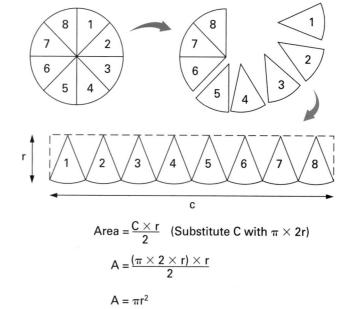

$$\text{Area} = \frac{C \times r}{2} \quad \text{(Substitute C with } \pi \times 2r)$$

$$A = \frac{(\pi \times 2 \times r) \times r}{2}$$

$$A = \pi r^2$$

FIGURE 6-13   The Circle.

6-2-6   *Other Polygons*

Figure 6-14(a) shows that if lines are drawn from the center of any polygon to each vertex, equivalent triangles are formed, which enables us to determine the polygon's area. In Figure 6-14(b), the formula given allows us to determine the number of degrees in the angles for each polygon type. From the examples shown we can see that each angle in an equilateral triangle is 60°, for a square each angle is 90° as expected, for a pentagon each angle is 108°, a hexagon has six 120° angles, and an octagon has eight 135° vertices.

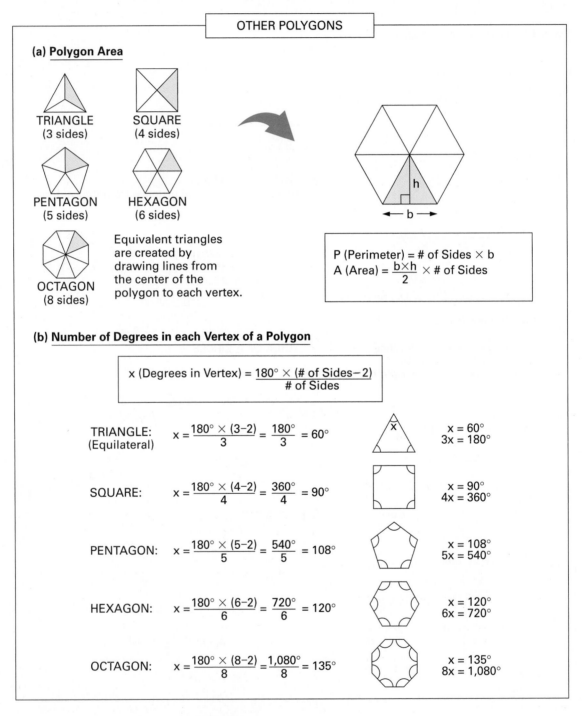

FIGURE 6-14   **Other Polygons.**

Now that you have completed this section, you should be able to:

- **Objective 3.** *Define the term* plane figure, *and identify the different types of polygons.*

- **Objective 4.** *List the formulas for determining the perimeter and area of plane figures.*

- **Objective 5.** *Explain how the Pythagorean theorem relates to right-angle triangles.*

- **Objective 6.** *Determine the length of one side of a right-angle triangle when the lengths of the other two sides are known.*

- **Objective 7.** *Describe how vectors are used to represent the magnitude and direction of physical quantities and how they are arranged in a vector diagram.*

- **Objective 8.** *Explain how the Pythagorean theorem can be applied to a vector diagram to calculate the magnitude of a resultant vector.*

Use the following questions to test your understanding of Section 6-2:

1. Define the terms:
  a. Plane figure
  b. Polygon

2. Sketch the following and list the formulas for perimeter and area:
  a. Square
  b. Rectangle
  c. Rhombus
  d. Parallelogram
  e. Trapezoid
  f. Right triangle
  g. Scalene triangle

  h. Equilateral triangle
  i. Isosceles triangle
  j. Circle

3. Calculate the lengths of the unknown sides in the following triangles.

4. Calculate the magnitude of the resultant vectors in the following vector diagrams.

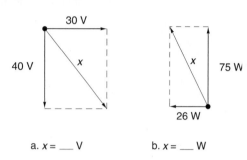

a. $x =$ ___ V   b. $x =$ ___ W

c.
$x =$ ___ mm
$y =$ ___ mm

6-3   TRIGONOMETRY

In the previous section you saw how Pythagoras's theorem could be used to find the length of one side of a right-triangle, when the length of the other two sides are known. **Trigonometry** is the study of the properties of triangles, and in this section we will examine the right-angle triangle in a little more detail, and see how angles and sides are related.

Until this point we have called the three sides of our right triangle *A, B,* and *C.* Figure 6-15 shows the more common names given to the three sides of a right-angle triangle. Side *C,* called the **hypotenuse,** is always the longest of the three sides. Side *B,* called the **adjacent side,** always extends between the hypotenuse and the vertical side. An angle called **theta** (θ, a Greek letter) is formed between the hypotenuse and the adjacent side. The side that is always opposite the angle θ is called the **opposite side.**

Angles are always measured in degrees because all angles are part of a circle, like the one shown in Figure 6-16(a). A circle is divided into 360 small sections called **degrees.** In Figure 6-16(b) our right triangle has been placed within the circle. In this example the triangle occupies a 45° section of the circle and therefore theta equals 45 degrees ($\theta = 45°$). Mov-

Trigonometry
The study of the properties of triangles.

Hypotenuse
The side of a right-angle triangle that is opposite the right angle.

Adjacent
The side of a right-angle triangle that has a common endpoint, in that it extends between the hypotenuse and the vertical side.

FIGURE 6-15   Names Given to the Sides of a Right Triangle.

A = Opposite Side

B = Adjacent Side

ing from left to right in Figure 6-16(c) you will notice that angle θ increases from 5° to 85°. The length of the hypotenuse (H) in all these examples remains the same; however, the adjacent side's length decreases ($A\downarrow$) and the opposite side's length increases ($O\uparrow$) as angle θ is increased ($\theta\uparrow$). This relationship between the relative length of a triangle's sides and theta means that we do not have to know the length of two sides to calculate a third. If we have the value of just one side and the angle theta, we can calculate the length of the other two sides.

6-3-1   *Sine of Theta (Sin θ)*

In the preceding section we discovered that a relationship exists between the relative length of a triangle's sides and the angle theta. **Sine** is a comparison between the length of the opposite side and the length of the hypotenuse. Expressed mathematically,

$$\text{Sine of theta } (\sin \theta) = \frac{\text{opposite side } (O)}{\text{hypotenuse } (H)}$$

Because the hypotenuse is always larger than the opposite side, the result will always be less than 1 (a decimal fraction). Let us use this formula in a few examples to see how it works.

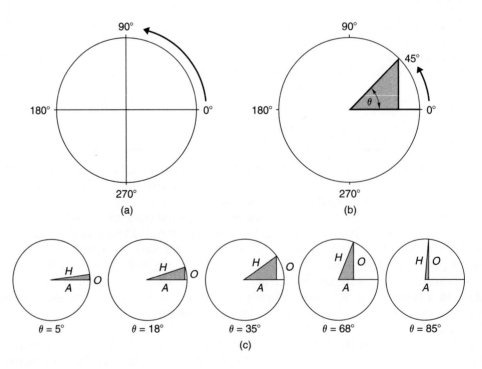

FIGURE 6-16   Angle Theta (θ).

CHAPTER 6 / GEOMETRY AND TRIGONOMETRY

CALCULATOR KEYS

Name: Sine key

Function: Instructs the calculator to find the sine of the displayed value (angle → value).

Example: sin 37° = ?

Press keys: | 3 | | 7 | | SIN |

Answer: 0.601815

Name: Arcsine (sin $^{-1}$) or inverse sine sequence.

Function: Calculates the smallest angle whose sine is in the display (value → angle).

Example: invsin 0.666 = ?

Press keys: | . | | 6 | | 6 | | 6 | | INV | | SIN |

Answer: 41.759°

■ **INTEGRATED MATH APPLICATION:**

In Figure 6-17(a), angle theta is equal to 41°, and the opposite side is equal to 20 centimeters in length. Calculate the length of the hypotenuse.

■ *Solution:*

Inserting these values in our formula, we obtain the following:

$$\sin \theta = \frac{O}{H}$$

$$\sin 41° = \frac{20 \text{ cm}}{H}$$

By looking up 41° in a sine trigonometry table or by using a scientific calculator that has all the trigonometry tables stored permanently in its memory, you will find that the sine of 41° is 0.656. This value describes the fact that when $\theta = 41°$, the opposite side will be 0.656, or 65.6%, as long as the hypotenuse side. By inserting this value into our formula and

(a)

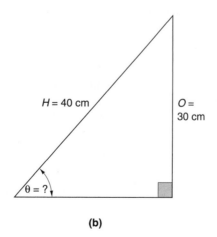

(b)

FIGURE 6-17   **Sine of Theta.**

transposing the formula according to the rules of algebra, we can determine the length of the hypotenuse.

$$\sin 41° = \frac{20 \text{ cm}}{H}$$

Calculator sequence: [4] [1] [SIN]

$$0.656 = \frac{20 \text{ cm}}{H}$$

$$0.656 \times H = \frac{20 \text{ cm} \times \cancel{H}}{\cancel{H}}$$

Multiply both sides by H.

$$\frac{\cancel{0.656} \times H}{\cancel{0.656}} = \frac{20 \text{ cm}}{0.656}$$

Divide both sides by 0.656.

$$H = \frac{20 \text{ cm}}{0.656} = 30.5 \text{ cm}$$

■   INTEGRATED MATH APPLICATION:

Figure 6-17(b) illustrates another example; however, in this case the lengths of sides H and O are known but θ is not.

■   *Solution:*

$$\sin \theta = \frac{O}{H}$$

$$\sin \theta = \frac{30 \text{ cm}}{40 \text{ cm}}$$

$$= 0.75$$

The ratio of side O to side H is 0.75, or 75%, which means that the opposite side is 75%, or 0.75, as long as the hypotenuse. To calculate angle θ we must isolate it on one side of the equation. To achieve this we must multiply both sides of the equation by **arcsine,** or **inverse sine** (invsin), which does the opposite of sine.

Arcsine or Inverse Sine

The inverse function to the sine. (If y is the sine of θ, then θ is the arcsine of y.)

$$\sin \theta = 0.75$$

$$\text{invsin} (\sin \theta) = \text{invsin } 0.75$$

$$\theta = \text{invsin } 0.75$$

$$= 48.6°$$

Take the inverse sine of 0.75.
Calculator sequence: [.] [7] [5] [INV] [SIN]

In summary, therefore, the sine trig functions take an angle θ and give you a number x. The inverse sine (arcsin) trig functions take a number x and give you an angle θ. In both cases, the number x is the ratio of the opposite side to the hypotenuse.

Sine: angle θ → number x
Inverse sine: number x → angle θ

6-3-2   *Cosine of Theta (Cos θ)*

Cosine

The trigonometric function that for an acute angle is the ratio between the leg adjacent to the angle when it is considered part of a right triangle, and the hypotenuse.

Sine is a comparison between the opposite side and the hypotenuse, and **cosine** is a comparison between the adjacent side and the hypotenuse.

$$\text{Cosine of theta (cos } \theta) = \frac{\text{adjacent } (A)}{\text{hypotenuse } (H)}$$

CHAPTER 6 / GEOMETRY AND TRIGONOMETRY

CALCULATOR KEYS

Name: Cosine key

Function: Instructs the calculator to find the cosine of the displayed value (angle → value).

Example: cos 26° = ?

Press keys: [2] [6] [COS]

Answer: 0.89879

Name: Arccosine (cos^{-1}) or inverse cosine sequence.

Function: Calculates the smallest angle whose cosine is in the display (value → angle).

Example: invcos 0.234 = ?

Press keys: [.] [2] [3] [4] [INV] [COS]

Answer: 76.467

■ **INTEGRATED MATH APPLICATION:**

Figure 6-18(a) illustrates a right triangle in which the angle θ and the length of the hypotenuse are known. From this information, calculate the length of the adjacent side.

■ *Solution:*

$$\cos \theta = \frac{A}{H}$$

$$\cos 30° = \frac{A}{40 \text{ cm}}$$

Calculator sequence: [3] [0] [COS]

$$0.866 = \frac{A}{40 \text{ cm}}$$

Looking up the cosine of 30° you will obtain the fraction 0.866. This value states that when θ = 30°, the adjacent side will always be 0.866, or 86.6%, as long as the

(a)   (b)

FIGURE 6-18   Cosine of Theta.

hypotenuse. By transposing this equation we can calculate the length of the adjacent side:

$$0.866 = \frac{A}{40 \text{ cm}}$$

$$40 \times 0.866 = \frac{A}{40 \text{ cm}} \times 40 \qquad \text{Multiply both sides by 40.}$$

$$A = 0.866 \times 40$$

$$= 34.64 \text{ cm}$$

■ **INTEGRATED MATH APPLICATION:**

Calculate the angle θ in Figure 6-18(b) if $A = 120$ centimeters and $H = 180$ centimeters.

■ *Solution:*

$$\cos \theta = \frac{A}{H}$$

$$= \frac{120 \text{ cm}}{180 \text{ cm}}$$

$$= 0.667 \qquad A \text{ is } 66.7\% \text{ as long as } H.$$

$$\text{invcos}(\cos \theta) = \text{invcos } 0.667 \qquad \text{Multiply both sides by invcos.}$$

$$\theta = \text{invcos } 0.667 \qquad \textit{Calculator sequence:} \quad \boxed{0}\;\boxed{.}\;\boxed{6}\;\boxed{6}\;\boxed{7}\;\boxed{\text{INV}}\;\boxed{\text{COS}}$$

$$= 48.2°$$

The inverse cosine trig function performs the reverse operation of the cosine function:

$$\text{Cosine:} \qquad \text{angle } \theta \longrightarrow \text{number } x$$

$$\text{Inverse cosine:} \quad \text{number } x \longrightarrow \text{angle } \theta$$

6-3-3   *Tangent of Theta (Tan θ)*

Tangent

The trigonometric function that for an acute angle is the ratio between the leg opposite the angle when it is considered part of a right triangle, and the leg adjacent.

Tangent is a comparison between the opposite side of a right triangle and the adjacent side.

$$\text{Tangent of theta (tan } \theta) = \frac{\text{opposite } (O)}{\text{adjacent } (A)}$$

CALCULATOR KEYS

Name: Tangent key

Function: Instructs the calculator to find the tangent of the displayed value (angle → value).

Example: sin 73° = ?

Press keys: $\boxed{7}\;\boxed{3}\;\boxed{\text{TAN}}$

Answer: 3.2709

Name: Arctangent $(\tan^{-1})$ or inverse tangent sequence.

Function: Calculates the smallest angle whose tangent is in the display (value → angle).

Example: invtan 0.95 = ?

Press keys: $\boxed{.}\;\boxed{9}\;\boxed{5}\;\boxed{\text{INV}}\;\boxed{\text{TAN}}$

Answer: 43.5312

■ INTEGRATED MATH APPLICATION:

Figure 6-19(a) illustrates a right triangle in which $\theta = 65°$ and the opposite side is 43 centimeters. Calculate the length of the adjacent side.

■ *Solution:*

$$\tan \theta = \frac{O}{A}$$

$$\tan 65° = \frac{43 \text{ cm}}{A}$$

$$2.14 = \frac{43 \text{ cm}}{A}$$

$$2.14 \times A = \frac{43 \text{ cm}}{A} \times A$$

$$\frac{2.14 \times A}{2.14} = \frac{43 \text{ cm}}{2.14}$$

$$A = \frac{43 \text{ cm}}{2.14}$$

$$= 20.1 \text{ cm}$$

Calculator sequence: ⬛6⬛ ⬛5⬛ ⬛TAN⬛

Whenever $\theta = 65°$, the opposite side will be 2.14 times longer than the adjacent side.

Multiply both sides by A.

Divide both sides by 2.14.

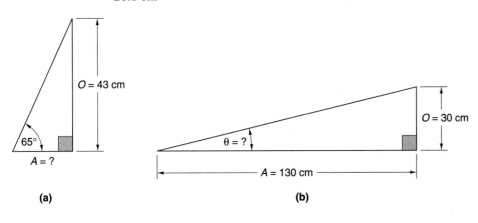

(a)   (b)

FIGURE 6-19   **Tangent of Theta.**

■ INTEGRATED MATH APPLICATION:

Calculate the angle θ in Figure 6-19(b) if $O = 30$ centimeters and $A = 130$ centimeters.

■ *Solution:*

$$\tan \theta = \frac{O}{A}$$

$$= \frac{30 \text{ cm}}{130 \text{ cm}}$$

$$= 0.231$$

$$\text{invtan} (\tan \theta) = \text{invtan } 0.231$$

$$\theta = \text{invtan } 0.231$$

$$= 12.99 \text{ or } 13°$$

O is 0.231 or 23.1% as long as *A*.

Multiply both sides by invtan.
Calculator sequence: ⬛0⬛ ⬛.⬛ ⬛2⬛⬛3⬛⬛1⬛ ⬛INV⬛ ⬛TAN⬛

6-3-4   *Summary*

As you have seen, trigonometry involves the study of the relationships among the three sides (O, H, A) of a right triangle and also the relationships among the sides of the right triangle and the number of degrees contained in the angle theta (θ).

If the lengths of two sides of a right triangle are known, and the length of the third side is needed, remember that

$$H^2 = O^2 + A^2$$

If the angle θ is known along with the length of one side, or if angle θ is needed and the lengths of the two sides are known, one of the three formulas can be chosen based on which variables are known and what is needed.

$$\sin \theta = \frac{O}{H} \quad \cos \theta = \frac{A}{H} \quad \tan \theta = \frac{O}{A}$$

When I was introduced to trigonometry, my mathematics professor spent 15 minutes having the whole class practice what he described as an old Asian war cry that went like this: "SOH CAH TOA." After he explained that it wasn't a war cry but in fact a memory aid to help us remember that SOH was in fact sin $\theta = O/H$, CAH was cos $\theta = A/H$, and TOA was tan $\theta = O/A$, we understood the method in his madness.

■ INTEGRATED MATH APPLICATION:

Sine Wave

A waveform that represents periodic oscillations in which the amplitude of displacement at each point is proportional to the sine of the phase angle of the displacement and that is visualized as a sine curve.

The **sine wave** is the most common type of electrical waveform. It is the natural output of a generator that converts a mechanical input, in the form of a rotating shaft, into an electrical output in the form of a sine wave. In fact, for one cycle of the input shaft, the generator will produce one sinusoidal ac voltage waveform, as shown in Figure 6-20. When the input shaft of the generator is at 0°, the ac output is 0 volts. As the shaft is rotated through 360° the ac output voltage will rise to a maximum or peak positive voltage at 90°, fall back to 0 volts at 180°, then reach a maximum or peak negative voltage at 270°, and finally return to 0 volts at 360°. If this ac voltage is applied across a closed circuit, it produces a current that continually reverses or alternates in direction. (It switches from being a positive voltage to a negative voltage, back to a positive voltage, and so on.) The sine wave is the most common type of waveform shape, and why the name *sine* was given to this wave needs to be explained further.

Figure 6-21(a) shows the correlation between the 360° of a circle and the 360° of a sine wave. Within the circle a triangle has been drawn to represent the right-angle triangle

FIGURE 6-20   **Degrees of a Sine Wave.**

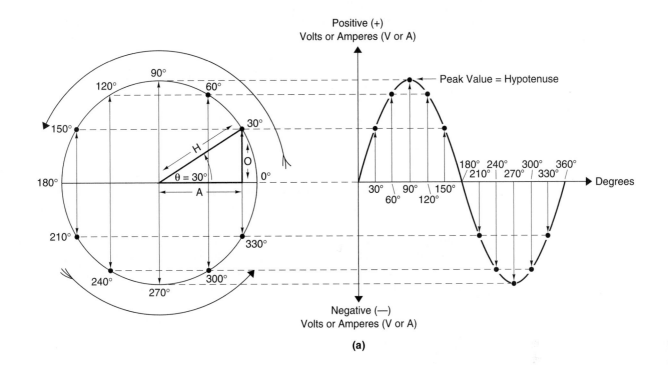

(a)

Angle	Sine
0...°	0.000
15	0.259
30	0.500
45	0.707
60	0.866
75	0.966
90	1.000

(b)

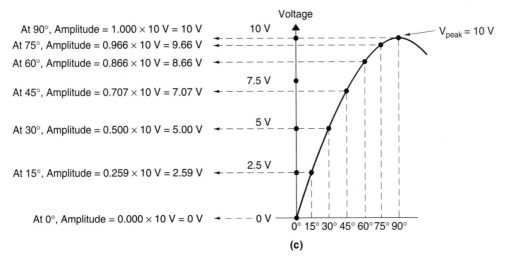

At 90°, Amplitude = 1.000 × 10 V = 10 V

At 75°, Amplitude = 0.966 × 10 V = 9.66 V

At 60°, Amplitude = 0.866 × 10 V = 8.66 V

At 45°, Amplitude = 0.707 × 10 V = 7.07 V

At 30°, Amplitude = 0.500 × 10 V = 5.00 V

At 15°, Amplitude = 0.259 × 10 V = 2.59 V

At 0°, Amplitude = 0.000 × 10 V = 0 V

(c)

FIGURE 6-21   **Meaning of Sine.**

formed at 30°. The hypotenuse side will always remain the same length throughout 360°, and will equal the maximum voltage or peak of the sine wave. The opposite side of the triangle is equal to the amplitude vector of the sine wave at 30°. To calculate the amplitude of the opposite side, and therefore the amplitude of the sine wave at 30°, we use the sine of theta formula discussed previously:

$$\sin \theta = \frac{\text{opposite}}{\text{hypotenuse}}$$

$$\sin 30° = \frac{O}{H}$$

$$0.500 = \frac{O}{H}$$

Calculator sequence: $\boxed{3}\ \boxed{0}\ \boxed{\text{SIN}}$

This equation tells us that at 30°, the opposite side is 0.500 or half the size of the hypotenuse. At 30°, therefore, the amplitude or magnitude of the sine wave will be 0.5 (50%) of the peak value.

Figure 6-21(b) lists the sine values at 15° increments. Figure 6-21(c) shows an example of a 10 volt peak sine wave. At 15°, a sine wave will always be at 0.259 (sine of 15°) of the peak value, which for a 10 volt sine wave will be 2.59 volts (0.259 × 10 volts = 2.59 volts). At 30°, a sine wave will have increased to 0.500 (sine of 30°) of the peak value. At 45°, a sine wave will be at 0.707 of the peak, and so on. The sine wave is called a sine wave because it changes in amplitude at the same rate as the sine trigonometric function.

SELF-TEST EVALUATION POINT FOR SECTION 6-3

Now that you have completed this section, you should be able to:

■ **Objective 9.** *Define the trigonometric terms opposite, adjacent, hypotenuse, theta, sine, cosine, and tangent.*

■ **Objective 10.** *Demonstrate how the sine, cosine, and tangent trigonometric functions can be used to* calculate (1) the length of an unknown side of a right-angle triangle if the length of another side and the angle theta are known or (2) the angle theta of a right-angle triangle if the lengths of two sides of the triangle are known.

Use the following questions to test your understanding of Section 6-3.

1. Calculate the length of the unknown side in the following triangles.

2. Calculate the angle θ for the following right-angle triangles.

6-4 SOLID FIGURES

As mentioned previously, a plane figure is a flat surface that has only two dimensions—length and width. A **solid figure**, on the other hand, is a three-dimensional figure having length, width and height.

6-4-1 *Prisms*

A **prism** is a solid figure made up of plane figure faces, and has a least one pair of parallel surfaces. Figure 6-22 details the facts and formulas for the prism. The visual breakdown in Figure 6-22(a) shows how a cube is made up of an upper base, lower base, and a set of sides. The **lateral surface area** (**L**) of the cube is the area of the sides only, as shown in Figure 6-22(b). The **total surface area** (**S**) is the lateral surface area of the cube plus the area of the two bases, as shown in Figure 6-22(c). The holding capacity or **volume** of the cube measures the space inside the prism, and is described visually in Figure 6-22(d). Figure 6-22(e) shows examples of other prism types.

6-4-2 *Cylinders*

A **cylinder** is a solid figure with curved side walls extending between two identical circular bases. Figure 6-23 details the facts and formulas for the cylinder. As with prisms, plane figure formulas are used to calculate the cylinder's lateral surface area and base area, and as expected, the volume of a cylinder is the product of its base area and height.

6-4-3 *Spheres*

A **sphere** is a globular solid figure in which any point on the surface is at an equal distance from the center. Figure 6-24 lists the formulas for a sphere along with a visual breakdown for a sphere's surface area and volume.

6-4-4 *Pyramids and Cones*

Figure 6-25(a) (p. 164) lists the formulas for a **pyramid,** which has only one base and three or more lateral surfaces that taper up to a single point called an **apex.** Figure 6-25(a) lists the formulas for a **cone,** which has a circular base and smooth surface that extends up to an apex. Figures 6-25(b) and 6-26(b) (p. 165), list the formulas for the frustum of a pyramid and the frustum of a cone. A **frustum** is the base section of a solid pyramid or cone, and is formed by cutting off the top of the solid figure so that the upper base and lower base are parallel to one another.

Solid Figure
A 3-dimensional figure having length, width, and height.

Prism
A solid figure made up of plane figure faces; having at least one pair of parallel surfaces.

Lateral Surface Area
The side area of a solid figure.

Total Surface Area
The lateral surface area of a solid figure plus the area of its two bases.

Volume
The space within a solid figure, or its holding capacity.

Cylinder
A solid figure with curved side walls extending between two identical circular bases.

Sphere
A globular solid figure in which any point on the surface is at equal distance from the center.

Pyramid
A solid figure having only one base and 3 or more lateral surfaces that taper up to a single point.

Apex
The uppermost point or tip of a solid figure.

Cone
A solid figure having a circular base and smooth surface that extends up to an apex.

Frustum
The base section of a solid pyramid or cone.

SELF-TEST EVALUATION POINT FOR SECTION 6-4

Now that you have completed this section, you should be able to:

■ *Objective 11.* *Define the term* solid figure, *and identify the different types.*

■ *Objective 12.* *List the formulas for determining the lateral surface area, total surface area, and volume of solid figures.*

Use the following questions to test your understanding of Section 6-4:

1. What is the difference between a plane figure and a solid figure?

2. What is the difference between lateral surface area and total surface area?

3. Sketch the following solid figures and list the formulas for lateral surface area, total surface area and volume.

   a. Prism   **d.** Pyramid
   b. Cylinder   **e.** Cone
   c. Sphere

4. What is a frustum?

FIGURE 6-22   **Prisms.**

FIGURE 6-23   Cylinders.

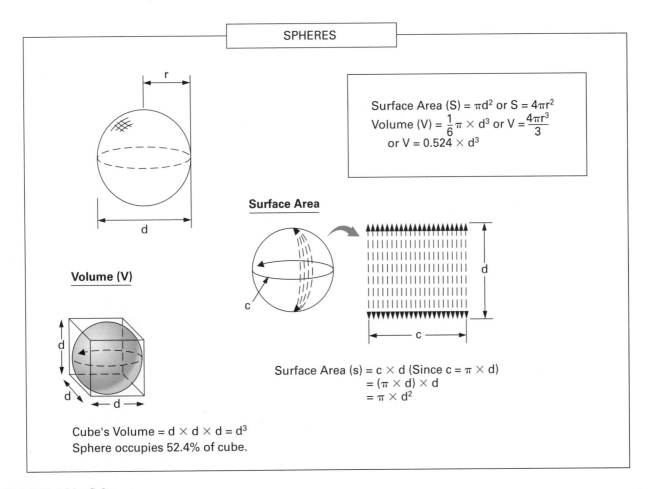

FIGURE 6-24   Spheres.

(a) PYRAMID

Lateral Surface Area (L) = $\frac{1}{2}$ ps

Volume (V) = $\frac{1}{3}$ Ah

p = perimeter of base   s = slant height
A = area of base   h = height

Lateral Surface Area (L)

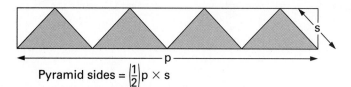

Pyramid sides = $\left|\frac{1}{2}\right|$ p × s

Volume (V)

Cube Volume = A × h.
Pyramid occupies $\frac{1}{3}$ of cube.

(b) FRUSTUM OF PYRAMID

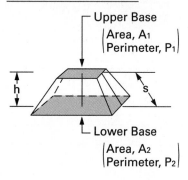

Upper Base
$\left|\begin{array}{l} \text{Area, } A_1 \\ \text{Perimeter, } P_1 \end{array}\right|$

Lower Base
$\left|\begin{array}{l} \text{Area, } A_2 \\ \text{Perimeter, } P_2 \end{array}\right|$

Lateral Surface Area (L) = $\frac{1}{2}$ (P_1 + P_2) s

Volume (V) = $\frac{1}{3}$ h (A_1 + A_2 + $\sqrt{A_1 \times A_2}$)

FIGURE 6-25   **Pyramids and Frustums of Pyramids.**

(a) CONE

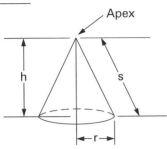

Apex

Lateral Surface Area (L) = πrs or L = $\frac{1}{2}\pi ds$

Volume (V) = $\frac{1}{3}\pi r^2 h$ or V = $\frac{1}{12}\pi d^2 h$

r = base radius   h = height
s = slant height   d = base diameter

(b) FRUSTUM OF CONE

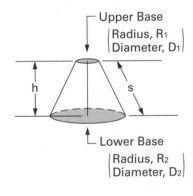

Upper Base
$\left(\begin{array}{l}\text{Radius, } R_1 \\ \text{Diameter, } D_1\end{array}\right)$

Lower Base
$\left(\begin{array}{l}\text{Radius, } R_2 \\ \text{Diameter, } D_2\end{array}\right)$

Lateral Surface Area (L) = $\pi s\,(r_1 + r_2)$

or L = $\frac{1}{2}\pi s\,(d_1 + d_2)$

Volume (V) = $\frac{1}{3}\pi h\,(r_1{}^2 + r_2{}^2 + (r_1 r_2))$

or V = $\frac{1}{12}\pi h\,(d_1{}^2 + d_2{}^2 + (d_1 d_2))$

FIGURE 6-26 Cones and Frustums of Cones.

REVIEW QUESTIONS

Multiple Choice Questions

1. What instrument can be used to measure an angle?

 a. Compass   **c.** Ruler
 b. Protractor   **d.** None of the above

2. A right angle is equal to:

 a. 45°   **c.** 180°
 b. 100°   **d.** 90°

3. A protractor can be used to:

 a. Measure an angle **c.** Both (a) and (b)
 b. Draw an angle   **d.** None of the above

4. A _____ is a quadrilateral parallelogram, that has four sides that are all of equal length.

 a. Square   **c.** Trapezoid
 b. Rectangle   **d.** Prism

5. Trigonometry is the study of:

 a. Triangles   **c.** Trigonometric functions
 b. Angles   **d.** All the above

6. The Pythagorean theorem states that the _____ of the length of the hypotenuse of a right-angle triangle equals the sum of the _____ of the lengths of the other two sides.

 a. Square, square roots **c.** Square, squares
 b. Square root, squares **d.** Square root, square roots

7. A vector is an arrow whose length is used to represent the _____ of a quantity and whose point is used to indicate the same quantity's _____.

 a. Magnitude, size   **c.** Direction, size
 b. Magnitude, direction **d.** Direction, magnitude

8. If two vectors in a vector diagram are not working together, or are out of phase with one another, they must be _____ to obtain a _____ vector.

 a. Added, resultant
 b. Multiplied by one another, parallel
 c. Vectorially added, resultant
 d. Both (a) and (b)

9. The _____ of a right-angle triangle is always the longest side.

   a. Opposite side
   b. Adjacent side
   c. Hypotenuse
   d. Angle theta

10. The _____ of a right-angle triangle is always across from the angle theta.

    a. Opposite side
    b. Adjacent side
    c. Hypotenuse
    d. Angle theta

11. The _____ of a right-angle triangle is always formed between the hypotenuse and the adjacent side.

    a. Opposite side   c. Hypotenuse
    b. Adjacent side   d. Angle theta

12. Which trigonometric function can be used to determine the length of the hypotenuse if the angle theta and the length of the opposite side are known?

    a. Tangent
    b. Sine
    c. Cosine

13. Which trigonometric function can be used to determine the angle theta when the lengths of the opposite and adjacent sides are known?

    a. Tangent
    b. Sine
    c. Cosine

14. Which trigonometric function can be used to determine the length of the adjacent side when the angle theta and the length of the hypotenuse are known?

    a. Tangent
    b. Sine
    c. Cosine

15. A plane figure has _____ dimensions, while a solid figure has _____ dimensions.

    a. 3, 4   c. 2, 3
    b. 2, 4   d. 3, 2

Communication Skill Questions

16. Define the following terms: (6-1 to 6-4)

    a. Geometry   c. Plane figure
    b. Trigonometry   d. Solid figure

17. Describe how to use a protractor to draw and measure an angle. (6-1)

18. Sketch the following plane figures and list their associated formulas for calculating perimeter and area: (6-2)

    a. Square   f. Right triangle
    b. Rectangle   g. Scalene triangle
    c. Rhombus   h. Equilateral triangle
    d. Parallelogram   i. Isosceles triangle
    e. Trapezoid   j. Circle

19. What are the number of degrees in each vertex of a: (6-2-5)

    a. Square   d. Hexagon
    b. Equilateral triangle   e. Octagon
    c. Pentagon

20. Describe the purpose of Pythagoras's theorem.   (6-2-2)

21. Demonstrate how the sine, cosine and tangent trigonometric functions can be used to calculate: (6-3)

    a. The length of an unknown side of a right triangle when the length of another side is known and the angle theta is known.
    b. The angle theta of a right triangle if the lengths of two sides of the triangle are known.

22. Sketch the following solid figures and list their associated formulas: (6-4)

    a. Prism   d. Pyramid
    b. Cylinder   e. Cone
    c. Sphere

23. Why is it necessary to have inverse trigonometric functions?   (6-3)

24. What is a sine wave?   (6-3)

25. What is a frustum?   (6-4)

Practice Problems

26. Sketch the following:

    a. A line segment.   e. Two perpendicular lines.
    b. Two parallel lines.   f. An angle of 45°.
    c. A right angle.   g. A vertical line.
    d. An acute angle.   h. An obtuse angle.

27. Calculate the length of the hypotenuse (C side) of the following right triangles.

    a. $A = 20$ mi, $B = 53$ mi
    b. $A = 2$ km, $B = 3$ km
    c. $A = 4$ in., $B = 3$ in.
    d. $A = 12$ mm, $B = 12$ mm

28. Calculate the length or magnitude of the resultant vectors in the following vector diagrams.

a.

V_L = 2.6 V, $V_{S=}$?, V_R = 5.5 V

b.
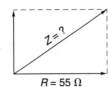
X_L = 26 Ω, Z = ?, R = 55 Ω

c.
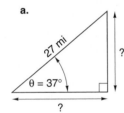
V_R = 12 V, V_C = 8 V, $V_{S=}$?

d.

P_R = 3.9 mw, P_X = 6.3 mw, $P_{A=}$?

29. Calculate the value of the following trigonometric functions.

a. sin 0°
b. sin 30°
c. sin 45°
d. sin 60°
e. sin 90°
f. cos 0°
g. cos 30°
h. cos 45°

i. cos 60°
j. cos 90°
k. tan 0°
l. tan 30°
m. tan 45°
n. tan 60°
o. tan 90°

30. Calculate angle θ from the function given.

a. sin θ = 0.707, θ = ?
b. sin θ = 0.233, θ = ?
c. cos θ = 0.707, θ = ?
d. cos θ = 0.839, θ = ?
e. tan θ = 1.25, θ = ?
f. tan θ = 0.866, θ = ?

31. Name each of the triangle's sides, and calculate the unknown values.

a.

27 mi, θ = 37°, ?, ?

b.

θ = 21°, ?, ?, 29 cm

c.

34 V, θ = ?, ?, 18.5 V

32. If a cylinder has a radius of 3 cm and a height of 10 cm, calculate its:

a. Lateral surface area
b. Total surface area
c. Volume

33. If a sphere has a diameter of 2 inches, what would be its surface area and volume?

34. What is the lateral surface area of a rectangular prism that has a base perimeter of 40 cm and a height of 12 cm?

35. Calculate the lateral surface area of a pyramid that has a base perimeter of 24 inches and a slant height of 8 inches?

Web Site Questions

Go to the web site http://www.prenhall.com/cook, select the textbook *Introductory Computer Mathematics, 2e,* select this chapter, and then follow the instructions when answering the multiple-choice practice problems.

Logarithms and Graphs

The First Pocket Calculator

During the seventeenth century, European thinkers were obsessed with any device that could help them with mathematical calculation. Scottish mathematician John Napier decided to meet this need, and in 1614 he published his new discovery of logarithms. In this book, consisting mostly of tediously computed tables, Napier stated that a logarithm is the exponent of a base number. For example:

The common logarithm (base 10) of 100 is 2 ($100 = 10^2$).
The common logarithm of 10 is 1 ($10 = 10^1$).
The common logarithm of 27 is 1.43136 ($27 = 10^{1.43136}$).
The common logarithm of 6 is 0.77815 ($6 = 10^{0.77815}$).

Any number, no matter how large or small, can be represented by or converted to a logarithm. Napier also outlined how the multiplication of two numbers could be achieved by simply adding the numbers' logarithms. For example, if the logarithm of 2 (which is 0.30103) is added to the logarithm of 4 (which is 0.60206), the result will be 0.90309, which is the logarithm of the number 8 ($0.30103 + 0.60206 = 0.90309$, $2 \times 4 = 8$). Therefore, the multiplication of two large numbers can be achieved by looking up the logarithms of the two numbers in a log table, adding them together, and then finding the number that corresponds to the sum in an antilog (reverse log) table. In this example, the antilog of 0.90309 is 8.

Napier's table of logarithms was used by William Oughtred, who developed, just 10 years after Napier's death in 1617, a handy mechanical device that could be used for rapid calculation. This device, considered the first pocket calculator, was the slide rule.

From *Portraits of Eminent Mathematicians*, by David Eugene Smith, published by Pictorial Mathematics, New York, 1936.

Outline and Objectives

Introduction

Logarithm

The exponent that indicates the power to which a number is raised to produce a given number.

In the first half of this chapter, we will discuss **logarithms.** Although this seems like a new topic, it is simply a continuation of our discussion of exponents from Chapter 4. For example, you know that a distance of 1 kilometer is equal to 1×10^3 meters (or 1000 meters). The power of 10, exponent 3 in this example, is the logarithm of 1000 (log 1000 = 3). Logarithms can be used to simplify multiplication to addition, division to subtraction, raising to a power to multiplication, and extracting square roots to division. Today, multiplication, division, raising to a power, and extracting square roots can be accomplished more easily with a calculator, and so the use of logarithms to simplify these operations has decreased. Logarithms have found new applications, however, in many areas of science and technology, and so it is important to have a clear understanding of them.

In the second half of this chapter, we will study graph interpretation, types, and plotting. A graph is a diagram used frequently in all areas of science and technology to show the relationship between two or more factors. Graphs actually make it easier for us to understand information because they convert long lists of recorded data into an easy-to-understand visual representation.

7-1   COMMON LOGARITHMS

Common Logarithm

A logarithm whose base is 10.

Stated simply, a **common logarithm** is a base 10 exponent. So far, you have already had considerable experience with base 10 exponents, or as they are officially called, *power of 10.* For example, you know that a computer having 1 megabyte of memory means that the memory can store 1×10^6 bytes (or 1,000,000 bytes), and that a distance of 1 kilometer is equal to 1×10^3 meters (or 1000 meters). The power-of-10 exponent used in these examples is the logarithm; for the first example, *the common logarithm (base 10) of 1,000,000 is 6,* and for the second example, *the common logarithm (base 10) of 1000 is 3.* If we were to state the first of these examples in mathematical form, it would appear as follows:

$$10^6 = 1,000,000 \qquad \log_{10} 1,000,000 = 6$$

Exponent · Number (for $10^6 = 1,000,000$); Logarithm · Base · Number (for $\log_{10} 1,000,000 = 6$)

The first expression ($10^6 = 1,000,000$) is an *exponential expression* because the quantity is expressed in *exponential form.* The second expression ($\log_{10} 1,000,000 = 6$) is a *logarithmic expression* because the quantity is expressed in *logarithmic form.*

Natural Logarithm

A logarithm with *e* as a base, in which *e* is a transcendental number having a value to eight decimal places of 2.71828183.

There are two basic types of logarithms: **natural logarithms,** which use a base of 2.71828, and *common logarithms,* which use a base of 10. The base of a logarithm should be indicated when there is any chance of confusion; however, when using only common logs, it is not necessary to indicate or say the base because it is assumed to be 10. Therefore, $\log_{10}$ 1,000,000 = 6 can be simplified to log 1,000,000 = 6.

As another example: What is the log of 1000? To find the answer, we simply convert 1000 to its power-of-10 equivalent, which in exponential form is 10^3. Because the common logarithm of any number is its base 10 exponent, in logarithmic form, log 1000 = 3. Here is a list of logs for numbers that are greater than 1:

Exponential Form	Logarithmic Form
$10^0 = 1$	$\log 1 = 0$
$10^1 = 10$	$\log 10 = 1$
$10^2 = 100$	$\log 100 = 2$
$10^3 = 1000$	$\log 1000 = 3$
$10^4 = 10,000$	$\log 10,000 = 4$
$10^5 = 100,000$	$\log 100,000 = 5$
$10^6 = 1,000,000$	$\log 1,000,000 = 6$

What do we do when we are dealing with numbers that are a fraction of 1, for example, 0.001, 0.000001, and so on? The answer is to follow exactly the same procedure. First, express the number or value in exponential form ($10^{-3} = 0.001$). Then, because the logarithm is the base 10 exponent, express the number in logarithmic form ($\log 0.001 = -3$). Here is a list of logs for numbers that are a fraction of 1:

Exponential Form	Logarithmic Form
$10^0 = 1$	$\log 1 = 0$
$10^{-1} = 0.1$	$\log 0.1 = -1$
$10^{-2} = 0.01$	$\log 0.01 = -2$
$10^{-3} = 0.001$	$\log 0.001 = -3$
$10^{-4} = 0.0001$	$\log 0.0001 = -4$
$10^{-5} = 0.00001$	$\log 0.00001 = -5$
$10^{-6} = 0.000001$	$\log 0.000001 = -6$

7-1-1   *The Calculator's Table of Common Logarithms*

Your calculator contains a table of common logarithms. It operates as follows.

CALCULATOR KEYS

Name: Common logarithm key

Function: Calculates the common logarithm (base 10) of the displayed number.

Example: log 35.5 = ?

Press keys: 3 5 . 5 LOG

Answer: 1.5502283

Therefore, $35.5 = 10^{1.5502283}$.

■ **EXAMPLE:**

Using a calculator, determine the log of 1000.

■ *Solution:*

To determine the log of 1000 using a calculator, you input the value 1000 and then press the log calculator key, as follows:

Log 1000 = ?
Press keys: $\boxed{1}\boxed{0}\boxed{0}\boxed{0}\boxed{\text{LOG}}$
Answer: 3

■ **EXAMPLE:**

Determine the log of 0.00001.

■ *Solution:*

To determine the log of 0.00001 using the calculator, you simply input the value 0.00001 and then press the log calculator key, as follows:

Log 0.00001 = ?
Press keys: $\boxed{0}\boxed{.}\boxed{0}\boxed{0}\boxed{0}\boxed{0}\boxed{1}\boxed{\text{LOG}}$
Answer: −5

7-1-2   *Numbers That Are Not Multiples of 10*

Only numbers that are powers of 10, for example, 0.001, 0.1, 100, 1,000,000,000, and so on, have an exponent or logarithm that is a whole number. In the following examples, you will see that any number between two multiples of 10 will have a decimal fraction. Using your calculator, check the logs (which have been rounded off to 4 decimal places) for the following numbers:

Logarithmic Form	**Exponential Form**
log 0.000089 = −4.0506	$10^{-4.0506} = 0.000089$
log 0.28 = −0.5528	$10^{-0.5528} = 0.28$

Because the example numbers are a fraction of 1, the log or exponent will be negative.

log 2.5 = 0.3979	$10^{0.3979} = 2.5$
log 6 = 0.7782	$10^{0.7782} = 6$
log 9.75 = 0.9890	$10^{0.9890} = 9.75$

Because the log of 1 is 0, and the log of 10 is 1, all the example numbers between 1 and 10 will have logs or exponents between 0 and 1.

log 12 = 1.0792	$10^{1.0792} = 12$
log 76.4 = 1.8831	$10^{1.8831} = 76.4$

Because the log of 10 is 1 and the log of 100 is 2, all the example numbers between 10 and 100 will have logs or exponents between 1 and 2.

log 135.54 = 2.132	$10^{2.132} = 135.54$
log 873 = 2.941	$10^{2.941} = 873$

Because the log of 100 is 2 and the log of 1000 is 3, all the example numbers between 100 and 1000 will have logs or exponents between 2 and 3.

7-1-3 *Using Logarithms to Simplify Multiplication*

Now that we know what logarithms are and how to find them, let us see what we can do with them. As mentioned previously, one reason for converting numbers to logarithms is to simplify calculations. For example, using logarithms, *we can reduce the multiplication of two numbers to simple addition.* This advantage was first shown in this chapter's opening vignette; however, let us study the steps involved in a little more detail. To find the product of two numbers using logs, you have to follow these four steps:

STEP 1.   Find the log of each number.
STEP 2.   Change each number to base 10.
STEP 3.   Add the base 10 exponents.
STEP 4.   Convert the log answer back to a number using the antilog.

■   **INTEGRATED MATH APPLICATION:**

Let us use logarithms to calculate the product of 26 × 789.

■   *Solution:*

STEP 1:   Find the log of each number:

$$\log 26 = 1.41497$$
$$\log 789 = 2.89708$$

STEP 2:   Change each number to base 10. Because the log of a number is the base 10 exponent, we have the following:

$$26 = 10^{1.41497}$$
$$789 = 10^{2.89708}$$

STEP 3:   The answer to 26 × 789 can be obtained by simply adding the exponents from step 2.

$$10^{1.41497} \times 10^{2.89708} = 10^{1.41497 + 2.89708} = 10^{4.31205}$$

Note: This law has always applied to exponents; for example, $100 \times 1000 = 10^2 \times 10^3 = 100{,}000 = 10^5$. When multiplying like bases, the exponents or number of zeros are added.

STEP 4:   Once you have converted all your numbers to logarithms and finished all your calculations, you then have to convert the log answer back to a number using a **common antilogarithm** table. This reverse log table is also stored in your calculator and operates as follows.

Antilogarithm
The number corresponding to a given logarithm.

CALCULATOR KEYS

Name: Common antilogarithm (10^x) key

Function: Calculates the common antilogarithm of the displayed value.

Example: $10^{1.5502283}$

Press keys: ☐ 1 ☐ . ☐ 5 ☐ 5 ☐ 0 ☐ 2 ☐ 2 ☐ 8 ☐ 3 ☐ 10^x

Answer: 35.5

Therefore, $10^{1.5502283} = 35.5$.

7-1-4 *Using Logarithms to Simplify Division*

Using logarithms, *we can reduce the division of one number by another number to simple subtraction.* The steps for calculating the result of a division using logs are:

STEP 1. Find the log of the dividend and divisor.
STEP 2. Change each number to base 10.
STEP 3. Subtract the log of the divisor from the log of the dividend.
STEP 4. Convert the log answer back to a number using the antilog.

■ INTEGRATED MATH APPLICATION:

Let us use logarithms to calculate the quotient of $8992.5 \div 165$.

■ *Solution:*

STEP 1: Find the log of both the dividend (8992.5) and the divisor (165):

$$\log 8992.5 = 3.95388$$
$$\log 165 = 2.21748$$

STEP 2: Change each number to base 10. Because the log of a number is the base 10 exponent, we have the following:

$$8992.5 = 10^{3.95388}$$
$$165 = 10^{2.21748}$$

STEP 3: The answer to $8992.5 \div 165$ can be obtained by simply subtracting the log or exponent of the divisor from the log of the dividend.

$$10^{3.95388} \div 10^{2.21748} = 10^{3.95388 - 2.21748} = 10^{1.7364}$$

Note: This law has always applied to exponents; for example, $10{,}000 \div 10 = 10^4 \div 10^1 = 10^{4-1} = 10^3 = 1000$. When like bases are divided, the exponents are subtracted.

STEP 4: Convert the log answer back to a number using the antilog table in your calculator.

$$\text{antilog } 1.7364 = 54.5$$

Using the calculator to double-check this answer, we find that $8992.5 \div 165 = 54.5$.

SELF-TEST EVALUATION POINT FOR SECTION 7-1

Now that you have completed this section, you should be able to:

■ *Objective 1. Define the terms* common logarithm, exponential expression, logarithmic expression, *and* natural logarithm.

■ *Objective 2. Describe how to use the calculator's stored table of common logarithms to find the log of any number.*

■ *Objective 3. Explain how logarithms can be used to simplify multiplication and division.*

7-2   GRAPHS

A **graph** is a diagram showing the relationship between two or more factors. As an example, the graph in Figure 7-1 shows how new technologies, such as electric light, radio, the telephone and television, have been adopted in homes in the United States. This type of graph is called a **line graph** because the data are represented by points joined by line segments. The vertical scale in this graph indicates U.S. households in percent (0% to 100%), while the horizontal scale indicates the year (1900 to 1980). The four plotted lines in this graph show how quickly these new technologies found their way into the home. The data used to create the four plotted lines was obtained by recording what percentage of homes in the United States had electric light, radio, telephone, and television every year from 1900 to 1980.

Graph
A diagram (as a series of one or more points, lines, line segments, curves, or areas) that represents the variation of a variable in comparison with that of one or more other variables.

Line Graph
A graph in which points representing values of a variable are connected by a broken line.

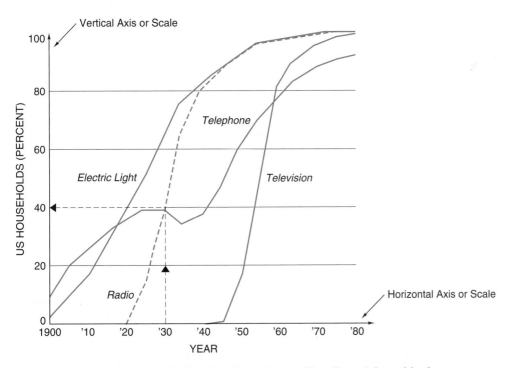

FIGURE 7-1   **A Graph Showing How New Technologies Have Been Adopted in the United States.**

7-2-1　　*Reading Graphs*

If the yearly facts for the graph shown in Figure 7-1 were listed in columns instead of illustrated graphically, it would be difficult to see the characteristics of each individual line graph, or to compare the characteristics of one technology with those of another. To explain this point, let us try an example.

■　**INTEGRATED MATH APPLICATION:**

From the graph shown in Figure 7-1, determine the following:

    a. Approximately what percentage of U.S. households had radio in 1930?

    b. Approximately what percentage of U.S. households had telephones in 1960?

    c. Which technology was the slowest to be adopted into the home, and which was the fastest?

■　*Solution:*

    a. To find this point on the graph, move along the horizontal scale until you reach 1930, and then proceed directly up until you connect with the plotted radio line graph (shown as a dashed line in Figure 7-1). Comparing this point with the vertical scale, you can see that it corresponds to about 40%. In 1930, therefore, approximately 40% of U.S. households had radio.

    b. To find this point on the graph, move along the horizontal scale until you reach 1960, and then proceed directly up until you connect with the plotted telephone line graph. Comparing this point with the vertical scale, you can see that it corresponds to about 80%. In 1960, therefore, approximately 80% of U.S. households had a telephone.

    c. Comparing the four line graphs shown in Figure 7-1, you can see that the telephone seemed to be the slowest technology to be adopted, while the television was the fastest.

Questions (a) and (b) in the previous example show how a graph can be used to analyze the characteristics of each individual line graph. Question (c) showed how a graph can be used to compare the characteristics of several line graphs.

■　**INTEGRATED MATH APPLICATION:**

From the graph shown in Figure 7-2, determine the following:

    a. Which camera type sells the most, and approximately how many were sold in 1992?

    b. Which camera type came on and then went off the market?

    c. Compare the sales of 35mm cameras in 1984 with sales in 1994.

■　*Solution:*

    a. Disposable or one-time-use cameras sell the most, and in 1992 approximately 22 million units were sold.

    b. Disc cameras.

    c. In 1984 approximately 5 million 35mm cameras were sold, whereas in 1994 approximately 11 million 35mm cameras were sold.

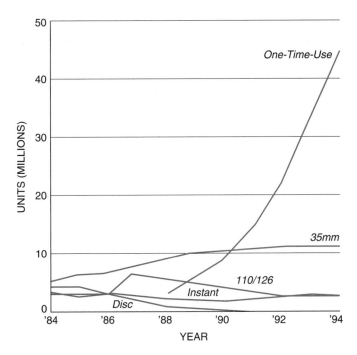

FIGURE 7-2   **A Graph Showing Camera Sales.**

7-2-2   *Types of Graphs*

There are three types of graphs: line graphs, bar graphs, and circle graphs. *Line graphs,* like the examples shown in Figures 7-1 and 7-2, are ideal for displaying historical data in which a factor is constantly changing.

Bar graphs use parallel bars of different sizes to compare one variable with others. For example, Figure 7-3(a) shows how a bar graph can be used to compare the sales records of four computer salespersons for the month of May.

Bar Graph or Bar Chart
A graphic means of quantitative comparison by rectangles with lengths proportional to the measure of the data being compared.

■ **INTEGRATED MATH APPLICATION:**

From the graph shown in Figure 7-3(a), determine the following:

a. How many sales representatives are there?

b. Which sales rep sold the most computers in the month of May, and which sales rep sold the least?

FIGURE 7-3   **Bar Graphs and Circle Graphs.**

c. How many computers were sold in May?

d. Calculate the number of computers Ann sold as a percentage of the total sold.

■ *Solution:*

a. There are four bars in this bar graph, one for each of the four sales reps.

b. Ann has the largest bar in the graph because she sold 23 computers, while Pete has the smallest bar because he sold only 12 computers.

c. Ann sold 23, Jim sold 15, Pete sold 12, and Sally sold 15. Therefore, a total of 65 computers were sold in May.

d. To calculate the number of computers Ann sold as a percentage of the total sold, use the following formula;

$$\text{Percentage} = \frac{\text{units sold by rep}}{\text{total units sold}} \times 100$$

$$\text{Percentage} = \frac{23}{65} \times 100 \qquad (23 \div 65 = 0.35)$$

$$= 0.35 \times 100 = 35\%$$

Pie Chart or Circle Graph

A circular chart cut by radii into segments illustrating relative magnitudes.

Figure 7-3(b) shows an example of a **circle graph,** which uses different size segments within a circle to compare one variable with others. These circle graphs are often called **pie charts** because the circle resembles a pie and the segments look like different slices in the pie. The size of the segments within a circle graph is usually given as a percentage, as shown in the example in Figure 7-3(b), with the whole circle equal to 100%. Like the bar graph in Figure 7-3(a), the circle graph in Figure 7-3(b) compares the sales records of four computer salespersons for the month of May. Because the data represented in these two graphs are the same, the only difference between these two graphs are the presentation format.

■ **INTEGRATED MATH APPLICATION:**

From the graph shown in Figure 7-3(b), determine the following:

a. How many sales representatives are there?

b. Which sales rep sold the most computers in the month of May, and which sales rep sold the least?

c. If Ann sold 23 units, Jim sold 15 units, Pete sold 12 units, and Sally sold 15 units, calculate each sales rep's percentage of total sales.

d. If a total of 87 computers were sold in the month of June and Jim again sold 23%, how many did he sell?

■ *Solution:*

a. There are four segments in this circle graph, one for each of the four sales reps.

b. The largest segment is assigned to Ann because she sold 23 computers, and the smallest segment is assigned to Pete, who sold only 12 computers.

c. Each sales rep's percentage was calculated as follows:

$$\text{Ann's percentage} = \frac{\text{units sold by Ann}}{\text{total units sold}} \times 100 = \frac{23}{65} \times 100 = 0.35 \times 100 = 35\%$$

$$\text{Jim's percentage} = \frac{\text{units sold by Jim}}{\text{total units sold}} \times 100 = \frac{15}{65} \times 100 = 0.23 \times 100 = 23\%$$

$$\text{Pete's percentage} = \frac{\text{units sold by Pete}}{\text{total units sold}} \times 100 = \frac{12}{65} \times 100 = 0.19 \times 100 = 19\%$$

$$\text{Sally's percentage} = \frac{\text{units sold by Sally}}{\text{total units sold}} \times 100 = \frac{15}{65} \times 100 = 0.23 \times 100 = 23\%$$

d. $23\% \times 87 = 0.23 \times 87 = 20$ Jim sold 20 computers in June.

7-2-3   *Plotting a Graph*

As the expression goes, "a picture is worth a thousand words," and graphs are pictures that can convert a thousand lines of recorded data into an easy-to-understand visual representation. To understand the graph better, let us collect data from an experiment and then plot these data on a graph.

It has been proven through experimentation that the human ear responds logarithmically to changes in sound level. To explain this in more detail, refer to the illustration in Figure 7-4(a), which shows a tone generator connected to a speaker. As the volume control of the generator is increased the power out of the speaker is gradually increased from 1 watt to 100 watts. The listener is asked to signal the points at which a change in volume is detected. These points are listed in the table in Figure 7-4(b). Studying this table, you can see that the power has to be increased from its start point of 1 watt (1.0 W) to approximately 1.26 W before the ear can detect an increase in volume. As the power was increased further, a change in volume was detected at 1.58 W, 2 W, 2.5 W, 3.17 W, and so on, as shown in the table in Figure 7-4(b).

Studying the table in Figure 7-4(b), you can see that the human ear is rather slow to respond to increased stimulation. For example, even when the sound intensity is doubled to twice its original value (1 W to 2 W), it is not interpreted by the listener as being twice as loud. In fact the listener seems to detect only a small change in loudness. To make the tone sound twice as loud to the listener, the power must be increased ten times (1 W to 10 W), and to make the tone sound three times as loud, the power must be increased one hundred times (1 W to 100 W). Let us see how these data can be put to some practical purpose.

■   **INTEGRATED MATH APPLICATION:**

You are considering the purchase of a new amplifier for your home music system and cannot decide whether to buy a 50 W or 100 W unit. If the 100 W unit is twice the price, will you get twice the benefit?

■   *Solution:*

Although the 100 W amplifier delivers twice the output power, your ears will perceive this as only a small difference in volume. For your ears to detect twice the volume, therefore, you would need an amplifier that can deliver 10 times the power. Compared with a 50 W amplifier, a 500 W amplifier will sound as though you are getting twice your money's worth.

The graph in Figure 7-4(c) shows the **logarithmic response** of the human ear by plotting the information from Figure 7-4(b). At point 10, the sound is sensed by the human ear as being twice (2 times) as loud, although power has actually increased from 1 W to 10 W (10 times). At point 20, the sound is sensed as being three times (3 times) as loud, although power has actually increased from 1 W to 100 W (100 times). This response is called logarithmic because a multiple-of-10 change was needed in the input sound level (power had to be increased from 1 W to 10 W to 100 W) to be detected as a change in the sensed output sound (twice as much, or three times as much, respectively).

Logarithmic Response
Having or being a response or output that is logarithmic.

Points at Which a Change in Volume Was Detected

	Sound reference
1.0W (log 1 = 0)	
1. 1.26W (log 1.26 = 0.1 or $10^{0.1}$)	
2. 1.58W (log 1.58 = 0.2 or $10^{0.2}$)	
3. 2.0W (log 2.0 = 0.3 or $10^{0.3}$)	
4. 2.5W (log 2.5 = 0.4 or $10^{0.4}$)	
5. 3.17W (log 3.17 = 0.5 or $10^{0.5}$)	
6. 3.98W (log 3.98 = 0.6 or $10^{0.6}$)	
7. 5.02W (log 5.02 = 0.7 or $10^{0.7}$)	
8. 6.31W (log 6.31 = 0.8 or $10^{0.8}$)	
9. 7.94W (log 7.94 = 0.9 or $10^{0.9}$)	
10. 10.0W (log 10 = 1.0 or 10^{1})	Sound sensed as twice as loud
11. 12.6W (log 12.6 = 1.1 or $10^{1.1}$)	
12. 15.8W (log 15.8 = 1.2 or $10^{1.2}$)	
13. 20.0W (log 20 = 1.3 or $10^{1.3}$)	
14. 25.1W (log 25.1 = 1.4 or $10^{1.4}$)	
15. 31.7W (log 31.7 = 1.5 or $10^{1.5}$)	
16. 39.8W (log 39.8 = 1.6 or $10^{1.6}$)	
17. 50.2W (log 50.2 = 1.7 or $10^{1.7}$)	
18. 63.1W (log 63.1 = 1.8 or $10^{1.8}$)	
19. 79.4W (log 79.4 = 1.9 or $10^{1.9}$)	
20. 100.0W (log 100 = 2.0 or 10^{2})	Sound sensed as three times as loud

TONE GENERATOR

SPEAKER

LISTENER

Pure Tone is applied to speaker—volume is adjusted to gradually increase power from 1 watt to 100 watts.

(a)

Listener is asked to sense change in power level.

MICROPHONE
Also senses change in power level.

(b)

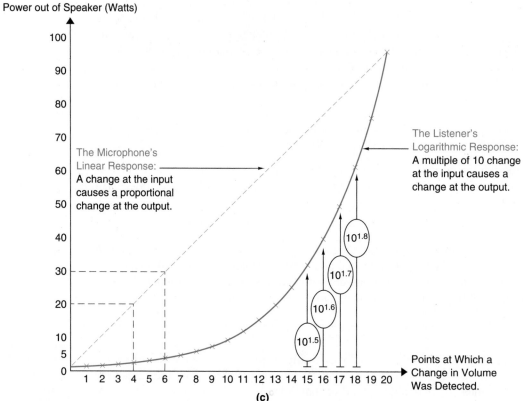

(c)

FIGURE 7-4   **The Human Ear's Logarithmic Response to Changes in Sound Level.**

FIGURE 7-5 A Graphing Calculator.

GRAPHING KEYS ———

EDITING KEYS ———

ADVANCED ———
FUNCTION KEYS

SCIENTIFIC ———
CALCULATOR KEYS

The logarithmic response of the human ear is very different from the **linear response** of the microphone, which is also shown in the graph in Figure 7-4(c). By definition, a linear change at the input causes a proportional change at the output. With a linear response, each change in the sound level applied to the microphone input will result in a proportional change in the microphone's output signal. For example, when the sound level is increased from 20 W to 30 W, the increase on the horizontal scale will be from point 4 to point 6 (a 10 W change at the input results in a 2-point change at the output). Similarly, when the sound level is increased from 70 W to 80 W, the increase on the horizontal scale will be from point 14 to point 16 (a 10 W change at the input will again result in a 2-point change at the output).

Linear Response

Having or being a response or output that is directly proportional to the input.

7-2-4   *The Graphing Calculator*

Up to this point we have been using a standard scientific calculator to perform mathematical operations. Graphing calculators possess all the functions of the scientific calculator, with the added ability that they can be used to graph functions. An example of a typical graphing calculator is shown in Figure 7-5.

The keyboard is divided into four zones: *graphing keys* access the calculator's interactive graphing features, *editing keys* allow you to edit expressions and values, *advanced function keys* display menus that access the advanced functions, and *scientific calculator keys* access the standard capabilities discussed previously.

To explain why we would need graphing capabilities, let us go straight to an application example.

■   **INTEGRATED MATH APPLICATION:**

Figure 7-6 shows a 20-cm × 25-cm sheet of cardboard with two $X \times X$ corners cut out of the left side, and two $X \times 12.5$-cm rectangles cut out of the right side. The cardboard can then be folded into a box with a lid that will have a volume that is equal to:

$$\boxed{V = A \times B \times X}$$

Volume = length × width × height

where $A = 20 - 2X$
$B = 25/2 - X$

FIGURE 7-6 A Cardboard Box.

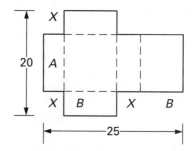

If a small value of X is chosen, the base of the box will be large, but the sides or height of the box would be small. On the other hand, if a large value of X is chosen, the base of the box will be small, but the sides or height of the box will be large. The question, therefore, is, What value of X will give the box the maximum value of volume?

■ *Solution:*

If $V = A \times B \times X$, and $A = 20 - 2X$, and $B = 25/2 - X$, then

$$V = (20 - 2X) \times (25/2 - X) \times (X)$$

To enter this equation into the graphing calculator, you would use the "$Y =$ Editor" function and input the formula as shown in Figure 7-7(a). Now, selecting the calculator's "Table" function will determine the volume of the box ($Y1$) for every possible value of X, as shown in Figure 7-7(b). Scanning down the table shown in Figure 7-7(c), you can see that the maximum value of $Y1$ or volume is 410.26, which occurs when $X = 3.7$. Pressing the calculator's "Graph" function will display the plotted data in graphical rather than tabular form, as seen in Figure 7-7(d).

This ability of the graphing calculator to display every possible output result for every possible input value can be extremely helpful in the analysis of a problem.

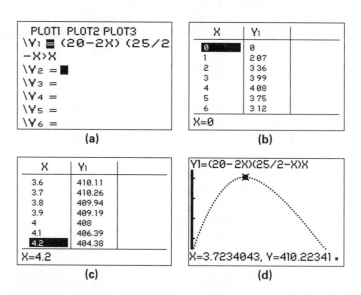

FIGURE 7-7 Display Results for Box Volume.

Now that you have completed this section, you should be able to:

- **Objective 5.** *Define the purpose of a graph.*
- **Objective 6.** *Describe how to interpret the data displayed in a graph.*
- **Objective 7.** *Explain the differences among a line graph, a bar graph, and a circle graph or pie chart.*
- **Objective 8.** *Show how the data collected from an experiment are plotted on a graph.*
- **Objective 9.** *Describe the difference between a logarithmic response and a linear response.*
- **Objective 10.** *Explain how the graphing calculator can be used to find the solution to a problem.*

Use the following questions to test your understanding of Section 7-2.

1. Plot a line graph to display the following data:

The Number of Indoor and Outdoor Movie Screens

YEAR	INDOOR	OUTDOOR
1975	12,000	4000
1980	14,000	4000
1985	18,000	3000
1990	23,000	2000

2. Draw a circle graph and bar graph to indicate the following data: Each 24 h day Tom works for 8 h, studies for 3 h, showers and dresses for 2 h, eats for 1.5 h, sleeps for 8 h, and exercises for 1.5 h.

3. What is the difference between a standard scientific calculator and a graphing calculator?

4. Describe the purpose for each of the following graphing calculator steps:
   - **a.** $Y =$ Editor
   - **b.** Table
   - **c.** Graph

REVIEW QUESTIONS

Multiple Choice Questions

1. Logarithms were first discovered by:
   - **a.** Carl Gauss
   - **b.** John Napier
   - **c.** Isaac Newton
   - **d.** Blaise Pascal

2. Natural logarithms use a base of _____, while common logarithms use a base of _____.
   - **a.** 10, 100
   - **b.** 10, 2.71828
   - **c.** 2.71828, 100
   - **d.** 2.71828, 10

3. What is the log of 10,000?
   - **a.** 9.21
   - **b.** 6
   - **c.** 4
   - **d.** 3

4. What is the log of 0.32?
   - **a.** −0.5
   - **b.** −1.1
   - **c.** −1.5
   - **d.** −0.15

5. What is the common antilogarithm of 10^6?
   - **a.** 1,000,000
   - **b.** 1000.000
   - **c.** 1000
   - **d.** 100,000

6. A_____ is a diagram showing the relationship between two or more factors.
   - **a.** Scale
   - **b.** Logarithm
   - **c.** Graph
   - **d.** Exponent

7. In a _____ _____, parallel columns of different sizes are used to compare one variable with others.
   - **a.** Line graph
   - **b.** Bar graph
   - **c.** Circle graph
   - **d.** All the above

8. In a _____ _____, the data are represented by points joined by line segments.
   - **a.** Line graph
   - **b.** Bar graph
   - **c.** Circle graph
   - **d.** All the above

9. In a _____ _____, different size segments within a circle are used to compare one variable with others.
   - **a.** Line graph
   - **b.** Bar graph
   - **c.** Pie chart
   - **d.** All the above

10. With a _____ response, a multiple-of-10 change at the input causes a change at the output, whereas with a _____ response, a change at the input causes a proportional change at the output.
    - **a.** Linear, logarithmic
    - **b.** Logarithmic, linear
    - **c.** Logarithmic, exponential
    - **d.** Linear, exponential

Communication Skill Questions

11. What is a common logarithm? (7–1)

12. What is the difference between a natural logarithm and a common logarithm? (7–1)

13. What is a common antilogarithm? (7–1)

14. State the four-step procedure you would use to simplify multiplication using logarithms. (7–1)

15. Arbitrarily choose values, and then describe the following:
   a. Multiplication using logarithms
   b. Division using logarithms

16. What is a graph? (7–2)

17. List some of the different types of graphs. (7–2)

18. Sketch a diagram and describe the different types of graphs. (7–2)

19. What is a linear response? (7–2)

20. What is a logarithmic response? (7–2)

Practice Problems

21. Determine the log of the following:

a. 10	**g.** 0.5
b. 10,000	**h.** 0.75
c. 1,234,567	**i.** 0.125
d. 50.5	**j.** 0.95
e. 1.5	**k.** 12.25
f. 0.25	**l.** 75.75

22. Determine the antilog of the following:

a. 1	**g.** 2.5
b. 2	**h.** 1.25
c. 3	**i.** 0.125
d. 4	**j.** 0.75
e. 5	**k.** 0.33
f. 1.5	**l.** 0.95

23. What steps should be followed to simplify multiplication to addition using logarithms?

24. Apply the four-step procedure from Problem 23 to the following:
   a. $2 \times 2 = ?$
   b. $244 \times 30 = ?$
   c. $1.25 \times 150 = ?$
   d. $0.5 \times 0.33 = ?$

25. What steps should be followed to simplify division to subtraction using logarithms?

26. Apply the four-step procedure from Problem 25 to the following:
   a. $12 \div 2 = ?$
   b. $244 \div 2 = ?$
   c. $250 \div 0.25 = ?$
   d. $1255 \div 0.5 = ?$

27. Identify the types of graphs shown in Figure 7-8.

28. Refer to Figure 7-8(a), and answer the following questions:
   a. Which music type has remained the most popular?
   b. Which music type is the least popular?
   c. Which music type became less popular as country music had an increase in popularity in 1992 and 1993?
   d. What was the approximate market share for each of the music types in 1995?

29. Refer to Figure 7-8(b), and answer the following questions:
   a. What is the main reason for the average person to connect to the Internet?
   b. In order, what are the top five reasons why people want to go online?

30. Figure 7-9 shows an electronic device called a *potentiometer*. This device has two terminals (*A* and *B*), a circular resistive track, and a wiper that can be placed at any point on the resistive track by turning a control shaft. (Potentiometers are often used as volume controls for music systems and televisions.) There are two basic types of potentiometers: those with linear resistive tracks, and those with tapered or nonlinear resistive tracks. Figure 7-9 lists the resistance values that were measured between terminals *A* and *B* for both a linear and a tapered potentiometer. When the control shaft was fully counterclockwise (CCW), the resistance between terminals *A* and *B* for both the linear and tapered potentiometer was 0 ohms (0 Ω). When the control shaft was turned one-quarter clockwise (¼ CW), the resistance between terminals *A* and *B* for the linear potentiometer was 250 Ω, and for the tapered potentiometer it was 350 Ω. When the control shaft was turned one-half clockwise (½ CW), the resistance between terminals *A* and *B* for the linear potentiometer was 500 Ω, and for the tapered potentiometer it was 625 Ω. When the control shaft was turned three-quarters clockwise (¾ CW), the resistance between terminals *A* and *B* for the linear potentiometer was 750 Ω, and for the tapered potentiometer it was 900 Ω. Finally, when the control shaft was turned fully clockwise (CW), the resistance between terminals *A* and *B* for the linear potentiometer was 1000 Ω (1 kΩ), and for the tapered potentiometer it was also 1000 Ω. Sketch a "resistance versus control shaft position" line graph, and then plot a linear line and tapered line.

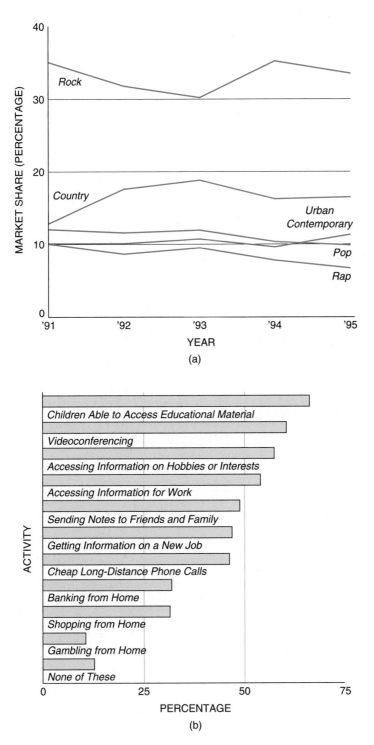

FIGURE 7-8   Graphs (a) Music Types. (b) Why People Connect to the Internet.

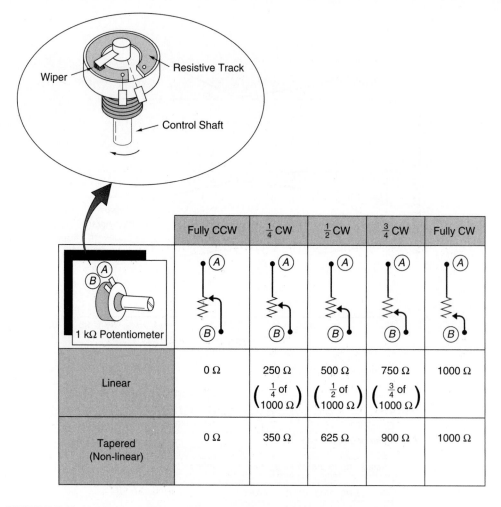

	Fully CCW	$\frac{1}{4}$ CW	$\frac{1}{2}$ CW	$\frac{3}{4}$ CW	Fully CW
1 kΩ Potentiometer	(A) ⚡ (B)	(A) ⚡ (B)	(A) ⚡ (B)	(A) ⚡ (B)	(A) ⚡ (B)
Linear	0 Ω	250 Ω $\left(\frac{1}{4} \text{ of } 1000\ \Omega\right)$	500 Ω $\left(\frac{1}{2} \text{ of } 1000\ \Omega\right)$	750 Ω $\left(\frac{3}{4} \text{ of } 1000\ \Omega\right)$	1000 Ω
Tapered (Non-linear)	0 Ω	350 Ω	625 Ω	900 Ω	1000 Ω

FIGURE 7-9 **A Linear and Tapered Resistive Track Within a Potentiometer.**

Web Site Questions

Go to the web site http://www.prenhall.com/cook, select the textbook *Introductory Computer Mathematics, 2e,* select this chapter, and then follow the instructions when answering the multiple-choice practice problems.

Analog to Digital

Moon Walk

In July of 1969 almost everyone throughout the world was caught at one stage or another gazing up to the stars and wondering. For U.S. astronaut Neil Armstrong, the age-old human dream of walking on the moon was close to becoming a reality. On earth, millions of people and thousands of newspapers and magazines waited anxiously to celebrate a successful moon landing. Then, the rather brief but eloquent message from Armstrong was transmitted 240,000 miles to Houston, Texas, where it was immediately retransmitted to a waiting world. This message was: "That's one small step for man, one giant leap for mankind." These words were received by many via television, but for magazines and newspapers this entire mission, including the speech from Armstrong, was converted into a special code made up of ON–OFF pulses that traveled from computer to computer. Every letter of every word was converted into a code that used the two symbols of the binary number system, 0 and 1. This code is still used extensively by all modern computers and is called *American Standard Code for Information Interchange,* or ASCII (pronounced "askey").

It is only fitting that these codes made up of zeros and ones conveyed the finale to this historic mission, since they played such an important role throughout. Commands encoded into zeros and ones were used to control almost everything, from triggering the takeoff to keeping the spacecraft at the proper angle for reentry into the earth's atmosphere.

No matter what its size or application, a digital electronic computer is quite simply a system that manages the flow of information in the form of zeros and ones. Referring to the ASCII code table in Figure 8-2 on p. 594, see if you can decode the following famous Armstrong message that reads from left to right:

0100010	1010100	1101000	1100001	1110100	0100111	1110011	0100000	1101111	1101110
1100101	0100000	1110011	1101101	1100001	1101100	1101100	0100000	1110011	1110100
1100101	1110000	0100000	1100110	1101111	1110010	0100000	1100001	0100000	1101101
1100001	1101110	0101100	0100000	1101111	1101110	1100101	0100000	1100111	1101101
1100001	1101110	1110100	0100000	1101100	1100101	1100001	1110000	0100000	1100110
1101111	1110010	0100000	1101101	1100001	1101110	1101011	1101001	1101110	1100100
0101110	0100010	0100000	0100000	0101101	1001110	1100101	1101001	1101100	0100000
1000001	1110010	1101101	1110011	1110100	1110010	1101111	1101110	1100111	0101100
0100000	1000001	1110000	1101111	1101100	1101100	1101111	0100000	0110001	0110001

Outline and Objectives

Introduction

Since World War II, no branch of science has contributed more to the development of the modern world than electronics. It has stimulated dramatic advances in the fields of communication, computing, consumer products, industrial automation, test and measurement, and health care. It has now become the largest single industry in the world, exceeding the automobile and oil industries, with annual sales of electronic systems exceeding $2 trillion.

One of the most important trends in this huge industry has been a gradual shift from *analog electronics* to *digital electronics.* This movement began in the 1960s and is almost complete today. In fact, a recent statistic stated that, on average, 90% of the circuitry within electronic systems is now digital and only 10% is analog. This digitalization of the electronics industry is bringing sectors that were once separate closer together. For example, two of the largest sectors or branches of electronics are *computing* and *communications.* Being able to communicate with each other using the common language of digital has enabled computers and communications to interlink so that computers can now function within communication-based networks, and communications networks can now function through computer-based systems. Industry experts call this merging *convergence* and predict that digital electronics will continue to unite the industry and stimulate progress in practically every field of human endeavor.

Needless to say, this part of the book you are about to read involving digital electronic mathematics and concepts is an essential element in your study of electronics.

8-1   ANALOG AND DIGITAL DATA AND DEVICES

Before we begin, let's first review analog electronics and then compare it to digital, so that we can clearly see the differences between the two.

8-1-1   *Analog Data and Devices*

Figure 8-1(a) shows an electronic circuit designed to amplify speech information detected by a microphone. One of the easiest ways to represent data or information is to have a voltage change in direct proportion to the information it is representing. In the example in Figure 8-1(a), the *pitch and loudness* of the sound waves applied to the microphone should control the *frequency and amplitude* of the voltage signal from the microphone. The output voltage signal from the microphone is said to be an analog of the input speech signal. The word **analog** means "similar to," and in Figure 8-1(a) the electronic signal produced by the microphone is an analog (or similar) to the speech signal, since a change in speech "loudness or pitch" will cause a corresponding change in signal voltage "amplitude or frequency."

In Figure 8-1(b), a light detector or solar cell converts light energy into an electronic signal. This information signal represents the amount of light present, since changes in voltage amplitude result in a change in light-level intensity. Once again, the output electronic signal is an analog (or similar) to the sensed light level at the input.

Figure 8-1, therefore, indicates two analog circuits. The microphone in Figure 8-1(a) generates an **ac analog signal** that is amplified by an **ac amplifier circuit.** The microphone is considered an **analog device or component** and the amplifier an **analog circuit.** The light detector in Figure 8-1(b) would also be an analog component; however, in this example it generates a **dc analog signal** that is amplified by a **dc amplifier circuit.**

Analog

The representation of physical properties by a proportionally varying signal.

AC Analog Signal

An analog signal that alternates positive and negative.

AC Amplifier

An amplifier designed to increase the magnitude of an ac signal.

Analog Circuit

A circuit designed to manage analog signals.

Analog Device or Component

Advice or component that makes up a circuit designed to manage analog signals.

DC Analog Signal

An analog signal that is always either positive or negative.

DC Amplifier

An amplifier designed to increase the magnitude of a dc signal.

190

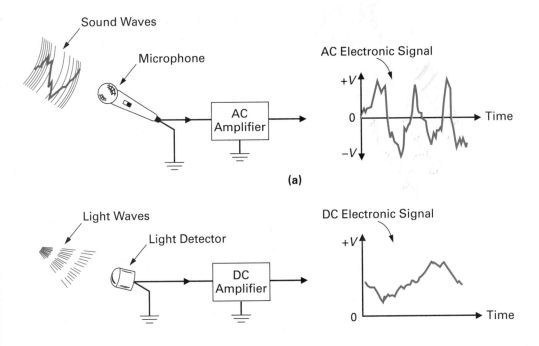

(a)

FIGURE 8-1   Analog Data and Devices.

Both of the information signals in Figure 8-1 vary smoothly and continuously, in accordance with the natural quantities they represent (sound and light). Analog circuits are often called **linear circuits,** since linear by definition is *the variation of an output in direct proportion to the input.* This linear circuit response is evident in Figure 8-1, where you can see that in both the dc and ac circuit, the output signal voltage varies in direct proportion to the sound or light signal input.

Linear Circuit
A circuit in which the output varies in direct proportion to the input.

8-1-2   *Digital Data and Devices*

In **digital electronic circuits,** information is first converted into a coded group of pulses. To explain this concept, let's take a closer look at the example shown in Figure 8-2(a). This code consists of a series of HIGH and LOW voltages, in which the HIGH voltages are called "1s" (ones) and the LOW voltages are called "0s" (zeros). Figure 8-2(b) lists the ASCII code, which is one example of a digital code. Referring to Figure 8-2(a), you will notice that the "1101001" information or data stream code corresponds to lowercase *i* in the ASCII table shown highlighted in Figure 8-2(b). Computer keyboards are one of many devices that make use of the digital ASCII code. In Figure 8-2(c), you can see how the lowercase *i* ASCII code is generated whenever the *i* key is pressed, encoding the information *i* into a group of pulses (1101001).

The next question you may have is: Why do we go to all this trouble to encode our data or information into these two-state codes? The answer can best be explained by examining history. The early digital systems constructed in the 1950s made use of a decimal code that used 10 levels or voltages, with each of these voltages corresponding to one of the 10 digits in the decimal number system ($0 = 0$ V, $1 = 1$ V, $2 = 2$ V, $3 = 3$ V, up to $9 = 9$ V). The circuits that had to manage these decimal codes, however, were very complex since they had to generate one of 10 voltages and sense the difference among all 10 voltage levels. This complexity led to inaccuracy, since some circuits would periodically confuse one voltage level for a different voltage level. *The solution to the problem of circuit complexity and inaccuracy was solved by a adopting a two-state system instead of a 10-state system.* Using a two-state or two-digit system, you can generate codes for any number, letter, or symbol, as we have seen

Digital Electonic Circuit
A circuit designed to manage digital information signals.

I C =
Integrated
Circut

FIGURE 8-2   Two-State Digital Information. (a) Information or Data Code for *i*. (b) ASCII Code. (c) Keyboard ASCII Generator.

in the ASCII table in Figure 8-2. The electronic circuits that manage these two-state codes are less complex, since they have to generate and sense only either a HIGH or LOW voltage. In addition, two-state circuits are much more accurate, since there is little room for error between the two extremes of ON and OFF, or HIGH voltage and LOW voltage.

Abandoning the 10-state system and adopting the two-state system for the advantages of circuit simplicity and accuracy meant that we were no longer dealing with the decimal number system. As a result, having only two digits (0 and 1) means that we are now operating in the two-state number system, which is called **binary.** Figure 8-3(a) shows how the familiar decimal system has 10 digits or levels labeled 0 through 9, and Figure 8-3(b) shows how we could electronically represent each decimal digit. With the base 2, or binary number

Binary

Having only two alternatives, two-state.

FIGURE 8-3   **Electronically Representing the Digits of a Number System.**

FIGURE 8-4   **Analog and Digital Readouts.**

system, we have only two digits in the number scale, and therefore only the digits 0 (zero) and 1 (one) exist in binary, as shown in Figure 8-3(c). These two states are typically represented in an electronic circuit as two different values of voltage (binary 0 = LOW voltage; binary 1 = HIGH voltage), as shown in Figure 8-3(d).

Using combinations of **binary digits** (abbreviated as **bits**) we can represent information as a binary code. This code is called a **digital signal** because it is an *information signal* that makes use of *binary digits*. Today, almost all information, from your telephone conversations to the music on your compact discs, is **digitized** or converted to binary data form.

You will probably hear the terms *analog* and *digital* used to describe the difference between the two types of meters shown in Figure 8-4. On the <u>analog readout multimeter</u> shown in Figure 8-4(a), the amount of pointer deflection across the scale is an analog (or similar) to the magnitude of the electrical property being measured. On the other hand, with the **digital readout multimeter** shown in Figure 8-4(b), the magnitude of the electrical property being measured is displayed using digits, which in this case are decimal digits.

SELF-TEST EVALUATION POINT FOR SECTION 8-1

Now that you have completed this section, you should be able to:

■ **Objective 1.** *Define the differences between analog and digital data and devices.*

■ **Objective 2.** *Explain why analog circuits are also referred to as linear circuits.*

■ **Objective 3.** *Describe the function of the digital ASCII code.*

■ **Objective 4.** *Explain why the 10-state digital system was replaced by the two-state digital system.*

Use the following questions to test your understanding of Section 8-1:

1. What does the word *analog* mean?
2. What is an analog signal?
3. What is the ASCII code?
4. What were the two key reasons for adopting the two-state binary system in place of the 10-state decimal system?
5. How many digits does the binary number system have?
6. Describe the difference between an analog readout watch and a digital readout watch.

Whenever you have two different forms, like analog and digital, there is always a need to convert back and forth between the two. Computers are used at the heart of almost every electronic system today because of their ability to quickly process and store large amounts of data, make systems more versatile, and perform many functions. Many of the input signals applied to a computer for storage or processing are analog in nature, and therefore a data conversion circuit is needed to interface these analog inputs with a digital system. Similarly, a data conversion circuit may also be needed to interface the digital computer system with an analog output device.

Figure 8-5 shows a simplified block diagram of a computer. No matter what the application, a computer has three basic blocks: a microprocessor unit, a memory unit, and an input/output unit.

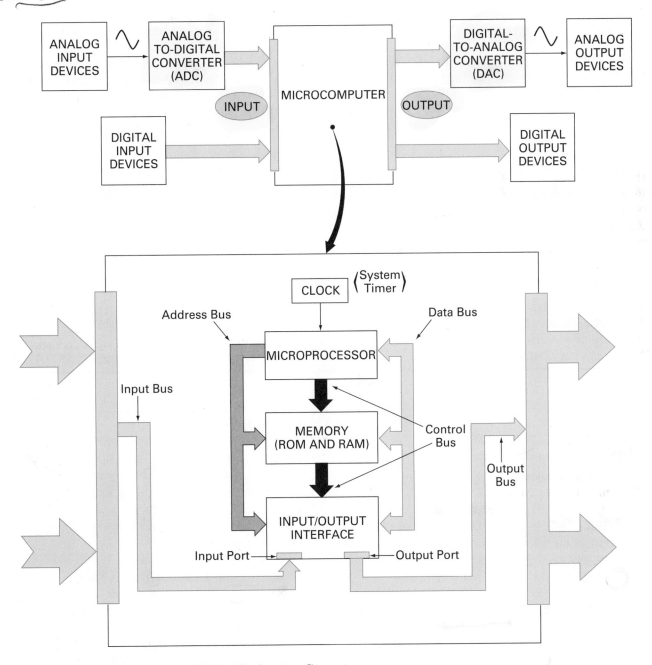

FIGURE 8-5   Connecting Analog and Digital Devices to a Computer.

Analog-to-Digital Converter

A circuit that converts analog input signals into equivalent digital output signals.

Input Port

An electronic doorway used for receiving information into a computer.

Output Port

An electronic doorway used for transmitting information out of a computer.

Digital-to-Analog Converter

A circuit that converts digital input signals into equivalent analog output signals.

Getting information into the computer is the job of input devices, and as you can see in Figure 8-5, there are two input paths into a computer: one path from analog input devices, and the other path from digital input devices. Information in the form of light, sound, heat, pressure, or any other real-world quantity is analog in nature and has an infinite number of input levels. Sensors or transducers of this type, such as photoelectric cells for light or microphones for sound, will generate analog signals. Since the computer operates on information in only digital or binary form, a translation or conversion is needed if the computer is going to be able to understand or interpret an analog signal. This signal processing is achieved by an **analog-to-digital converter (ADC),** which transforms the varying analog input voltage into equivalent digital codes that the computer can understand. Once the analog information has been encoded into digital form, the information can then enter the computer through an electronic doorway called the **input port.** Some input devices, such as a computer keyboard, automatically generate a digital code for each key that is pressed (ASCII codes). The codes generated by these digital input devices can be connected directly to the computer, without having to pass through a converter.

In contrast, the digital output information from a computer's **output port** can be used to drive either an analog device or a digital device. If an analog output is desired, as in the case of a voice signal for a speaker, the digital computer's output will have to be converted into a corresponding equivalent analog signal by a **digital-to-analog converter (DAC). If** only a digital output is desired, as in the case of a printer, which translates the binary codes into printed characters, the digital computer output can be connected directly to the output device without the need for a converter.

SELF-TEST EVALUATION POINT FOR SECTION 8-2

Now that you have completed this section, you should be able to:

- **Objective 5.** *List the three basic blocks within a computer.*

- **Objective 6.** *Describe how analog and digital devices are connected to a computer.*

- **Objective 7.** *Define the function of the analog-to-digital converter (ADC) and the digital-to-analog converter (DAC).*

Use the following questions to test your understanding of Section 8-2:

1. What are the three basic blocks of a computer?
2. Give the full names of the following acronyms:
   a. ADC   b. DAC
3. An ADC will convert a/an _____ input into an equivalent _____ output.
4. A DAC will convert a/an _____ input into an equivalent _____ output.

REVIEW QUESTIONS

Multiple Choice Questions

1. Which of the following is an example of a digital electronic circuit?
   a. Amplifier
   b. MPU
   c. Power supply
   d. ON/OFF lighting circuit

2. Which of the following is an example of an analog electronic circuit?
   a. Amplifier
   b. CPU
   c. ASCII keyboard
   d. ON/OFF lighting circuit

3. Digital circuits based on the binary number system are often referred to as _____ circuits. On the other hand, the output of analog circuits varies in direct proportion to the input, which is why analog circuits are often called _linear_ circuits.

   a. Ten-state, linear
   b. Linear, two-state
   c. Digital, 10-state
   d. Two-state, linear

4. A digital multimeter displays the measured quantity using
   a. Binary digits
   b. A pointer and scale
   c. Decimal digits
   d. An abacus

5. Why do digital electronic systems use the two-state system instead of the 10-state system?
   a. Circuit simplicity
   b. Accuracy
   c. Both (a) and (b)
   d. None of the above

6. A _____ converts an analog input voltage into a proportional digital output code.
   a. ROM
   b. DAC
   c. ADC
   d. RWM

7. A _____DAC_____ converts a digital input code into a proportional analog output voltage.

   a. ROM   c. ADC
   b. DAC *(circled)*   d. RWM

8. What are the three basic blocks of a computer?

   a. ADC, I/O, memory   c. MPU, I/O, memory
   b. DAC, ADC, I/O   d. ROM, MPU, I/O

Communication Skill Questions

9. Define the following: (8-1)

   a. Analog   c. Binary
   b. Convergence   d. Linear

10. Why do digital systems convert information into binary codes instead of decimal codes? (8-1-2)

11. Why are analog circuits also called linear circuits? (8-1-1)

12. Describe the differences between an analog and digital readout display. (8-1-1)

13. What are the three basic blocks of a computer? (8-2)

14. Give the full names of the following abbreviations and acronyms:

    a. Bit   c. ADC
    b. ASCII   d. DAC

15. What is the function of the ADC and the DAC? (8-2)

Web Site Questions

Go to the web site http://www.prenhall.com/cook, select the textbook *Introductory Computer Mathematics, 2e*, select this chapter, and then follow the instructions when answering the multiple-choice practice problems.

(handwritten annotations):
micro processor unit, I/O, memory
American Standard Code
Analog to Digital Conversion
Digital to Analog Conversion
Analog to Digital Conversion
Digital to Analog Conversion
Digital readout display in decimal digits, electrical prop display in digits
Variation of output is direct proportion to input
Integrated circuit is a chip
Analog displays amount of pointer deflection across scale is an analog to the electrical property being measured.

Number Systems and Codes

Leibniz's Language of Logic

Gottfried Wilhelm Leibniz was born in Leipzig, Germany, in 1646. His father was a professor of moral philosophy and spent much of his time discussing thoughts and ideas with his son. Tragically, his father died when Leibniz was only six, and from that time on, Leibniz spent hour upon hour in his late father's library reading all his books.

By the age of 12, Leibniz had taught himself history, Latin, and Greek. At 15 he entered the University of Leipzig, where he came across the works of scholars such as Johannes Kepler and Galileo. The new frontiers of science fascinated him, so he added mathematics to his curriculum.

In 1666, while finishing his university studies, the 20-year-old Leibniz wrote what he modestly called a schoolboy essay, titled "DeArte Combinatoria," which translated means "On the Art of Combination." In this work, he described how all thinking of any sort on any subject could be reduced to exact mathematical statements. Logic, or as he called it, the laws of thought, could be converted from the verbal realm, which is full of ambiguities, into precise mathematical statements. In order to achieve this, however, Leibniz stated that a "universal language" would be needed. Most of his professors found the paper either baffling or outrageous, and this caused Leibniz not to pursue the idea any further.

After graduating from the university, Leibniz was offered a professorship at the University of Nuremberg, which he turned down for a position as an international diplomat. This career proved not as glamorous as he imagined, since most of his time was spent in uncomfortable horse-drawn coaches traveling between European capitals.

In 1672 his duties took him to Paris, where he met Dutch mathematician and astronomer Christian Huygens. After seeing the hours that Huygens spent on endless computations, Leibniz set out to develop a mechanical calculator. A year later Leibniz unveiled the first machine that could add, subtract, multiply, and divide decimal numbers.

In 1676 Leibniz began to concentrate more on mathematics. It was at this time that he invented calculus, which was also independently discovered by Isaac Newton in England. Leibniz's focus, however, was on the binary number system, which preoccupied him for years. He worked tirelessly to document the long combinations of 1s and 0s that make up the modern binary number system, and perfected binary arithmetic. Ironically, for all his genius, Leibniz failed to make the connection between his 1666 essay and binary, which was the universal language of logic that he was seeking.

This discovery would remain hidden until another self-taught mathematician named George Boole uncovered it a century and a quarter after Leibniz's death in 1716.

Outline and Objectives

Introduction

One of the best ways to understand anything new is to compare it with something that you are already familiar with so that the differences are highlighted. In this chapter we will be examining the *binary number system,* which is the language used by digital computer electronic circuits. To best understand this new number system, we will compare it with the number system most familiar to you, the *decimal* (base 10) *number system.* Although this system is universally used to represent quantities, many people do not fully understand its weighted structure. This review of decimal has been included so that you can compare the base 10 number system with the system used internally by digital electronics, the binary (base 2) number system.

In addition to binary, two other number systems are widely used in conjunction with digital electronics: the octal (base 8) number system, and the hexadecimal (base 16) number system. These two number systems are used as a type of shorthand for large binary numbers and enable us to represent a large group of binary digits with only a few octal or hexadecimal digits.

The last section in this chapter covers many of the different binary codes, which use a combination of 1s and 0s to represent letters, numbers, symbols, and other information.

Keep in mind throughout this chapter that different number systems are just another way to count, and as with any new process, you will become more proficient and more at ease with practice as we proceed through the chapters that follow.

9-1   THE DECIMAL NUMBER SYSTEM

The decimal system of counting and keeping track of items was first created by Hindu mathematicians in India in A.D. 400. Since it involved the use of fingers and thumbs, it was natural that this system would have 10 digits. The system found its way to all the Arab countries by A.D. 800, where it was named the Arabic number system, and from there it was eventually adopted by nearly all the European countries by A.D. 1200, where it was called the **decimal number system.**

Decimal Number System

A base 10 number system.

Base

With number systems, it describes the number of digits used.

The key feature that distinguishes one number system from another is the number system's **base** or *radix.* This base indicates the number of digits that will be used. The decimal number system, for example, is a base 10 number system, which means that it uses 10 digits (0 through 9) to communicate information about an amount. A subscript is sometimes included after a number when different number systems are being used, to indicate the base of the number. For example, $12,567_{10}$ is a base 10 number, whereas 10110_2 is a base 2 number.

9-1-1   *Positional Weight*

The position of each digit of a decimal number determines the weight of that digit. A 1 by itself, for instance, is worth only 1, whereas a 1 to the left of three 0s makes the 1 worth 1000.

In decimal notation, each position to the left of the decimal point indicates an increased positive power of 10, as seen in Figure 9-1(a). The total quantity or amount of the number is therefore determined by the size and the weighted position of each digit. For ex-

$$\text{Value of number} = (6 \times 10^3) + (0 \times 10^2) + (1 \times 10^1) + (9 \times 10^0)$$
$$= (6 \times 1000) + (0 \times 100) + (1 \times 10) + (9 \times 1)$$
$$= 6000 + 0 + 10 + 9 = 6019_{10}$$

(a)

(b)   (c)

FIGURE 9-1   **The Decimal (Base 10) Number System.**

ample, the value shown in Figure 9-1(a) has six thousands, zero hundreds, one ten, and nine ones, which combined makes a total of 6019_{10}.

In the decimal number system, the leftmost digit is called the **most significant digit (MSD),** and the rightmost digit is called the **least significant digit (LSD).** Applying these definitions to the example in Figure 9-1(a), we see that the 6 is the MSD, since its position carries the most weight, and the 9 is the LSD, since its position carries the least weight.

9-1-2   *Reset and Carry*

Before proceeding to the binary number system, let us review one other action that occurs when counting in decimal. This action, which is familiar to us all, is called **reset and carry.** Referring to Figure 9-1(b), you can see that a reset and carry operation occurs after a count of 9. The units column, which has reached its maximum count, resets to 0 and carries a 1 to the tens column, resulting in a final count of 10.

This reset and carry action will occur in any column that reaches its maximum count. For example, Figure 8-1(c) shows how two reset and carry operations will take place after a count of 99. The units column resets and carries a 1 to the tens column, which in turn resets and carries a 1 to the hundreds column, resulting in a final count of 100.

Most Significant Digit (MSD)

The leftmost, largest-weight digit in a number.

Least Significant Digit (LSD)

The rightmost, smallest-weight digit in a number.

Reset and Carry

An action that occurs when a column has reached its maximum count.

Now that you have completed this section, you should be able to:

■ **Objective 1.** *Explain how different number systems operate.*

■ **Objective 2.** *Describe the differences between the decimal and binary number systems.*

■ **Objective 3.** *Explain the positional weight and the reset and carry action for decimal.*

Use the following questions to test your understanding of Section 9-1:

1. What is the difference between the Arabic number system and the decimal number system?
2. What is the base or radix of the decimal number system?
3. Describe the positional weight of each of the digits in the decimal number 2639.
4. What action occurs when a decimal column advances beyond its maximum count?

9-2 THE BINARY NUMBER SYSTEM

As in the decimal system, the value of a binary digit is determined by its position relative to the other digits. In the decimal system, each position to the left of the decimal point increases by a power of 10. Similarly, in the binary number system, since it is a base 2 (bi) number system, each place to the left of the **binary point** increases by a power of 2. Figure 9-2 shows how columns of the binary and decimal number systems have different weights. For example, with binary, the columns are weighted so that 2^0 is one, 2^1 is two, 2^2 is four, 2^3 is eight ($2 \times 2 \times 2 = 8$), and so on.

As we know, the base or radix of a number system also indicates the maximum number of digits used by the number system. The base 2 binary number system uses only the first two digits on the number scale, 0 and 1. The 0s and 1s in binary are called binary digits, or *bits,* for short.

Binary Point

A symbol used to separate the whole from the fraction in a binary number.

9-2-1 *Positional Weight*

As in the decimal system, each column in binary carries its own weight, as seen in Figure 9-3(a). With the decimal number system, each position to the left increases 10 times. With binary, the weight of each column to the left increases 2 times. The first column, therefore, has a weight of 1, the second column has a weight of 2, the third column 4, the fifth 8, and so on. The value or quantity of a binary number is determined by the digit in each column and the

FIGURE 9-2 **A Comparison of Decimal and Binary.**

weight of each position

DECIMAL			BINARY			
10^2	10^1	10^0	2^3	2^2	2^1	2^0
100	10	1	8	4	2	1
		0				0
		1				1
		2			1	0
		3			1	1
		4		1	0	0
		5		1	0	1
		6		1	1	0
		7		1	1	1
		8	1	0	0	0
		9	1	0	0	1
	1	0	1	0	1	0
	1	1	1	0	1	1
	1	2	1	1	0	0

2^6	2^5	2^4	2^3	2^2	2^1	2^0
	32	16	8	4	2	1
	1	0	1	1	0	1

MSB   LSB

$$
\begin{aligned}
\text{Value of number} &= (1 \times 2^5) + (0 \times 2^4) + (1 \times 2^3) + (1 \times 2^2) + (1 \times 2^1) + (1 \times 2^0) \\
&= (1 \times 32) + (0 \times 16) + (1 \times 8) + (1 \times 4) + (0 \times 2) + (1 \times 1) \\
&= 32 + 0 + 8 + 4 + 0 + 1 \\
&= 45_{10}
\end{aligned}
$$

(a)

DECIMAL		BINARY			
10^1	10^0	2^3	2^2	2^1	2^0
10	1	8	4	2	1
	0				0
	1				1
	2			1	0
	3			1	1
	4		1	0	0
	5		1	0	1
	6		1	1	0
	7		1	1	1
	8	1	0	0	0
	9	1	0	0	1
1	0	1	0	1	0
1	1	1	0	1	1
1	2	1	1	0	0
1	3	1	1	0	1
1	4	1	1	1	0
1	5	1	1	1	1

(b)

FIGURE 9-3   **The Binary (Base 2) Number System.**

positional weight of the column. For example, in Figure 9-3(a), the binary number 101101_2 is equal in decimal to 45_{10}, since we have 1×32, 1×8, 1×4, and 1×1 ($32 + 8 + 4 + 1 = 45$). The 0s in this example are not multiplied by their weights (16 and 2), since they are "0" and of no value. The leftmost binary digit is called the most significant bit (MSB) since it carries the most weight, and the rightmost digit is called the least significant bit (LSB), since it carries the least weight. Applying these definitions to the example in Figure 9-3(a), we see that the 1 in the thirty-twos column is the MSB, and the 1 in the units column is the LSB.

9-2-2   *Reset and Carry*

The reset and carry action occurs in binary in exactly the same way as does in decimal; however, since binary has only two digits, a column will reach its maximum digit much sooner, and therefore the reset and carry action in the binary number system will occur much more frequently.

Referring to Figure 9-3(b), you can see that the binary counter begins with 0 and advances to 1. At this stage, the units column has reached its maximum, and therefore the next count forces the units column to reset and carry into the next column, producing a count of 0010_2 (2_{10}). The units column then advances to a count of 0011_2 (3_{10}). At this stage, both the units column and twos column have reached their maximums. As the count advances by 1 it will cause the units column to reset and carry a 1 into the twos column, which will also have to reset and carry a 1 into the fours column. This will result in a final count of 0100_2 (4_{10}). The count will then continue to 0101_2 (5_{10}), 0110_2 (6_{10}), 0111_2 (7_{10}), and then 1000_2 (8_{10}), which, as you can see in Figure 9-3(b), is a result of three reset and carries.

Comparing the binary reset and carry with the decimal reset and carry in Figure 9-3(b), you can see that since the binary number system uses only two digits, binary numbers quickly turn into multidigit figures. For example, a decimal eight (8) uses only one digit, whereas a binary eight (1000) uses four digits.

9-2-3   *Converting Binary Numbers to Decimal Numbers*

Binary numbers can easily be converted to their decimal equivalent by simply adding together all the column weights that contain a binary 1, as we did previously in Figure 9-3(a).

■   **INTEGRATED MATH APPLICATION:**

Convert the following binary numbers to their decimal equivalents.

   a. 1010

   b. 101101

■   *Solution:*

   a. Binary column weights   32   16   8   4   2   1
        Binary number   1   0   1   0

$$\text{Decimal equivalent} = (1 \times 8) + (0 \times 4) + (1 \times 2) + (0 \times 1)$$
$$= 8 + 0 + 2 + 0 = 10_{10}$$

   b. Binary column weights   32   16   8   4   2   1
        Binary number   1   0   1   1   0   1

$$\text{Decimal equivalent} = (1 \times 32) + (0 \times 16) + (1 \times 8) + (1 \times 4)$$
$$+ (0 \times 2) + (1 \times 1)$$
$$= 32 + 0 + 8 + 4 + 0 + 1 = 45_{10}$$

■   **INTEGRATED MATH APPLICATION:**

The LEDs in Figure 9-4 are being used as a 4-bit (4-binary-digit) display. When the LED is OFF, it indicates a binary 0, and when the LED is ON, it indicates a binary 1. Determine the decimal equivalent of the binary displays shown in Figure 9-4(a), (b), and (c).

■   *Solution:*

   a. $0101_2 = 5_{10}$

   b. $1110_2 = 14_{10}$

   c. $1001_2 = 9_{10}$

FIGURE 9-4   **Using LEDs to Display Binary Numbers.**

Handwritten notes:

03FF
11 AD

16^3	16^2	16^1	16^0
4096	256	16	1
0	3	F	F
0	768	240	15

= 1023

16^3	16^2	16^1	16^0
4096	256	16	1
1	1	A	D
4096	256	160	13

4525

9-2-4   *Converting Decimal Numbers to Binary Numbers*

To convert a decimal number to its binary equivalent, continually subtract the largest possible power of two until the decimal number is reduced to zero, placing a binary 1 in columns that are used and a binary 0 in the columns that are not used. To see how simple this process is, refer to Figure 9-5, which shows how decimal 53 can be converted to its binary equivalent. As you can see in this example, the first largest power of two that can be subtracted from decimal 53 is 32, and therefore a 1 is placed in the thirty-twos column, and 21 remains. The largest power of two that can be subtracted from the remainder 21 is 16, and therefore a 1 is placed in the sixteens column, and 5 remains. The next largest power of two that can be subtracted from 5 is 4, and therefore a 1 is placed in the fours column, and 1 remains. The final 1 is placed in the units column, and therefore decimal 53 is represented in binary as 110101, which indicates that the value is the sum of 1×32, 1×16, 1×4, and 1×1.

■   **INTEGRATED MATH APPLICATION:**

Convert the following decimal numbers to their binary equivalents.

   a. 25
   b. 55

■   *Solution:*

a.
Binary column weights	32	16	8	4	2	1
Binary number		1	1	0	0	1

    Decimal equivalent $25 - 16 = 9, 9 - 8 = 1, 1 - 1 = 0$

b.
Binary column weights	32	16	8	4	2	1
Binary number	1	1	0	1	1	1

    Decimal equivalent $55 - 32 = 23, 23 - 16 = 7, 7 - 4 = 3,$
    $3 - 2 = 1, 1 - 1 = 0$

Handwritten notes:
16 of Binary
= 10 of Decimal

10 = 16

DECIMAL

10	1
5	3

BINARY

64	32	16	8	4	2	1
	1	1	0	1	0	1

$$-3\ 2$$
$$2\ 1$$
$$-1\ 6$$
$$5$$
$$-\quad 4$$
$$1$$
$$-\quad 1$$
$$0$$

FIGURE 9-5 **Decimal-to-Binary Conversion.**

■ **INTEGRATED MATH APPLICATION:**

Figure 9-6 shows a strip of magnetic tape containing binary data. The shaded dots indicate magnetized dots (which are equivalent to binary 1s), and the unshaded circles on the tape represent unmagnetized points (which are equivalent to binary 0s). What binary values should be stored in rows 4 and 5 so that decimal 9_{10} and 33_{10} are recorded?

FIGURE 9-6 **Binary Numbers on Magnetic Tape.**

MSB —— LSB

Row 0 = 21_{10}
Row 1 = 30_{10}
Row 2 = 9_{10}
Row 3 = 7_{10}
? Row 4 = 19_{10}
? Row 5 = 33_{10}

Solution:

a.
Binary column weights	32	16	8	4	2	1
Binary number		1	0	0	1	1
Decimal equivalent	$19 - 16 = 3, 3 - 2 = 1, 1 - 1 = 0$					
Stored magnetic data	○ ● ○ ○ ● ●					

b.
Binary column weights	32	16	8	4	2	1
Binary number	1	0	0	0	0	1
Decimal equivalent	$33 - 32 = 1, 1 - 1 = 0$					
Stored magnetic data	● ○ ○ ○ ○ ●					

9-2-5   Converting Information Signals

Figure 9-7(a) shows how analog data can be converted into digital data, and Figure 9-7 (b) shows how digital data can be converted into analog data. To give this process purpose, let us consider how the music information stored on a compact disc (CD) is first recorded and then played back. Like all sound, music is made up of waves of compressed air. When these waves strike the diaphragm of a microphone, an analog voltage signal is generated. In the past, this analog data was recorded on magnetic tapes or as a grooved track on a record. These data storage devices were susceptible to wear and tear, temperature, noise, and age. Using digital recording techniques, binary codes are stored on a compact disc to achieve near-perfect fidelity to live sound.

Figure 9-7(a) shows how an analog-to-digital converter (ADC) is used during the recording process to convert the analog music input into a series of digital output codes. These digital codes are used to control the light beam of a recording laser so that it will engrave the binary 0s and 1s onto a compact disc (CD) in the form of pits or spaces. The ADC is triggered into operation by a sampling pulse that causes it to measure the input voltage of the analog signal at that particular time and generate an equivalent digital output code. For example, on the active edge of sampling pulse 1, the analog input voltage is at 2 V, so the binary code 010 (decimal 2) is generated at the output of the ADC. On the active edge of sampling pulse 2, the analog voltage has risen to 4 V, so the binary code 100 (decimal 4) is generated at the output of the ADC. On the active edge of sampling pulses 3, 4, and 5, the binary codes 110 (decimal 6), 111 (decimal 7), and 110 (decimal 6) are generated at the output, representing the analog voltages 6 V, 7 V, and 6 V, respectively, and so on.

Figure 9-7(b) shows how a digital-to-analog converter (DAC) is used during the music playback process to convert the digital codes stored on a CD to an analog output signal. Another laser is used during playback to read the pits and spaces on the compact disc as 0s and 1s. These codes are then applied to the DAC, which converts the digital input codes into discrete voltages. The DAC is triggered into operation by a strobe pulse that causes it to convert the code currently being applied at the input into an equivalent output voltage.

For example, on the active edge of strobe pulse 1, the digital code 010 (decimal 2) is being applied to the DAC, so it will generate 2 V at its output. On the active edge of strobe pulse 2, the digital code 100 (decimal 4) is being applied to the DAC, so it will generate 4 V at its output. On the active edge of strobe pulse 3, the digital code 110 (decimal 6) is being applied to the DAC, so it will generate 6 V at its output. On the active edge of strobe pulse 4, the digital code 111 (decimal 7) is being applied to the DAC, so it will generate 7 V at its output. On the active edge of strobe pulse 5, the digital code 110 (decimal 6) is being applied to the DAC, so it will generate 6 V at its output, and so on. If the output of a DAC is then applied to a low-pass filter, the discrete voltage steps can be blended into a smooth wave that closely approximates the original analog wave, as shown by the dashed line in Figure 9-7(b).

FIGURE 9-7   Converting Information Signals.

DECIMAL			HEXADECIMAL				BINARY					
10^2	10^1	10^0	16^3	16^2	16^1	16^0	2^5	2^4	2^3	2^2	2^1	2^0
100	10	1	4096	256	16	1	32	16	8	4	2	1
		0				0	0	0	0	0	0	0
		1				1	0	0	0	0	0	1
		2				2	0	0	0	0	1	0
		3				3	0	0	0	0	1	1
		4				4	0	0	0	1	0	0
		5				5	0	0	0	1	0	1
		6				6	0	0	0	1	1	0
		7				7	0	0	0	1	1	1
		8				8	0	0	1	0	0	0
		9				9	0	0	1	0	0	1
	1	0				A	0	0	1	0	1	0
	1	1				B	0	0	1	0	1	1
	1	2				C	0	0	1	1	0	0
	1	3				D	0	0	1	1	0	1
	1	4				E	0	0	1	1	1	0
	1	5				F	0	0	1	1	1	1
	1	6			1	0	0	1	0	0	0	0
	1	7			1	1	0	1	0	0	0	1
	1	8			1	2	0	1	0	0	1	0
	1	9			1	3	0	1	0	0	1	1
	2	0			1	4	0	1	0	1	0	0
	2	1			1	5	0	1	0	1	0	1
	2	2			1	6	0	1	0	1	1	0
	2	3			1	7	0	1	0	1	1	1
	2	4			1	8	0	1	1	0	0	0
	2	5			1	9	0	1	1	0	0	1
	2	6			1	A	0	1	1	0	1	0
	2	7			1	B	0	1	1	0	1	1
	2	8			1	C	0	1	1	1	0	0
	2	9			1	D	0	1	1	1	0	1
	3	0			1	E	0	1	1	1	1	0
	3	1			1	F	0	1	1	1	1	1
	3	2			2	0	1	0	0	0	0	0
	3	3			2	1	1	0	0	0	0	1
	3	4			2	2	1	0	0	0	1	0

(a)

A9	DE7	78	F9	10C
AA	DE8	79	FA	10D
AB	DE9	7A	FB	10E
AC	DEF	7B	FC	10F
AD	DF0	7C	FD	110
AE	DF1	7D	FE	111
AF	DF2	7E	FF	112
B0	DF3	7F	100	
B1		80	101	
		81		

(b)

FIGURE 9-8   Hexadecimal Reset and Carry (a) Number System Comparison (b) Hex Counting Examples.

Now that you have completed this section, you should be able to:

■ **Objective 4.** *Explain the positional weight, the reset and carry action, and conversion for binary.*

Use the following questions to test your understanding of Section 9-2:

1. What is the decimal equivalent of 11010_2?
2. Convert 23_{10} to its binary equivalent.
3. What are the full names for the acronyms LSB and MSB?
4. Convert 110_{10} to its binary equivalent.

9-3 THE HEXADECIMAL NUMBER SYSTEM

If digital electronic circuits operate using binary numbers and we operate using decimal numbers, why is there any need for us to have any other system? The hexadecimal, or hex, system is used as a sort of shorthand for large strings of binary numbers, as will be explained in this section. To begin, let us examine the basics of this number system and then look at its application.

Hexadecimal

A base 16 number system.

Hexadecimal means "sixteen," and this number system has 16 different digits, as shown in Figure 9-8(a), which shows a comparison between decimal, hexadecimal, and binary. Looking at the first 10 digits in the decimal and hexadecimal columns, you can see that there is no difference between the two columns; however, beyond 9, hexadecimal makes use of the letters A, B, C, D, E, and F. Digits 0 through 9 and letters A through F make up the 16 total digits of the hexadecimal number system. Comparing hexadecimal with decimal once again, you can see that $A_{16} = 10_{10}$, $B_{16} = 11_{10}$, $C_{16} = 12_{10}$, $D_{16} = 13_{10}$, $E_{16} = 14_{10}$ and $F_{16} = 15_{10}$. Having these extra digits means that a column reset and carry will not occur until the count has reached the last and largest digit, F. The hexadecimal column in Figure 9-8(a) shows how reset and carry will occur whenever a column reaches its maximum digit, F, and this action is further illustrated in the examples in Figure 9-8(b).

9-3-1 *Converting Hexadecimal Numbers to Decimal Numbers*

To find the decimal equivalent of a hexadecimal number, simply multiply each hexadecimal digit by its positional weight. For example, referring to Figure 9-9, you can see that the positional weight of each of the hexadecimal columns, as expected, is progressively 16 times larger as you move from right to left. The hexadecimal number 4C, therefore, indicates that the value is 4×16 ($4 \times 16 = 64$) and $C \times 1$. Since C is equal to 12 in decimal (12×1), the result of $12 + 64$ is 76, so hexadecimal 4C is equivalent to decimal 76.

■ **INTEGRATED MATH APPLICATION:**

Convert hexadecimal 8BF to its decimal equivalent.

■ *Solution:*

Hexadecimal column weights

4096	256	16	1
	8	B	F

Decimal equivalent

$$= (8 \times 256) + (B \times 16) + (F \times 1)$$
$$= 2048 + (11 \times 16) + (15 \times 1)$$
$$= 2048 + 176 + 15$$
$$= 2239$$

16^4	16^3	16^2	16^1	16^0
65,536	4096	256	16	1
			4	C

Value of Number $= (4 \times 16) + (C \times 1)$
$\qquad\qquad\qquad = (4 \times 16) + (12 \times 1)$
$\qquad\qquad\qquad = \quad 64 \quad + \quad 12 \; = 76_{10}$

FIGURE 9-9 **Positional Weight of the Hexadecimal Number System.**

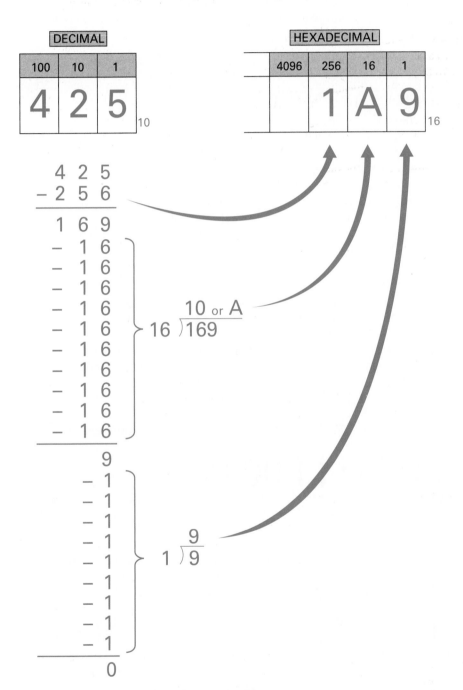

FIGURE 9-10 **Decimal-to-Hexadecimal Conversion.**

9-3-2 Converting Decimal Numbers to Hexadecimal Numbers

Decimal-to-hexadecimal conversion is achieved in the same way as decimal to binary. First, subtract the largest possible power of 16, and then keep subtracting the largest possible power of 16 from the remainder. Each time a subtraction takes place, add a 1 to the respective column until the decimal value has been reduced to zero. Figure 9-10 illustrates this procedure with an example showing how decimal 425 is converted to its hexadecimal equivalent. To begin, the largest possible power of sixteen (256) is subtracted once from 425 and therefore a 1 is placed in the 256s column, leaving a remainder of 169. The next largest power of 16 is 16, which can be subtracted 10 times from 169, and therefore the hexadecimal equivalent of 10, which is A, is placed in the sixteens column, leaving a remainder of 9. Since nine 1s can be subtracted from the remainder of 9, the units column is advanced nine times, giving us our final hexadecimal result, 1A9.

■ **INTEGRATED MATH APPLICATION:**

Convert decimal 4525 to its hexadecimal equivalent.

■ *Solution:*

Hexadecimal column weights	4096	256	16	1
	1	1	A	D

Decimal equivalent
$$= 4525 - 4096 = 429, \; 429 - 256 = 173,$$
$$173 - 16 - 16 - 16 - 16 - 16 - 16 -$$
$$16 - 16 - 16 - 16 = 13, \; 13 - 1 -$$
$$1 - 1 - 1 - 1 - 1 - 1 - 1 - 1 -$$
$$1 - 1 - 1 - 1 = 0$$

9-3-3 Converting between Binary and Hexadecimal

As mentioned at the beginning of this section, hexadecimal is used as a shorthand for representing large groups of binary digits. To illustrate this, Figure 9-11(a) shows how a 16-bit bi-

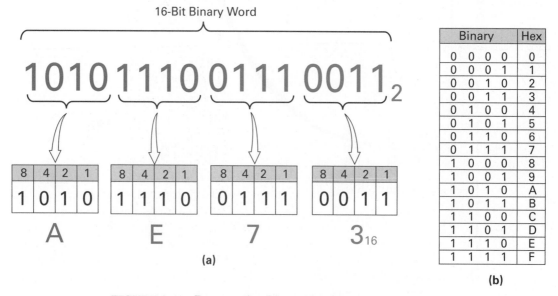

FIGURE 9-11 Representing Binary Numbers in Hexadecimal.

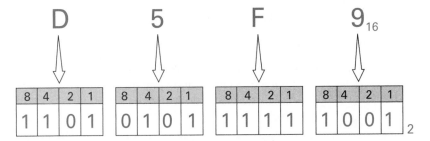

FIGURE 9-12   **Hexadecimal-to-Binary Conversion.**

nary number, which is more commonly called a 16-bit binary **word,** can be represented by 4 hexadecimal digits. To explain this, Figure 9-11(b) shows how a 4-bit binary word can have any value from 0_{10} (0000_2) to 15_{10} (1111_2), and since hexadecimal has the same number of digits (0 through F), we can use one hexadecimal digit to represent 4 binary bits. As you can see in Figure 9-11(a), it is much easier to work with a number like $AE73_{16}$ than with 1010111001110011_2.

To convert from hexadecimal to binary, we simply do the opposite, as shown in Figure 9-12. Since each hexadecimal digit represents 4 binary digits, a 4-digit hexadecimal number will convert to a 16-bit binary word.

Word

An ordered set of characters that is treated as a unit.

■   **INTEGRATED MATH APPLICATION:**

Convert hexadecimal 2BF9 to its binary equivalent.

■   *Solution:*

Hexadecimal number	2	B	F	9
Binary equivalent	0010	1011	1111	1001

CALCULATOR KEYS

Name: Entering and converting number bases

Function and Examples: Some calculators allow you to enter and convert number bases.

To enter a binary number, use the form:

0b*binaryNumber*   (for example: 0b11100110)
     └── Binary number with up to 32 digits
     └── Zero, not the letter *O*, and the letter *b*

To enter a hexadecimal number, use the form:

0h*hexadecimalNumber*   (for example: 0h89F2C)
     └── Hexadecimal number with up to 8 digits
     └── Zero, not the letter *O*, and letter *h*

If you enter without the 0b or 0h prefix, such as 11, it is always treated as a decimal number. If you omit the 0h prefix on a hexadecimal number containing A–F, all or part of the entry is treated as a variable.

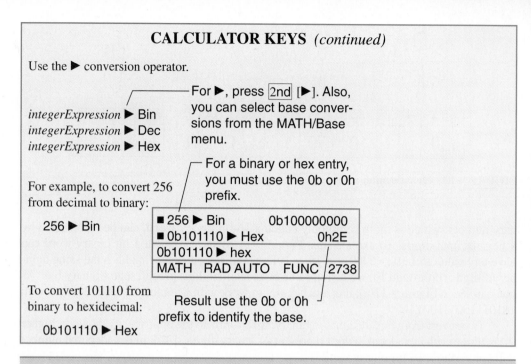

CALCULATOR KEYS *(continued)*

Use the ▶ conversion operator.

integerExpression ▶ Bin
integerExpression ▶ Dec
integerExpression ▶ Hex

For ▶, press 2nd [▶]. Also, you can select base conversions from the MATH/Base menu.

For example, to convert 256 from decimal to binary:

256 ▶ Bin

For a binary or hex entry, you must use the 0b or 0h prefix.

■ 256 ▶ Bin	0b100000000
■ 0b101110 ▶ Hex	0h2E
0b101110 ▶ hex	
MATH RAD AUTO FUNC	2738

To convert 101110 from binary to hexidecimal:

0b101110 ▶ Hex

Result use the 0b or 0h prefix to identify the base.

■ **INTEGRATED MATH APPLICATION:**

Convert binary 110011100001 to its hexadecimal equivalent.

■ *Solution:*

Binary number	1100	1110	0001
Hexadecimal equivalent	C	E	1

SELF-TEST EVALUATION POINT FOR SECTION 9-3

Now that you have completed this section, you should be able to:

■ *Objective 5.* *Explain the positional weight, the reset and carry action, and conversion for hexadecimal.*

Use the following questions to test your understanding of Section 9-3:

1. What is the base of each?
   a. Decimal number system
   b. Binary number system
   c. Hexadecimal number system
2. What are the decimal and hexadecimal equivalents of 10101_2?
3. Convert 1011 1111 0111 1010_2 to its hexadecimal equivalent.
4. What are the binary and hexadecimal equivalents of 33_{10}?

9-4 THE OCTAL NUMBER SYSTEM

Although not as frequently used as hexadecimal, the *octal number system* is also used as a shorthand for large binary words. It is very much like hexadecimal in that it allows for easy conversion from binary to octal and from octal to binary. Before we look at its application, however, let us examine the basics of this number system.

Octal

A base 8 number system.

Octal means "eight," and the octal number system has eight digits: 0 through 7. Figure 9-13(a) shows a comparison among decimal, octal, binary, and hexadecimal. Having only eight possible digits means that a column reset and carry will occur when the count has reached the last and largest digit, 7. At this point, the first octal column will reset and carry, as shown in Figure 9-13(a). This action occurs when any of the octal columns reaches the maximum digit, 7, as illustrated in the additional counting examples shown in Figure 9-13(b).

DECIMAL				OCTAL				BINARY					HEXADECIMAL			
1000	100	10	1	512	64	8	1	16	8	4	2	1	4096	256	16	1
			0				0	0	0	0	0	0				0
			1				1	0	0	0	0	1				1
			2				2	0	0	0	1	0				2
			3				3	0	0	0	1	1				3
			4				4	0	0	1	0	0				4
			5				5	0	0	1	0	1				5
			6				6	0	0	1	1	0				6
			7				7	0	0	1	1	1				7
			8			1	0	0	1	0	0	0				8
			9			1	1	0	1	0	0	1				9
		1	0			1	2	0	1	0	1	0				A
		1	1			1	3	0	1	0	1	1				B
		1	2			1	4	0	1	1	0	0				C
		1	3			1	5	0	1	1	0	1				D
		1	4			1	6	0	1	1	1	0				E
		1	5			1	7	0	1	1	1	1				F
		1	6			2	0	1	0	0	0	0			1	0
		1	7			2	1	1	0	0	0	1			1	1
		1	8			2	2	1	0	0	1	0			1	2
		1	9			2	3	1	0	0	1	1			1	3
		2	0			2	4	1	0	1	0	0			1	4

(a)

44	75	474
45	76	475
46	77	476
47	100	477
50	101	500
51	102	501

(b)

FIGURE 9-13   Octal Reset and Carry (a) Number System Comparison (b) Octal Counting Examples.

9-4-1   *Converting Octal Numbers to Decimal Numbers*

To find the decimal equivalent of an octal number, simply multiply each octal digit by its positional weight. For example, referring to Figure 9-14, you can see that the positional weight of each of the octal columns is as expected, progressively eight times larger as you move from right to left. The octal number 126 therefore indicates that the value is the sum of 1×64, 2×8, and 6×1. The result of $64 + 16 + 6$ is decimal 86.

8^5	8^4	8^3	8^2	8^1	8^0
32,768	4096	512	64	8	1
			1	2	6

$$\text{Value of Number} = (1 \times 64) + (2 \times 8) + (6 \times 1)$$
$$= 64 + 16 + 6 = 86_{10}$$

FIGURE 9-14 **Positional Weight of the Octal Number System.**

■ **INTEGRATED MATH APPLICATION:**

Convert octal 2437 to its decimal equivalent.

■ *Solution:*

Octal column weights

$$\frac{512 \quad 64 \quad 8 \quad 1}{2 \quad\quad 4 \quad\; 3 \quad 9}$$

Decimal equivalent
$$= (2 \times 512) + (4 \times 64) + (3 \times 8) + (7 \times 1)$$
$$= 1024 + 256 + 24 + 7$$
$$= 1311$$

9-4-2 *Converting Decimal Numbers to Octal Numbers*

Decimal-to-octal conversion is achieved in the same way as decimal to binary and decimal to hex. First, subtract the largest possible power of eight, and then keep subtracting the largest possible power of eight from the remainder. Each time a subtraction takes place, add 1 to the respective column until the decimal value has been reduced to zero. Figure 9-15 illustrates this procedure with an example showing how decimal 139 is converted to its octal equivalent. To begin, the largest possible power of eight (64) can be subtracted twice from

FIGURE 9-15 **Decimal-to-Octal Conversion.**

139, and therefore a 2 is placed in the 64 column. The next largest power of eight is 8 itself, and therefore the octal 8 column is advanced to 1, leaving a remainder of 3. Since 3 units can be subtracted from this final remainder of 3, a 3 is placed in the octal units column, giving us our final octal result, 213.

■ **INTEGRATED MATH APPLICATION:**

Convert decimal 3724 to its octal equivalent.

■ *Solution:*

Octal column weights

$$\frac{512 \quad 64 \quad 8 \quad 1}{7 \qquad 2 \qquad 1 \qquad 4}$$

Decimal equivalent
$$\begin{aligned} &= 3724 - 512 = 3212, 3212 - 512 = 2700, \\ &2700 - 512 = 2188, 2188 - 512 = 1676, \\ &1676 - 512 = 1164, 1164 - 512 = 652, \\ &652 - 512 = 140, 140 - 64 = 76, 76 - 64 = 12, \\ &12 - 8 = 4, 4 - 1 = 3, 3 - 1 = 2, 2 - 1 = 1, \\ &1 - 1 = 0 \end{aligned}$$

9-4-3   *Converting between Binary and Octal*

Like hexadecimal, octal is used as a shorthand for representing large groups of binary digits. To illustrate this, Figure 9-16(a) shows how a 16-bit binary number, which is more commonly called a 16-bit binary word, can be represented by six octal digits. To explain this, Figure 9-16(b) shows how a 3-bit binary word can have any value from 0_{10} (000_2) to 7_{10} (111_2), and since octal has the same number of digits (0 through 7), we can use one octal digit to represent three binary bits. As you can see in the example in Figure 9-16(a), it is much easier to work with a number like 127163 than with 1010111001110011.

To convert from octal to binary, we simply do the opposite, as shown in the example in Figure 9-17. Since each octal digit represents three binary digits, a 4-digit octal number will convert to a 12-bit binary word.

■ **INTEGRATED MATH APPLICATION:**

Convert octal number 2635 to its binary equivalent.

■ *Solution:*

Octal Number	2	6	3	5
Binary Equivalent	010	110	011	101

BINARY	OCTAL
0 0 0	0
0 0 1	1
0 1 0	2
0 1 1	3
1 0 0	4
1 0 1	5
1 1 0	6
1 1 1	7

(a)   (b)

FIGURE 9-16   **Representing Binary Numbers in Octal.**

FIGURE 9-17 **Octal-to-Binary Conversion.**

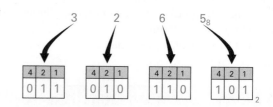

■ INTEGRATED MATH APPLICATION:

Convert binary 101111001000010110 to its octal equivalent.

■ *Solution:*

Binary Number	101	111	001	000	010	110
Octal Equivalent	5	7	1	0	2	6

SELF-TEST EVALUATION POINT FOR SECTION 9-4

Now that you have completed this section, you should be able to:

■ **Objective 6.** *Explain the positional weight, the reset and carry action, and conversion for octal.*

Use the following questions to test your understanding of Section 9-4:

1. The octal number system has a base of _____.
2. Convert 154_{10} into octal.
3. A 12-bit binary word will convert into _____ octal digits.
4. What is the octal equivalent of 100010110111_2?

9-5 BINARY CODES

Pure Binary

Uncoded binary.

The process of converting a decimal number to its binary equivalent is called *binary coding*. The result of the conversion is a binary number or code that is called **pure binary.** There are, however, binary codes used in digital circuits other than pure binary, and in this section we will examine some of the most frequently used.

9-5-1 *The Binary Coded Decimal (BCD) Code*

Binary-Coded Decimal (BCD)

A code in which each decimal digit is represented by a group of 4 binary bits.

No matter how familiar you become with binary, it will always be less convenient to work with than the decimal number system. For example, it will always take a short time to convert 1111000_2 to 120_{10}. Designers realized this disadvantage early on and developed a binary code that had decimal characteristics and that was appropriately named **binary coded decimal (BCD).** Being a binary code, it has the advantages of a two-state system, and since it has a decimal format, it is also much easier for an operator to interface via a decimal keypad or decimal display with systems such as pocket calculators and wristwatches.

The BCD code expresses each decimal digit as a 4-bit word, as shown in Figure 9-18(a). In this example, decimal 1753 converts to a BCD code of 0001 0111 0101 0011, with the first 4-bit code ($0001_2 = 1_{10}$) representing the 1 in the thousands column, the second 4-bit code ($0111_2 = 7_{10}$) representing the 7 in the hundreds column, the third 4-bit code ($0101_2 = 5_{10}$) representing the 5 in the tens column, and the fourth 4-bit code ($0011_2 = 3_{10}$) representing the 3 in the units column. As can be seen in Figure 9-18(a), the subscript BCD is often used after a BCD code to distinguish it from a pure binary number ($1753_{10} = 0001\ 0111\ 0101\ 0011_{BCD}$).

FIGURE 9-18 **Binary Coded Decimal (BCD) (a) Decimal-to-BCD Conversion Example (b) A Comparison of Decimal, Binary, and BCD.**

Figure 9-18(b) compares decimal, binary, and BCD. As you can see, the reset and carry action occurs in BCD at the same time it does in decimal. This is because BCD was designed to have only ten 4-bit binary codes, 0000, 0001, 0010, 0011, 0100, 0101, 0110, 0111, 1000, and 1001 (0 through 9), to make it easy to convert between this binary code and decimal. Binary codes 1010, 1011, 1100, 1101, 1110, and 1111 (A_{16} through F_{16}), are invalid codes that are not used in BCD because they are not used in decimal.

■ **INTEGRATED MATH APPLICATION:**

Convert the BCD code 0101 1000 0111 0000 to decimal, and the decimal number 369 to BCD.

■ *Solution:*

BCD code	0101	1000	0111	0000
Decimal equivalent	5	8	7	0
Decimal number	3	6	9	
BCD code	0011	0110	1001	

9-5-2 *The Excess-3 Code*

Excess-3 Code

A code in which the decimal digit *n* is represented by the 4-bit binary code *n* + 3.

The **excess-3 code** is similar to BCD and is often used in some applications because of certain arithmetic operation advantages. Referring to Figure 9-19(a), you can see that, like BCD, each decimal digit converted into a 4-bit binary code. The difference with the excess-3 code, however, is that a value of 3 is added to each decimal digit before it is converted into a 4-bit binary code. Although this code has an offset or excess of 3, it still uses only ten 4-bit binary codes (0011 through 1100) like BCD. The invalid 4-bit values in this code are therefore 0000, 0001, 0010 (decimal 0 through 2), and 1101, 1110, 1111 (decimal 13 through 15). Figure 9-19(b) shows a comparison between decimal, BCD, and excess-3.

■ **INTEGRATED MATH APPLICATION:**

Convert decimal 408 to its excess-3 code equivalent.

■ *Solution:*

Decimal number	4	0	8
+3 =	7	3	11
Excess-3 binary	0111	0011	1011

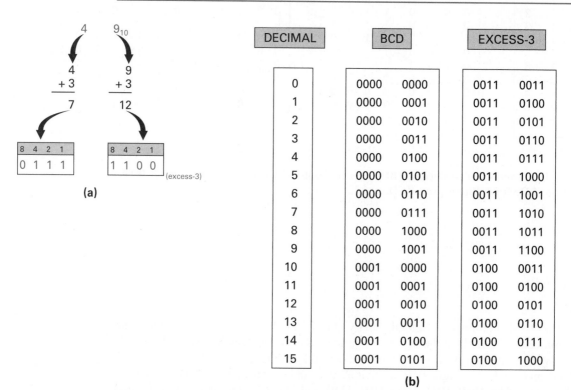

(a)

DECIMAL	BCD		EXCESS-3	
0	0000	0000	0011	0011
1	0000	0001	0011	0100
2	0000	0010	0011	0101
3	0000	0011	0011	0110
4	0000	0100	0011	0111
5	0000	0101	0011	1000
6	0000	0110	0011	1001
7	0000	0111	0011	1010
8	0000	1000	0011	1011
9	0000	1001	0011	1100
10	0001	0000	0100	0011
11	0001	0001	0100	0100
12	0001	0010	0100	0101
13	0001	0011	0100	0110
14	0001	0100	0100	0111
15	0001	0101	0100	1000

(b)

FIGURE 9-19   The Excess-3 Code (a) Example (b) Code Comparison.

9-5-3   The Gray Code

The **gray code,** shown in Figure 9-20, is a nonweighted binary code, which means that the position of each binary digit carries no specific weight or value. This code, named after its inventor, was developed so that only one of the binary digits changes as you step from one code group to the next code group in sequence. For example, in pure binary, a change from 3 (0011) to 4 (0100) results in a change in 3 bits, whereas in the gray code only 1 bit changes.

    The minimum-change gray code is used to reduce errors in digital electronic circuitry. To explain this in more detail, when a binary count changes, it takes a very small amount of time for the bits to change from 0 to 1, or from 1 to 0. This transition time can produce an incorrect intermediate code. For example, if when changing from pure binary 0011 (3) to 0100 (4), the LSB were to switch slightly faster than the other bits, an incorrect code would be momentarily generated, as shown in the following example.

Binary	Decimal
0 0 1 1	3
0 0 1 0	Error
0 1 0 0	4

This momentary error could trigger an operation that should not occur, resulting in a system malfunction. Using the gray code eliminates these timing errors, since only 1 bit changes at a time. The disadvantage of the gray code is that the codes must be converted to pure binary if numbers need to be added, subtracted, or used in other computations.

9-5-4   The American Standard Code for Information Interchange (ASCII)

To this point we have discussed only how we can encode numbers into binary codes. Digital electronic computers must also be able to generate and recognize binary codes that rep-

DECIMAL	BINARY	GRAY CODE
0	0000	0000
1	0001	0001
2	0010	0011
3	0011	0010
4	0100	0110
5	0101	0111
6	0110	0101
7	0111	0100
8	1000	1100
9	1001	1101
10	1010	1111
11	1011	1110
12	1100	1010
13	1101	1011
14	1110	1001
15	1111	1000

FIGURE 9-20   **The Gray Code.**

Gray Code
A nonweighted binary code in which sequential numbers are represented by codes that differ by one bit.

Alphanumeric Code

A code used to represent
the letters of the alphabet
and decimal numbers.

resent letters of the alphabet and symbols. The ASCII code is the most widely used **alphanumeric code** (alphabet and numeral code). As I write this book the ASCII codes for each of these letters is being generated by my computer keyboard, and the computer is decoding these ASCII codes and then displaying the alphanumeric equivalent (letter, number, or symbol) on my computer screen.

Figure 9-21 lists all the 7-bit ASCII codes and the full names for the abbreviations used. This diagram also shows how each of the 7-bit ASCII codes is made up of a 4-bit group that indicates the row of the table and a 3-bit group that indicates the column of the table. For example, the uppercase letter K is in column 100 (4_{10}) and in row 1011 (11_{10}) and is therefore represented by the ASCII code 1001011.

Column		0	1	2	3	4	5	6	7
Row Bits 4321 765 ➡		000	001	010	011	**100**	101	110	111
0	0000	NUL	DLE	SP	0	@	P	\	p
1	0001	SOH	DC1	!	1	A	Q	a	q
2	0010	STX	DC2	"	2	B	R	b	r
3	0011	ETX	DC3	#	3	C	S	c	s
4	0100	EOT	DC4	$	4	D	T	d	t
5	0101	ENQ	NAK	%	5	E	U	e	u
6	0110	ACK	SYN	&	6	F	V	f	v
7	0111	BEL	ETB	'	7	G	W	g	w
8	1000	BS	CAN	(	8	H	X	h	x
9	1001	HT	EM	)	9	I	Y	i	y
10	1010	LF	SUB	*	:	J	Z	j	z
11	**1011**	VT	ESC	+	;	**K**	[	k	{
12	1100	FF	FS	,	<	L	\	l	¦
13	1101	CR	GS	-	=	M	]	m	}
14	1110	SO	RS	.	>	N	⌒	n	~
15	1111	SI	US	/	?	O	—	o	DEL

Example: Upper Case "K" = Column 4   (100), Row 11 (1011)

Bits	7 6 5	4 3 2 1
ASCII Code	1 0 0	1 0 1 1

NUL Null	FF Form Feed	CAN Cancel
SOH Start of Heading	CR Carriage Return	EM End of Medium
STX Start of Text	SO Shift Out	SUB Substitute
ETX End of Text	SI Shift In	ESC Escape
EOT End of Transmission	SP Space (blank)	FS File Separator
ENQ Enquiry	DLE Data Link Escape	GS Group Separator
ACK Acknowledge	DC1 Device Control 1	RS Record Separator
BEL Bell (audible signal)	DC2 Device Control 2	US Unit Separator
BS Backspace	DC3 Device Control 3	DEL Delete
HT Horizontal Tabulation (punched card skip)	DC4 Device Control 4	
	NAK Negative Acknowledge	(ASCII
LF Line Feed	SYN Synchronous Idle	Abbreviations)
VT Vertical Tabulation	ETB End of Transmission Block	

FIGURE 9-21   **The ASCII Code.**

CHAPTER 9 / NUMBER SYSTEMS AND CODES

■ INTEGRATED MATH APPLICATION:

List the ASCII codes for the message "Digital."

■ *Solution:*

$$D = 1000100$$
$$i = 1101001$$
$$g = 1100111$$
$$i = 1101001$$
$$t = 1110100$$
$$a = 1100001$$
$$l = 1101100$$

■ INTEGRATED MATH APPLICATION:

Figure 9-22 illustrates a 7-bit register. A register is a circuit that is used to store or hold a group of binary bits. If a switch is closed, the associated "Q output" is grounded and 0 V (binary 0) is applied to the output. On the other hand, if a switch is open, the associated "Q output" is pulled HIGH due to the connection to +5 V via the 10 kΩ resistor, and therefore +5 V (binary 1) is applied to the output. What would be the pure binary value, hexadecimal value, and ASCII code stored in this register?

■ *Solution:*

	Q_6	Q_5	Q_4	Q_3	Q_2	Q_1	Q_0
Pure binary	1	0	0	0	0	1	1_2
Hexadecimal				43_{16}			
ASCII			C (uppercase C)				

FIGURE 9-22 A 7-Bit Switch Register.

SELF-TEST EVALUATION POINT FOR SECTION 9-5

Now that you have completed this section, you should be able to:

■ *Objective 7. Describe the following binary codes: binary coded decimal (BCD), the excess-3 code, the gray code, and the American Standard Code for Information Exchange (ASCII).*

Use the following questions to test your understanding of Section 9-5:

1. Convert 7629_{10} to BCD.
2. What is the decimal equivalent of the excess-3 code 0100 0100?
3. What is a minimum-change code?
4. What does the following string of ASCII codes mean?

   1000010   1100001   1100010   1100010   1100001
   1100111   1100101

Multiple Choice Questions

1. Binary is a base _____ number system, decimal a base _____ number system, and hexadecimal a base _____ number system.
   - **a.** 1, 2, 3
   - **b.** 2, 10, 16
   - **c.** 10, 2, 3
   - **d.** 8, 4, 2

2. What is the binary equivalent of decimal 11?
   - **a.** 1010
   - **b.** 1100
   - **c.** 1011
   - **d.** 0111

3. Which of the following number systems is used as a shorthand for binary?
   - **a.** Decimal
   - **b.** Hexadecimal
   - **c.** Binary
   - **d.** Both a and b

4. What would be displayed on a hexadecimal counter if its count was advanced by 1 from $39FF_{16}$?
   - **a.** 4000
   - **b.** 3A00
   - **c.** 3900
   - **d.** 4A00

5. What is the base of 101?
   - **a.** 10
   - **b.** 2
   - **c.** 16
   - **d.** Could be any base

6. Which of the following is an invalid BCD code?
   - **a.** 1010
   - **b.** 1000
   - **c.** 0111
   - **d.** 0000

7. What is the decimal equivalent of $1001\ 0001\ 1000\ 0111_{BCD}$?
   - **a.** 9187
   - **b.** 9A56
   - **c.** 8659
   - **d.** 254,345

8. What is the pure binary equivalent of 15_{16}?
   - **a.** 0000 1111
   - **b.** 0001 0101
   - **c.** 0101 0101
   - **d.** 0000 1101

9. Which of the following is an invalid excess-3 code?
   - **a.** 1010
   - **b.** 1000
   - **c.** 0111
   - **d.** 0000

10. Since the lower 4 bits of the ASCII code for the numbers 0 through 9 are the same as their pure binary equivalent, what is the 7-bit ASCII code for the number 6?
    - **a.** 0111001
    - **b.** 0110111
    - **c.** 0110011
    - **d.** 0110110

Communication Skill Questions

11. Give the full name of the following abbreviation and acronyms.
    - **a.** Bit
    - **b.** BCD
    - **c.** ASCII

12. What is the Arabic number system? (9-1)

13. Why does reset and carry occur after 9 in decimal and after 1 in binary? (9-1)

14. Briefly describe how to: (9-2)
    - **a.** convert a binary number to a decimal number.
    - **b.** convert a decimal number to a binary number.

15. Why is the hexadecimal number system used in conjunction with digital circuits? (9-3)

16. Briefly describe how to: (9-3)
    - **a.** convert a hexadecimal number to a decimal number.
    - **b.** convert a decimal number to a hexadecimal number.
    - **c.** convert between binary and hexadecimal.

17. In what application is the octal number system used? (9-4)

18. Briefly describe how to: (9-4)
    - **a.** convert an octal number to a decimal number.
    - **b.** convert a decimal number to an octal number.
    - **c.** convert between binary and octal.

19. Why is the BCD code easier to decode than pure binary? (9-5)

20. What is binary coded decimal and why is it needed? (9-5)

21. Briefly describe how to convert a BCD code to its decimal equivalent. (9-5)

22. What is the excess-3 code? (9-5)

23. What is the advantage of the gray code? (9-5)

24. What is the ASCII code and in what applications is it used? (9-5)

Practice Problems

25. What is the pure binary output of the switch register shown in Figure 9-23 and what is its decimal equivalent?

26. Convert the following into their decimal equivalents:
    - **a.** 110111_2
    - **b.** $2F_{16}$
    - **c.** 10110_{10}

27. What would be the decimal equivalent of the LED display in Figure 9-24 if it were displaying each of the following:
    - **a.** Pure binary
    - **b.** BCD
    - **c.** Excess-3

FIGURE 9-23   Four-Bit Switch Register.

MSB   LSB

 = LED Off = 0

= LED On = 1

FIGURE 9-24 **8-Bit LED Display.**

28. Convert the following binary numbers to octal and hexadecimal:

   a. 111101101001   **c.** 111000

   b. 1011   **d.** 1111111

29. Convert the following decimal numbers to BCD:

   a. 2,365   **b.** 24

30. Give the ASCII codes for the following:

   a. ?   **c.** 6

   b. $

31. Identify the following codes.

DECIMAL	(A)		(B)	(C)		(D)	
0	0000	0000	0000	0011	0011	011	0000
1	0000	0001	0001	0011	0100	011	0001
2	0000	0010	0011	0011	0101	011	0010
3	0000	0011	0010	0011	0110	011	0011
4	0000	0100	0110	0011	0111	011	0100
5	0000	0101	0111	0011	1000	011	0101

32. Convert the decimal number 23 into each of the following:

   a. ASCII   **d.** Hexadecimal

   b. Pure binary   **e.** Octal

   c. BCD   **f.** Excess-3

33. If a flashlight is used to transmit ASCII codes, with a dot (short flash) representing a 0 and a dash (long flash) representing a 1, what is the following message?

0000010 1010000 1110010 1100001 1100011 1110100 1101001
1100011 1100101 0100000 1001101 1100001 1101011 1100101
1110011 0100000 1010000 1100101 1110010 1100110 1100101
1100011 1110100 0000011

34. Convert the following:

   a. $1110110_2 =$ _____ $_{16}$

   b. $175_{10} =$ _____ $_2$

   c. $ABC_{16} =$ _____ $_2$

   d. $00110110_2 =$ _____ $_8$

Web Site Questions

Go to the web site http://prenhall.com/cook, select the textbook *Introductory Computer Mathematics, 2e,* select this chapter, and then follow the instructions when answering the multiple-choice practice problems.

Logic Gates

The Great Experimenter

Michael Faraday was born to James and Margaret Faraday on September 22, 1791. James Faraday, a blacksmith in failing health, had brought his family to London in hopes of finding work; however, his poor health meant that the Faradays were often penniless and hungry. At the age of 12, the young Faraday left school to work as an errand boy for a local bookseller and was later apprenticed to the bookbinder's trade. At 21, the gifted and engaging Faraday was at the right place at the right time, and with the right talents. He impressed the brilliant professor Humphry Davvy, who made him his research assistant at Count Rumford's Royal Institution. After only two years, Faraday was given a promotion and an apartment at the Royal Institution, and in 1821 he married. He lived at the Royal Institution for the rest of his active life, working in his laboratory, giving very dynamic lectures, and publishing over 450 scientific papers. Unlike many scientific papers, Faraday's papers never used a calculus formula to explain a concept or idea. Instead, Faraday would explain all his findings using logic and reasoning, so that a person trying to understand science did not need to be a scientist. It was this gift of being able to make even the most complex areas of science easily accessible to the student, coupled with his motivational teaching style, that made him so popular.

By 1855 he had written three volumes of papers on electromagnetism, the first dynamo, the first transformer, and the foundations of electrochemistry, a large amount on dielectrics, and even some papers on plasma. The unit of capacitance is the *farad,* in honor of his work in these areas of science.

Faraday began two series of lectures at the Royal Institution, which have continued to this day. Michael and Sarah Faraday were childless, but they both loved children, and in 1860 Faraday began a series of Christmas lectures expressly for children, the most popular of which was called "The Chemical History of a Candle." The other lecture series was the "Friday Evening Discourses," of which he himself delivered over a hundred. These very dynamic, enlightening, and entertaining lectures covered areas of science or technology for the layperson and were filled with demonstrations. On one evening in 1846, an extremely nervous and shy speaker ran off just moments before he was scheduled to give the Friday Evening Discourse. Faraday had to fill in, and to this day a tradition is still enforced whereby the lecturer for the Friday Evening Discourse is locked up for half an hour before the presentation with a single glass of whiskey.

Faraday was often referred to as "the great experimenter," and it was this consistent experimentation that led to many of his findings. He was fascinated by science and technology, and was always exploring new and sometimes dangerous horizons. In fact, in one of his reports he states, "I have escaped, not quite unhurt, from four explosions." When asked to comment on experimentation, his advice was to "Let your imagination go, guiding it by judgment and principle, but holding it in and directing it by experiment. Nothing is so good as an experiment which, while it sets an error right, gives you as a reward for your humility an absolute advance in knowledge."

Outline and Objectives

Introduction

Within any digital electronic system you will find that diodes and transistors are used to construct **logic gate circuits.** Logic gates are in turn used to construct flip-flop circuits, and flip-flops are used to construct register, counter, and a variety of other circuits. These logic gates, therefore, are used as the basic building blocks for all digital circuits, and their purpose is to control the movement of binary data and binary instructions. Logic gates, and all other digital electronic circuits, are often referred to as **hardware** circuits. By definition, the hardware of a digital electronic system includes all the electronic, magnetic, and mechanical devices of a digital system. In contrast, the **software** of a digital electronic system includes the binary data (like pure binary and ASCII codes) and binary instructions that are processed by the digital electronic system hardware. To use an analogy, we could say that a compact disc player is hardware, and the music information stored on a compact disc and processed by the player is software. Just as a CD player is useless without CDs, a digital system's hardware is useless without software. In other words, the information on the CD determines what music is played, and similarly, the digital software determines the actions of the digital electronic hardware.

Every digital electronic circuit uses logic-gate circuits to manipulate the coded pulses of binary language. These logic-gate circuits are constructed using diodes and transistors, and they are the basic decision-making elements in all digital circuits.

Logic Gate Circuits

Circuits containing predictable gate functions that either open or close their outputs.

Hardware

A term used to describe the electronic, magnetic, and mechanical devices of a digital system.

Software

A term used to describe the binary instructions and data processed by digital electronic system hardware.

10-1   HARDWARE FOR BINARY SYSTEMS

Digital circuits are often referred to as *switching circuits* because their control devices (diodes and transistors) are switched between the two extremes of ON and OFF. These digital circuits are also called **two-state circuits** because their control devices are driven into one of two states: either into the saturation state (fully ON), or cutoff state (fully OFF). These two modes of operation are used to represent the two binary digits of 1 and 0.

To develop a digital electronic system that could manipulate binary information, inventors needed a two-state electronic switch. Early machines used mechanical switches and electromechanical relays—like the examples in Figure 10-1(a) and (b)—to represent binary data by switching current ON or OFF. These mechanical devices were eventually replaced by the vacuum tube, shown in Figure 10-1(c), which, unlike switches and relays, had the advantage of no moving parts. The vacuum tube, however, was bulky, fragile, had to warm up, and consumed an enormous amount of power. Finally, compact and low-power digital electronic circuits and systems became a reality with the development of semiconductor diode and transistor switches, which are shown in Figures 10-1(d) and 10-1(e).

A logic gate accepts inputs in the form of HIGH or LOW voltages, judges these input combinations based on a predetermined set of rules, and then produces a single output in the form of a HIGH or LOW voltage. The term *logic* is used because the output is predictable or logical, and the term *gate* is used because only certain input combinations will "unlock the gate." For example, any HIGH input to an OR gate will unlock the gate and allow the HIGH at the input to pass through to the output.

Two-State Circuits

Switching circuits that are controlled to be either ON or OFF.

SELF-TEST EVALUATION POINT FOR SECTION 10-1

Now that you have completed this section, you should be able to:

■ *Objective 1. Describe the evolution of the two-state switch and how it is used to represent binary data.*

■ *Objective 2. Explain how the diode and transistor semiconductor switches can be used to construct digital electronic gates.*

Use the following questions to test your understanding of Section 10-1:

1. The _____ includes the electronic, magnetic, and mechanical devices of a digital electronic system.

2. The _____ includes the binary data and binary instructions that are processed by the digital electronic system.

3. Which two-state switch was the first to have no moving parts?

4. Which two semiconductor switches are used to implement digital electronic circuits?

10-2   BASIC LOGIC GATES *input to Oart*

10-2-1   *The OR Gate*

The **OR gate** can have two or more inputs but will always have a single output. Figure 10-2(a) shows a table listing all the input possibilities for this two-input OR gate. This table is often referred to as a **truth table** or **function table,** since it details the "truth" or the way in which

OR Gate

A logic gate that will give a HIGH output if either of its inputs are HIGH.

Truth Table or Function Table

A table used to show the action of a device as it reacts to all possible input combinations.

FIGURE 10-1   The Evolution of the Switch for Digital Electronic Circuits (a) A Nineteenth-Century Mechanical Turn Switch (b) An Electromechanical Relay (c) A 1906 Vacuum Tube (d) A 1939 Semiconductor Diode (e) A 1948 Semiconductor Transistor.

OR Gate

A	B	Y
0	0	0
0	1	1
1	0	1
1	1	1

0 = 0 V
1 = +5 V

(a)

A	B	C	Y
0	0	0	0
0	0	1	1
0	1	0	1
0	1	1	1
1	0	0	1
1	0	1	1
1	1	0	1
1	1	1	1

Operation Expressions

Any 1 gives a 1.

All inputs at 0; gives an output of 0.

(b)

(c)

FIGURE 10-2   **The OR Gate (a) Two-Input Truth Table (b) Three-Input Truth Table (c) OR Gate Symbols and Operation Expressions.**

this circuit will function. We could summarize the operation of this gate, therefore, by saying that *if either* A *OR* B *is HIGH, the output* Y *will be HIGH, and only when both inputs are LOW will the output be LOW.*

The truth table for a three-input OR gate is shown in Figure 10-2(b). Although another input has been added to this circuit, the action of the gate still remains the same, since any HIGH input will still produce a HIGH output, and only when all the inputs are LOW will the output be LOW.

Figure 10-2(c) shows the logic schematic symbols for a two- and three-input OR gate and the abbreviated operation expressions for the OR gate. Logic gates with a large number of inputs will have a large number of input combinations; however, the gate's operation will still be the same. To calculate the number of possible input combinations, you can use the following formula:

$$n = 2^x$$

where n = number of input combinations
2 is included because we are dealing with a base 2 number system
x = number of inputs

$$x = \text{# of inputs}$$

■ INTEGRATED MATH APPLICATION:

Construct a truth table for a four-input OR gate, and show the output logic level for every input combination.

■ *Solution:*

Figure 10-3(a) shows the truth table for a four-input OR gate. The number of possible input combinations will be

$$x =$$
$$n = 2^x$$
$$n = 2^4 = 16$$

Calculator sequence: $\boxed{2}\,\boxed{y^x}\,\boxed{4}\,\boxed{=}$

With only two possible digits (0 and 1) and four inputs, we can have a maximum of 16 different input combinations. Whether the OR gate has 2, 3, 4, or 444 inputs, it will still operate in the same predictable or logical way: when all inputs are LOW, the output will be LOW, and when any input is HIGH, the output will be HIGH. This result is shown in the Y column in Figure 10-3(a).

Alternate after every eight digits.
Alternate after every four digits.
Alternate after every two digits.
Alternate after every one digit.

$A_{(8)}$	$B_{(4)}$	$C_{(2)}$	$D_{(1)}$	Y
0	0	0	0	0
0	0	0	1	1
0	0	1	0	1
0	0	1	1	1
0	1	0	0	1
0	1	0	1	1
0	1	1	0	1
0	1	1	1	1
1	0	0	0	1
1	0	0	1	1
1	0	1	0	1
1	0	1	1	1
1	1	0	0	1
1	1	0	1	1
1	1	1	0	1
1	1	1	1	1

(a)

$A_{(8)}$	$B_{(4)}$	$C_{(2)}$	$D_{(1)}$
0	0	0	0
0	0	0	1
0	0	1	0
0	0	1	1
0	1	0	0
0	1	0	1
0	1	1	0
0	1	1	1
1	0	0	0
1	0	0	1
1	0	1	0
1	0	1	1

(b)

FIGURE 10-3   A Four-Input OR Gate.

An easy way to construct these truth tables is first to calculate the maximum number of input combinations and then start in the units (1s) column (D column) and move down, alternating the binary digits after every single digit (0101010101, and so on) up to the maximum count, as shown in Figure 10-3(b). Then go to the twos (2s) column (C column), and move down, alternating the binary digits after every 2 digits (001100110011, and so on) up to the maximum count. Then go to the fours (4s) column (B column), and move down, alternating the binary digits after every 4 digits (000011110000, and so on) up to the maximum count. Finally, go to the eights (8s) column (A column), and move down, alternating the binary digits after every 8 digits (0000000011111111, and so on) up to the maximum count.

These input combinations—or binary words—will start at the top of the truth table with binary 0 and then count up to a maximum value that is always 1 less than the maximum number of combinations. For example, with the four-input OR gate in Figure 10-3(a), the truth table begins with a count of 0000_2 (0_{10}) and then counts up to a maximum of 1111_2 (15_{10}). The maximum count within a truth table (1111_2 or 15_{10}) is always 1 less than the maximum number of combinations (16_{10}) because 0000_2 (0_{10}) is one of the input combinations (0000, 0001, 0010, 0011, 0100, 0101, 0110, 0111, 1000, 1001, 1010, 1011, 1100, 1101, 1110, 1111 is a total of 16 different combinations, with 15 being the maximum count). Stated in a formula:

$$Count_{max} = 2^x - 1$$

For example, in Figure 10-3: $Count_{max} = 2^4 - 1 = 16 - 1 = 15$
Calculator sequence: $\boxed{2}\ \boxed{y^x}\ \boxed{4}\ \boxed{-}\ \boxed{1}\ \boxed{=}$

10-2-2   The AND Gate

Like the OR gate, an **AND gate** can have two or more inputs but will always have a single output. Figure 10-4(a) shows the truth table for a two-input AND gate. We can summarize the operation of this gate by saying that *any LOW input will cause a LOW output, and only when both A AND B inputs are HIGH will the output Y be HIGH.*

AND Gate

A logic gate that will give a HIGH output only if all inputs are HIGH.

A	B	Y
0	0	0
0	1	0
1	0	0
1	1	1

(a)

A	B	C	Y
0	0	0	0
0	0	1	0
0	1	0	0
0	1	1	0
1	0	0	0
1	0	1	0
1	1	0	0
1	1	1	1

(b)

Operation Expressions

Any 0 gives a 0.

All inputs at 1; gives an output of 1.

(c)

FIGURE 10-4   The AND Gate (a) Two-Input Truth Table (b) Three-Input Truth Table (c) AND Gate Symbols.

Figure 10-4(b) shows the truth table for a three-input AND gate. Figure 10-4(c) shows the logic schematic symbols for a two- and three-input AND gate and the abbreviated operation expressions for the AND gate.

■ INTEGRATED MATH APPLICATION:

Develop a truth table for a five-input AND gate. Show the output logic level for every input combination, and indicate the range of values within the truth table.

■ *Solution:*

Figure 10-5 shows the truth table for a five-input AND gate. The number of possible input combinations will be

$$n = 2^x$$
$$n = 2^5 = 32$$

A five-input AND gate will still operate in the same predictable or logical way: when any input is LOW, the output will be LOW, and when all inputs are HIGH, the output will be HIGH. This result is shown in the Y column in Figure 10-5.

The range of values within the truth table will be

$$\text{Count}_{max} = 2^5 - 1 = 32 - 1 = 31$$

With the five-input OR gate in Figure 10-5, the truth table begins with a count of 00000_2 (0_{10}) and then counts up to a maximum of 11111_2 (31_{10}).

■ INTEGRATED MATH APPLICATION:

Referring to the circuit in Figure 10-6(a), describe what would be present at the output of the AND gate for each of the CONTROL switch positions.

■ *Solution:*

When the CONTROL switch is put in the ENABLE position, a HIGH is applied to the lower input of the AND gate, as shown in Figure 10-6(b). In this mode, the output Y will follow the square-wave input, since when the square-wave is HIGH, the AND gate inputs are both HIGH, and Y is HIGH. When the square-wave is LOW, the AND gate will have a LOW input, and Y will be LOW. The Y output, therefore, follows the square-wave input, and the AND gate is said to be equivalent to a closed switch, as shown in Figure 10-6(b). On the other hand, when the CONTROL switch is put in the DISABLE position, a LOW is applied to the lower input of the AND gate, as shown in Figure 10-6(c). In this mode, the output Y will always remain LOW, since any LOW input to an AND gate will always result in a LOW

FIGURE 10-5   A Five-Input AND Gate.

	A	B	C	D	E	Y
0	0	0	0	0	0	0
1	0	0	0	0	1	0
2	0	0	0	1	0	0
3	0	0	0	1	1	0
4	0	0	1	0	0	0
5	0	0	1	0	1	0
6	0	0	1	1	0	0
7	0	0	1	1	1	0
8	0	1	0	0	0	0
9	0	1	0	0	1	0
10	0	1	0	1	0	0
11	0	1	0	1	1	0
12	0	1	1	0	0	0
13	0	1	1	0	1	0
14	0	1	1	1	0	0
15	0	1	1	1	1	0
16	1	0	0	0	0	0
17	1	0	0	0	1	0
18	1	0	0	1	0	0
19	1	0	0	1	1	0
20	1	0	1	0	0	0
21	1	0	1	0	1	0
22	1	0	1	1	0	0
23	1	0	1	1	1	0
24	1	1	0	0	0	0
25	1	1	0	0	1	0
26	1	1	0	1	0	0
27	1	1	0	1	1	0
28	1	1	1	0	0	0
29	1	1	1	0	1	0
30	1	1	1	1	0	0
31	1	1	1	1	1	1

32 Possible Input Combinations

output. In this situation, the AND gate is said to be equivalent to an open switch, as shown in Figure 10-6(c).

    When used in applications such as this, the AND gate is said to be acting as a **controlled switch.**

Controlled Switch
An electronically controlled switch.

SELF-TEST EVALUATION POINT FOR SECTION 10-2

Now that you have completed this section, you should be able to:

■ **Objective 3.**   *Describe the discrete circuit construction, truth table, operation, and an application for the following basic logic-gate types:*

    a. The OR Gate
    b. The AND Gate

Use the following questions to test your understanding of Section 10-2:

1. The two basic logic gates are the _____ gate and the _____ gate.

2. With an OR gate, any binary 1 input will give a binary _____ output.

3. Which logic gate can be used as a controlled switch?

4. Only when all inputs are LOW will the output of an _____ gate be LOW?

5. With an AND gate, any binary 0 input will give a binary _____ output.

6. Only when all inputs are HIGH will the output of an _____ gate be HIGH.

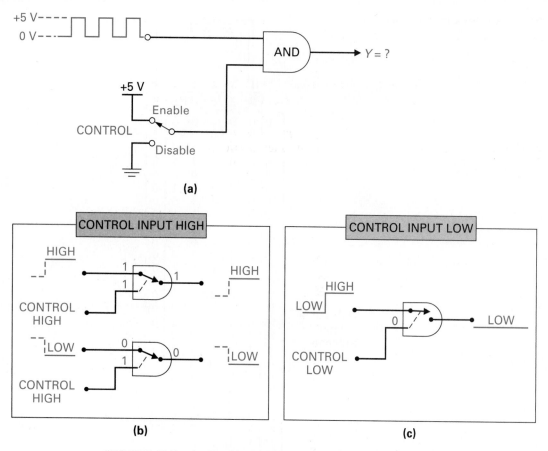

FIGURE 10-6   An Enable/Disable Control Circuit Using an AND Gate.

10-3   INVERTING LOGIC GATES

Although the basic OR and AND logic gates can be used to construct many digital circuits, in some circuit applications other logic operations are needed. For this reason, semiconductor manufacturers make other logic-gate types available. In this section, we will examine three inverting-type logic gates, called the NOT gate, the NOR gate, and the NAND gate. We will begin with the NOT or INVERTER gate.

10-3-1   *The NOT Gate*

The NOT, or logic INVERTER gate, is the simplest of all the logic gates because it has only one input and one output. Figure 10-7(a) shows the truth table for this gate. We can summarize

A	Y
0	1
1	0

(a)   (b)   (c)

FIGURE 10-7   The NOT Gate or Logic Inverter (a) Truth Table (b) Logic Symbols for a NOT Gate (c) Timing Diagram Showing Input-to-Output Inversion.

the operation of this gate by saying that *the output logic level is "NOT" the same as the input logic level* (a 1 input is NOT the same as the 0 output, or a 0 input is NOT the same as the 1 output), hence the name NOT gate. The output is, therefore, always the **complement,** or opposite, of the input.

Figure 10-7(b) shows the schematic symbols normally used to represent the NOT gate. The triangle is used to represent the logic circuit, and the **bubble,** or small circle either before or after the triangle, is used to indicate the complementary or inverting nature of this circuit. Figure 10-7(c) shows a typical example of a NOT gate's input/output waveforms. As you can see from these waveforms, the output waveform is an inverted version of the input waveform.

10-3-2 *The NOR Gate*

The name NOR is a combination of the two words NOT and OR and is used to describe a type of logic gate that contains an OR gate followed by a NOT gate, as shown in Figure 10-8(a).

Figure 10-8(b) shows the standard symbol used to represent the **NOR gate.** Studying the NOR gate symbol, you can see that the OR symbol is used to show how the two inputs are first ORed, and the bubble is used to indicate that the result from the OR operation is inverted before being applied to the output. As a result, the NOR gate's output is simply the opposite, or complement, of the OR gate, as can be seen in the truth table for a two-input NOR gate shown in Figure 10-8(c). If you compare the NOR gate truth table with the OR gate truth table, you can see that the only difference is that the output is inverted:

OR gate: When all inputs are 0, the output is **0,** and any 1 input gives a **1** output.

NOR gate: When all inputs are 0, the output is **1,** and any 1 input gives a **0** output.

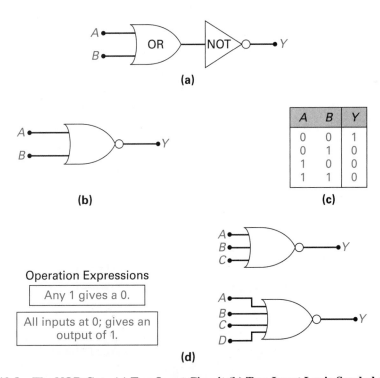

(a)

(b)

A	B	Y
0	0	1
0	1	0
1	0	0
1	1	0

(c)

Operation Expressions

Any 1 gives a 0.

All inputs at 0; gives an output of 1.

(d)

FIGURE 10-8 **The NOR Gate (a) Two-Input Circuit (b) Two-Input Logic Symbol (c) Two-Input Truth Table (d) NOR Gate Operation Expressions and Three- and Four-Input Logic Symbols.**

The operation expressions for the NOR gate are listed in Figure 10-8(d), along with the standard symbols for a three-input and a four-input NOR gate.

10-3-3 *The NAND Gate*

Like NOR, the name NAND is a combination of the two words NOT and AND, and it describes another type of logic gate that contains an AND gate followed by a NOT gate, as shown in Figure 10-9(a).

Figure 10-9(b) shows the standard symbol used to represent the **NAND gate.** The AND symbol is used to show how the two inputs are first "ANDed," and the triangle followed by the bubble is used to indicate that the result from the AND operation is inverted before being applied to the output. As a result, the NAND gate's output is simply the opposite, or complement, of the AND gate, as can be seen in the truth table for a two-input NAND gate shown in Figure 10-9(c). If you compare the NAND gate truth table with the AND gate truth table, you can see that the only difference is that the output is inverted. To compare:

> AND Gate: Any 0 input gives a **0** output, and only when both inputs are 1 will the output be **1.**
>
> NAND Gate: Any 0 input gives a **1** output, and only when both inputs are 1 will the output be **0.**

The operation expressions for the NAND gate are listed in Figure 10-9(d), along with the standard symbols for a three-input and four-input NAND gate.

(a)

A	B	Y
0	0	1
0	1	1
1	0	1
1	1	0

(c)

Operation Expressions

Any 0 gives a 1.

All inputs at 1; gives an output of 0.

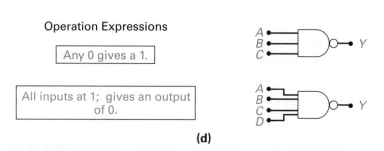

(d)

FIGURE 10-9   The NAND Gate (a) Two-Input Circuit (b) Two-Input Logic Symbol (c) Two-Input Truth Table (d) NAND Gate Operation Expressions and Three- and Four-Input Logic Symbols.

Now that you have completed this section, you should be able to:

■ **Objective 4.** *Describe the discrete circuit construction, truth table, operation, and an application for the following inverting logic-gate types:*
  a. The NOT Gate
  b. The NOR Gate
  c. The NAND Gate

Use the following questions to test your understanding of Section 10-3:

1. Which of the logic gates has only one input?
2. With a NOT gate any binary 1 input will give a binary _____ output.
3. Which logic gate is a combination of an AND gate followed by a logic INVERTER?
4. Only when all inputs are LOW, will the output of a(n) _____ gate be HIGH.
5. With a NAND gate, any binary 0 input will give a binary _____ output.
6. Only when all inputs are HIGH, will the output of a(n) _____ gate be LOW.

10-4 EXCLUSIVE LOGIC GATES

In this section, we will discuss the final two logic-gate types: the exclusive-OR (XOR) gate and exclusive-NOR (XNOR) gate. Although these two logic-gates are not used as frequently as the five basic OR, AND, NOT, NOR, and NAND gates, their function is ideal in some applications.

10-4-1 *The XOR Gate*

The **exclusive-OR (XOR) gate** logic symbol is shown in Figure 10-10(a), and its truth table is shown in Figure 10-10(b). As with the basic OR gate, the operation of the XOR is dependent

> **Exclusive-OR Gate**
>
> A logic-gate circuit that will give a HIGH output if any odd number of binary 1s are applied to the input.

(a)

	A	B	Y
Even	0	0	0
Odd	0	1	1
Odd	1	0	1
Even	1	1	0

(b)

low output
high output

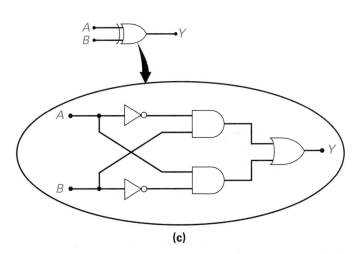

(c)

FIGURE 10-10   The Exclusive-OR (XOR) Gate. (a) Two-Input Logic Symbol. (b) Two-Input Truth Table. (c) Basic Gate Circuit.

on the HIGH or binary 1 inputs. With the basic OR, any 1 input will cause a 1 output. With the XOR, any odd number of binary 1s at the input will cause a binary 1 at the output. Looking at the truth table in Figure 10-10(b), you can see that when the binary inputs are 01 or 10, there is one binary 1 at the input (odd), and the output is 1. When the binary inputs are 00 or 11, however, there are two binary 0s or two binary 1s at the input (even), and the output is 0. To distinguish the XOR gate symbol from the basic OR gate symbol, you may have noticed that an additional curved line is included across the input.

Exclusive-OR logic gates are constructed by combining some of the previously discussed basic logic gates. Figure 10-10(c) shows how two NOT gates, two AND gates, and an OR gate can be connected to form an XOR gate. If inputs A and B are both HIGH or both LOW (even input), the AND gates will both end up with a LOW at one of their inputs, so the OR gate will have both of its inputs LOW, and therefore the final output will be LOW. On the other hand, if A is HIGH and B is LOW or if B is HIGH and A is LOW (odd input), one of the AND gates will have both of its inputs HIGH, giving a HIGH to the OR gate and, therefore, a HIGH to the final output. We can summarize the operation of this gate therefore by saying that *the output is HIGH only when there is an odd number of binary 1s at the input.*

■   INTEGRATED MATH APPLICATION:

List the outputs you would expect for all the possible inputs to a three-input XOR gate.

■   *Solution:*

$$n = 2^x$$
$$n = 2^3 = 8$$

A	B	C	Y	
0	0	0	0	Even
0	0	1	1	Odd
0	1	0	1	Odd
0	1	1	0	Even
1	0	0	1	Odd
1	0	1	0	Even
1	1	0	0	Even
1	1	1	1	Odd

The output is 1 only when there is an odd number of binary 1s at the input.

10-4-2   *The XNOR Gate*

The **exclusive-NOR (XNOR) gate** is simply an XOR gate followed by a NOT gate, as shown in Figure 10-11(a). Its logic symbol, which is shown in Figure 10-11(b), is the same as the XOR symbol except for the bubble at the output, which indicates that the result of the XOR operation is inverted before it appears at the output. Like the NAND and NOR gates, the XNOR is not a new gate but simply a previously discussed gate with an inverted output.

CHAPTER 10 / LOGIC GATES

The truth table for the XNOR gate is shown in Figure 10-11(c). Comparing the XOR gate with the XNOR gate, you can see that the XOR gate's output is HIGH only when there is an odd number of binary 1s at the input, whereas with the XNOR gate, the output is HIGH only when there is an even number of binary 1s at the input. The operation expression for the XNOR gate is listed in Figure 10-11(d).

CALCULATOR KEYS

Name: Comparing or manipulating bits

Function: Some calculators let you compare or manipulate bits in a binary number. You can enter an integer in any number base. Your entries are converted to binary automatically for the operation, and results are displayed according to the Base mode.

Example:

OPERATOR WITH SYNTAX	DESCRIPTION
not *integer*	Returns the one's complement, in which each bit is flipped.
(-) *integer*	Returns the two's complement, which is the one's complement +1.
integer1 **and** *integer2*	In a bit-by-bit **and** comparison, the result is 1 if both bits are 1; otherwise, the result is 0. The returned value represents the bit results.
integer1 **or** *integer2*	In a bit-by-bit **or** comparison, the result is 1 if either bit is 1; the result is 0 only if both bits are 0. The returned value represents the bit results.
integer1 **xor** *integer2*	In a bit-by-bit **xor** comparison, the result is 1 if either bit (but not both) is 1; the result is 0 if both bits are 0 or both bits are 1. The returned value represents the bit results.

Suppose you enter:

0h7AC36 and 0h3D5F

If Base mode = HEX:

```
■0h7AC36 and 0h3D5F
                      0h2C16
0h7ac36 and 0h3d5f
MAIN  RAD AUTO       FUNC      1/30
```

Internally, the hexadecimal integers are converted to a signed, 32-bit binary number.

If Base mode = BIN:

```
■0h7AC36 and 0h3D5F
              0b0110000010110
0h7ac36 and 0h3d5f
MAIN  RAD AUTO       FUNC      1/30
```

Then corresponding bits are compared.

0h7AC36 = 0b00000000000001111010110000110110
  and   **and**

$$0\text{h3D5F} = \frac{0b00000000000000000011110101011111}{0b00000000000000000010110000010110} = 0\text{h2C16}$$

Leading zeros are not shown in the result.

The result is displayed according to the Base mode.

(a)

(b)

A	B	Number of 1s	Y
0	0	Even	1
0	1	Odd	0
1	0	Odd	0
1	1	Even	1

Low output
high output

(c)

Operation Expression

Even number of 1s at the input
gives a 1 at the output.

(d)

FIGURE 10-11   The Exclusive-NOR (XNOR) Gate (a) Circuit (b) Logic Symbol (c) Truth
Table (d) Operation Expression.

SELF-TEST EVALUATION POINT FOR SECTION 10-4

Now that you have completed this section, you should be
able to:

■ **Objective 5.**   *Describe the discrete circuit
construction, truth table, operation, and an
application for the following exclusive logic-gate
types.*
   a. *The XOR Gate*
   b. *The XNOR Gate*

Use the following questions to test your understanding of

Section 10-4:

1. Which type of logic gate will always give a HIGH output
when any of its inputs are HIGH?

2. Only when an odd number of 1s is applied to the input, will
the output of an _____ gate be HIGH.

3. Only when an even number of 1s is applied to the input, will
the output of an _____ gate be HIGH.

4. Which type of logic gate will always give a LOW output
when an odd number of 1s is applied at the input?

10-5   IEEE/ANSI SYMBOLS FOR LOGIC GATES

The logic symbols presented so far in this chapter have been used for many years in the
digital electronics industry. In 1984 the *Institute of Electrical and Electronic Engineers
(IEEE)* and the *American National Standards Institute (ANSI)* introduced a new standard
for logic symbols, which is slowly being accepted by more and more electronics compa-

nies. The advantage of this new standard is that instead of using distinctive shapes to represent logic gates, it uses a special **dependency notation system.** Simply stated, the new standard uses a *notation* (or note) within a rectangular or square block to indicate how the output is *dependent* on the input. Figure 10-12 compares traditional logic symbols with the newer IEEE/ANSI logic symbols and describes the meaning of the dependency notations.

Dependency Notation System

A coding system used on schematic diagrams that uses notations to indicate how an output is dependent on inputs.

Traditional Logic Gate Symbol	IEEE/ANSI Logic Gate Symbols
NOT Gate	A "1" notation indicates gate has only one input. A triangle replaces the bubble at the output to indicate logic inversion.
OR Gate	The "≥1" (greater than or equal to 1) notation is used to indicate that the output will go HIGH whenever one or more of the inputs is HIGH.
AND Gate	The "&" (and) notation is used to indicate that A AND B (all inputs) must be HIGH for the output to go HIGH.
NOR Gate	The OR Notation with Inversion at the Output (NOR)
NAND Gate	The AND Notation with Inversion at the Output (NAND)
XOR Gate	The "=1" notation is used to indicate that the output will be HIGH when only one input equals 1 (odd number of 1s).
XNOR Gate	The XOR Notation with Inversion at the Output (XNOR)

FIGURE 10-12   Traditional and IEEE/ANSI Symbols for Logic Gates.

SECTION 10-5 / IEEE/ANSI SYMBOLS FOR LOGIC GATES

241

Now that you have completed this section, you should be able to:

- **Objective 6.** *Explain the major differences between the traditional and IEEE/ANSI logic-gate symbols.*

Use the following questions to test your understanding of Section 10-5:

1. Which type of logic-gate symbols make use of rectangular blocks?
2. What advantage does the IEEE/ANSI standard have over the traditional symbol standard?

REVIEW QUESTIONS

Multiple Choice Questions

1. Which device is currently being used extensively as a two-state switch in digital electronic circuits?
   - **a.** Transistor
   - **b.** Vacuum tube
   - **c.** Relay
   - **d.** Toggle switch

2. How many input combinations would a four-input logic gate have?
   - **a.** 8
   - **b.** 16
   - **c.** 32
   - **d.** 4

3. What would be the maximum count within the truth table for a four-input logic gate?
   - **a.** 7
   - **b.** 15
   - **c.** 17
   - **d.** 31

4. What would be the output from a three-input OR gate if its inputs were 101?
   - **a.** 1
   - **b.** 0
   - **c.** Unknown
   - **d.** None of the above

5. What would be the output from a three-input AND gate if its inputs were 101?
   - **b.** 0
   - **c.** Unknown
   - **d.** None of the above

6. What would be the output from a three-input NAND gate if its inputs were 101?
   - **a.** 1
   - **b.** 0
   - **c.** Unknown
   - **d.** None of the above

7. What would be the output from a three-input NOR gate if its inputs were 101?
   - **a.** 1
   - **b.** 0
   - **c.** Unknown
   - **d.** None of the above

8. What would be the output from a three-input XOR gate if its inputs were 101?
   - **a.** 1
   - **b.** 0
   - **c.** Unknown
   - **d.** None of the above

9. What would be the output from a three-input XNOR gate if its inputs were 101?
   - **a.** 1
   - **b.** 0
   - **c.** Unknown
   - **d.** None of the above

10. What would be the output from a NOT gate if its input was 1?
    - **a.** 1
    - **b.** 0
    - **c.** Unknown
    - **d.** None of the above

11. Which of the following logic gates is ideal as a controlled switch?
    - **a.** OR
    - **b.** AND
    - **c.** NOR
    - **d.** NAND

12. Which of the following logic gates will always give a LOW output whenever a LOW input is applied?
    - **a.** OR
    - **b.** AND
    - **c.** NOR
    - **d.** NAND

13. Which of the following logic gates will always give a HIGH output whenever a LOW input is applied?
    - **a.** OR
    - **b.** AND
    - **c.** NOR
    - **d.** NAND

14. Which of the following logic gates will always give a HIGH output whenever a HIGH input is applied?
    - **a.** OR
    - **b.** AND
    - **c.** NOR
    - **d.** NAND

15. Which of the following logic gates will always give a LOW output whenever a HIGH input is applied?
    - **a.** OR
    - **b.** AND
    - **c.** NOR
    - **d.** NAND

16. An XOR gate will always give a HIGH output whenever an _____ number of 1s is applied to the input.
    - **a.** odd
    - **b.** even
    - **c.** unknown
    - **d.** none of the above

17. An XNOR gate will always give a HIGH output whenever an _____ number of 1s is applied to the input.
    - **a.** odd
    - **b.** even
    - **c.** unknown
    - **d.** none of the above

18. Which of the dependency notation logic symbols contains an "&" within the square block and no triangle at the output?
    - **a.** OR
    - **b.** AND
    - **c.** NOR
    - **d.** NAND

19. Which of the dependency notation logic symbols contains a "≥1" within the square block and no triangle at the output?
    - **a.** OR
    - **b.** AND
    - **c.** NOR
    - **d.** NAND

20. Which of the dependency notation logic symbols contains a "≥1" within the square block and a triangle at the output?
    - **a.** OR
    - **b.** AND
    - **c.** NOR
    - **d.** NAND

Communication Skill Questions

21. List some of the two-state devices that have been used in digital electronic switches. (10-1)

22. What were the reasons for using two-state switches instead of 10-state switches? (10-1)

23. Sketch and describe how the semiconductor diode can be used to construct a logic-gate circuit. (10-1)

24. What is the meaning of the term *logic gate?* (10-1)

25. Sketch and describe how the semiconductor transistor can be used to construct a logic-gate circuit. (10-1)

26. Sketch the traditional logic symbols used for the seven logic-gate types. (10-2, 10-3)

27. Describe the differences between the truth tables for the OR gate and the AND gate. (10-2)

28. Describe the basic differences between the following: (10-2 and 10-3)
- **a.** The OR and the NOR gate.
- **b.** The AND and the NAND gate.

29. Give the full names for the following abbreviations. (10-3, 10-4, 10-5)

a. XOR	**e.** NAND
b. &	**f.** XNOR
c. = 1	**g.** NOR
d. ≥1	

30. Sketch the truth table and the IEEE/ANSI logic gate symbols for the OR, AND, NOT, NOR, NAND, XOR, and XNOR gates. (10-5)

Practice Problems

31. In the home security system shown in Figure 10-13(a), a two-input logic gate is needed to actuate an alarm (output *Y*) if either the window is opened (input *A*) or the door is opened (input *B*). Which decision-making logic gate should be used in this application?

32. In the office temperature-control system shown in Figure 10-13(b), a thermostat (input *A*) is used to turn a fan (output *Y*) ON and OFF. A thermostat-enable switch is connected to the other input of the logic gate (input *B*), and it will either enable thermostat control of the fan or disable thermostat control of the fan. Which decision-making logic gate should be used in this application?

33. Which logic gate would produce the output waveform shown in Figure 10-14?

34. Choose one of the six logic gates shown in Figure 10-15(a) and then sketch the output waveform that would result for the input waveforms shown in Figure 10-15(b).

35. Develop truth tables for each of the logic circuits shown in Figure 10-16.

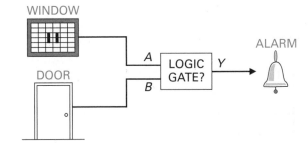

WINDOW	DOOR	ALARM
Closed (0)	Closed (0)	OFF (0)
Closed (0)	Open (1)	ON (1)
Open (1)	Closed (0)	ON (1)
Open (1)	Open (1)	ON (1)

(a)

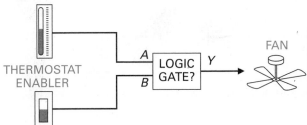

ENABLE	THERMOSTAT	FAN
OFF (0)	COLD (0)	OFF (0)
OFF (0)	HOT (1)	OFF (0)
ON (1)	COLD (0)	OFF (0)
ON (1)	HOT (1)	ON (1)

(b)

FIGURE 10-13 **Decision-Making Logic Gates.**

FIGURE 10-14   Timing Analysis of a Logic Gate.

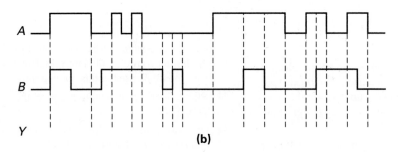

(a)

(b)

FIGURE 10-15   Timing Waveforms.

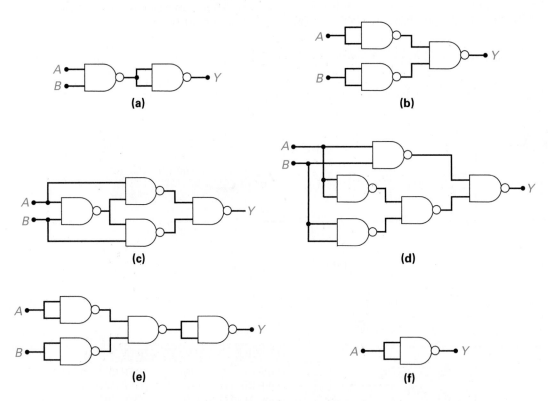

(a)

(b)

(c)

(d)

(e)

(f)

FIGURE 10-16   Using NAND Gates as Building Blocks for Other Gates.

Web Site Questions

Go to the web site http://www.prenhall.com/cook, select the textbook *Introductory Computer Mathematics, 2e,* se-lect this chapter, and then follow the instructions when answering the multiple-choice practice problems.

Boolean Expressions and Algebra

From Folly to Foresight

George Boole was born in the industrial town of Lincoln in eastern England in 1815. His parents were poor trade people, and even though there was a school for boys in Lincoln, there is no record of his ever having attended. In those hard times, children of the working class had no hope of receiving any form of education, and their lives generally followed the familiar pattern of their parents'. George Boole, however, was to break the mold. He would rise up from these humble beginnings to become one of the most respected mathematicians of his day.

Boole's father had taught himself a small amount of mathematics, and since his son of six seemed to have a thirst for learning, he began to pass on all his knowledge. At 8 years old, George had surpassed his father's understanding and craved more. George quickly realized that his advancement was heavily dependent on his understanding Latin. Luckily, a family friend who owned a local bookshop knew enough about the basics of Latin to get him on his way, and once he had taught Boole all he knew, Boole continued with the books at his disposal. By the age of 12 he had conquered Latin, and by 14 he had added Greek, French, German, and Italian to his repertoire.

At the age of 16, however, poverty stood in his way. Since his parents could no longer support him, he was forced to take a job as a poorly paid teaching assistant. After studying the entire school system, he left 4 year laters and opened his own school in which he taught all subjects. It was in this role that he discovered that his mathematics was weak, so he began studying the mathematical journals at the local library in an attempt to stay ahead of his students. He quickly discovered that he had a talent for mathematics. As well as mastering all the present-day ideas, he began to develop some of his own, which were later accepted for publication. After a stream of articles, he became so highly regarded that he was asked to join the mathematics faculty at Queens College in 1849.

After accepting the position, Boole concentrated more on his ideas, one of which was to develop a system of symbolic logic. In this system he created a form of algebra that had its own set of symbols and rules. Using this system, Boole could encode any statement that had to be proven (a proposition) into his symbolic language and then manipulate it to determine whether it was true or false. Boole's algebra had three basic operations that are often called logic functions—AND, OR, and NOT. Using these three operations, Boole could perform such operations as add, subtract, multiply, divide, and compare. These logic functions were binary in nature, and therefore dealt with only two entities—TRUE or FALSE, YES or NO, OPEN or CLOSED, ZERO or ONE, and so on. Boole's theory was that if all logical arguments could be reduced to one of two basic levels, the questionable middle ground would be removed, making it easier to arrive at a valid conclusion.

At the time, Boole's system, which was later called Boolean algebra, was either ignored or criticized by colleagues, who called it a folly with no practical purpose. Almost a century later, however, scientists would combine George Boole's Boolean algebra with binary numbers and make possible the digital electronic computer.

Outline and Objectives

Introduction

Boolean Algebra

A form of algebra invented by George Boole that deals with classes, propositions, and ON/OFF circuit elements associated with operators such as NOT, OR, NOR, AND, NAND, XOR, and XNOR.

In 1854 George Boole invented a symbolic logic that linked mathematics and logic. Boole's logical algebra, which today is known as **Boolean algebra,** states that each variable (input or output) can assume one of two values or states—true or false.

Boolean algebra had no practical application until Claude Shannon used it in 1938 to analyze telephone switching circuits in his MIT thesis. In his paper he described how the two variables of the Boolean algebra (true and false) could be used to represent the two states of the switching relay (open and closed).

Today the mechanical relays have been replaced by semiconductor switches; however, Boolean algebra is still used to express both simple and complex two-state logic functions in a convenient mathematical format. These mathematical expressions of logic functions make it easier for technicians to analyze digital circuits, and are a primary design tool for engineers. Boolean algebra allows circuits to be made simpler, less expensive, and more efficient.

In this chapter we will discuss the basics of Boolean algebra and then see how it can be applied to logic-gate circuit simplification.

11-1   BOOLEAN EXPRESSIONS FOR LOGIC GATES

The operation of each of the seven basic logic gates can be described with a Boolean expression. To explain this in more detail, let us first consider the most basic of all the logic gates, the NOT gate.

11-1-1   *The NOT Expression*

In Figure 11-1 you can see the previously discussed NOT gate, and because of the inversion that occurs between input and output, Y is always equal to the opposite, or complement, of input A.

$$Y = \text{NOT } A$$

Therefore, if the input is 0, the output will be 1.

$$Y = \text{NOT } 0 = 1$$

If, on the other hand, the A input is 1, the output will be 0.

$$Y = \text{NOT } 1 = 0$$

In Boolean algebra, this inversion of the A input is indicated with a bar over the letter A as follows:

$$Y = \overline{A} \text{ (pronounced "Y equals not A")}$$

Boolean Expression
$Y = \overline{A}$

FIGURE 11-1   The Boolean Expression for a NOT Gate.

248   CHAPTER 11 / BOOLEAN EXPRESSIONS AND ALGEBRA

Using this Boolean expression, we can easily calculate the output Y for either of the two input A conditions. For example, if $A = 0$,

$$Y = \overline{A} = \overline{0} = 1$$

If, on the other hand, the A input is 1,

$$Y = \overline{A} = \overline{1} = 0$$

11-1-2   *The OR Expression*

The operation of the OR gate can also be described with a Boolean expression, as seen in Figure 11-2(a). In Boolean algebra, a + sign is used to indicate the OR operation.

$$Y = A + B \text{ (pronounced "Y equals A or B")}$$

The expression $Y = A + B$ (Boolean equation) is the same as $Y = A$ OR B (word equation). If you find yourself wanting to say "plus" instead of "OR," don't worry; it will require some practice to get used to using this traditional symbol in a new application. Just as the word "wind" can be used in two different ways ("a gusty wind" or "to wind a clock"), then so can the + symbol be used to mean either addition or the OR operation. With practice you will get more comfortable with this symbol's dual role.

If OR gates have more than two inputs, as seen in Figure 11-2(b) or (c), the OR expression is simply extended, as follows.

For a three-input OR gate: $Y = A + B + C$

For a four-input OR gate: $Y = A + B + C + D$

The truth table for a two-input OR gate has been repeated in Figure 11-2(d). Using the Boolean expression for this gate, we can determine the output Y by substituting all the possible 1 and 0 input combinations for the input variables A and B. Studying this truth table you can see that the output of the OR gate is a 1 when either A OR B is 1. George Boole described the OR function as *Boolean addition*, which is why the "plus" (addition) symbol was used. Boolean addition, or *logical addition*, is different from normal addition, and the OR gate is said to produce at its output the *logical sum* of the inputs.

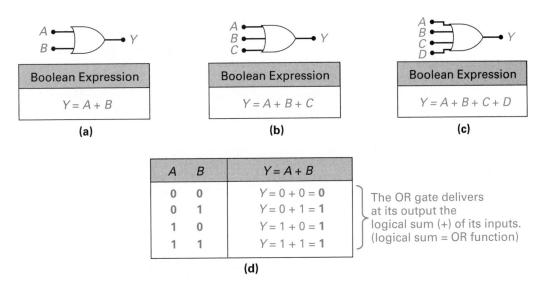

(a)   (b)   (c)

(d)

FIGURE 11-2   **The Boolean Expression for an OR Gate.**

Boolean Expression

$Y = A \cdot B$

(a)

Boolean Expression

$Y = ABC$

(b)

Boolean Expression

$Y = ABCD$

(c)

A	B	$Y = A \cdot B$
0	0	$Y = 0 \cdot 0 = 0$
0	1	$Y = 0 \cdot 1 = 0$
1	0	$Y = 1 \cdot 0 = 0$
1	1	$Y = 1 \cdot 1 = 1$

The AND gate delivers at its output the logical product (•) of its inputs. (logical product = AND function)

(d)

FIGURE 11-3   The Boolean Expression for an AND Gate.

11-1-3   *The AND Expression*

In Boolean algebra, the AND operation is indicated with a ·, and therefore the Boolean expression for the AND gate shown in Figure 11-3(a) would be

$Y = A \cdot B$ (pronounced "Y equals A and B")

This · is also used to indicate multiplication, and therefore like the + symbol, the · symbol has two meanings. In multiplication, the · is often dropped; however, the meaning is still known to be multiplication. For example:

$V = I \times R$ is the same as $V = I \cdot R$ or $V = IR$

In Boolean algebra the same situation occurs, in that

$Y = A \cdot B$ is the same as $Y = AB$

If AND gates have more than two inputs, as seen in Figure 11-3(b) and (c), the AND expression is simply extended.

The truth table for a two-input AND gate is shown in Figure 11-3(d). Using the Boolean expression, we can calculate the output Y by substituting all the possible 1 and 0 input combinations for the input variables A and B. Looking at the truth table, you can see that the output of the AND gate is a 1 only when both A and B are 1. George Boole described the AND function as *Boolean multiplication* or *logical multiplication,* which is why the period (multiplication) symbol was used. Boolean multiplication is different from standard multiplication as you have seen, and the AND gate is said to produce at its output the *logical product* of its inputs.

■   INTEGRATED MATH APPLICATION:

What would be the Boolean expression for the circuit shown in Figure 11-4, and what would be the output if $A = 0$ and $B = 1$?

FIGURE 11-4   **Single Bubble Input AND Gate.**

The *A* input in Figure 11-4 is first inverted before it is ANDed with the *B* input. The Boolean equation therefore, would be

$$Y = \overline{A} \cdot B$$

Using this expression, we can substitute the *A* and *B* for the 0 and 1 inputs, and then determine the output *Y*.

$$Y = \overline{A} \cdot B = \overline{0} \cdot 1 = 1 \cdot 1 = 1$$
(*Y* equals NOT *A* and *B* = NOT 0 AND 1 = 1 AND 1 = 1)

INTEGRATED MATH APPLICATION:

Give the Boolean expression and truth table for the logic circuit shown in Figure 11-5(a).

(a)

A	B	C	Y = AB + C
0	0	0	$Y = (0 \bullet 0) + 0 = 0 + 0 = 0$
0	0	1	$Y = (0 \bullet 0) + 1 = 0 + 1 = 1$
0	1	0	$Y = (0 \bullet 1) + 0 = 0 + 0 = 0$
0	1	1	$Y = (0 \bullet 1) + 1 = 0 + 1 = 1$
1	0	0	$Y = (1 \bullet 0) + 0 = 0 + 0 = 0$
1	0	1	$Y = (1 \bullet 0) + 1 = 0 + 1 = 1$
1	1	0	$Y = (1 \bullet 1) + 0 = 1 + 0 = 1$
1	1	1	$Y = (1 \bullet 1) + 1 = 1 + 1 = 1$

(b)

FIGURE 11-5   *(A* AND *B)* or *(C)* **Logic Circuit.**

Solution:

First, there are three inputs or variables to this circuit: *A, B,* and *C.* Inputs *A* and *B* are ANDed together $(A \cdot B)$ and the result is then ORed with input *C* $(+ C)$. The resulting equation is

$$Y = (A \cdot B) + C \quad \text{or} \quad Y = AB + C$$

As you can see in the truth table in Figure 11-5(b), the *Y* output is HIGH when *A* AND *B* are HIGH, OR if *C* is HIGH.

In the two previous examples, we have gone from a logic circuit to a Boolean equation. Now let us reverse the procedure and see how easy it is to go from a Boolean equation to a logic circuit.

■ **INTEGRATED MATH APPLICATION:**

Sketch the logic-gate circuit for the following Boolean equation.

$$Y = AB + CD$$

■ *Solution:*

Studying the equation, we can see that there are four variables: A, B, C, and D. Input A is ANDed with input B, giving $A \cdot B$ or AB, as seen in Figure 11-6(a). Input C is ANDed with input D, giving CD, as seen in Figure 11-6(b). The ANDed AB output and the ANDed CD output are then ORed to produce a final output Y, as seen in Figure 11-6(c).

FIGURE 11-6 **(A AND B) or (C AND D) Logic Circuit.**

$$Y = \overline{A + B} \text{ (pronounced "Y equals not A or B")}$$

11-1-4   *The NOR Expression*

The Boolean expression for the two-input NOR gate shown in Figure 11-7(a) is

Boolean Expression
$Y = \overline{A + B}$

(a)

A	B	$Y = \overline{A + B}$
0	0	$Y = \overline{0 + 0} = \overline{0} = 1$
0	1	$Y = \overline{0 + 1} = \overline{1} = 0$
1	0	$Y = \overline{1 + 0} = \overline{1} = 0$
1	1	$Y = \overline{1 + 1} = \overline{1} = 0$

(b)

FIGURE 11-7   **The Boolean Expression for a NOR Gate.**

To explain why this expression describes a NOR gate, let us examine each part in detail. In this equation the two input variables A and B are first ORed (indicated by the $A + B$ part of the equation), and then the result is complemented, or inverted (indicated by the bar over the whole OR output expression).

The truth table in Figure 11-7(b) tests the Boolean expression for a NOR gate for all possible input combinations.

While studying Boolean algebra Augustus DeMorgan discovered two important theorems. The first theorem stated that a bubbled input AND gate was logically equivalent to a NOR gate. These two logic gates are shown in Figure 11-8(a) and (b), respectively, with their associated truth tables. Comparing the two truth tables you can see how any combination of inputs to either gate circuit will result in the same output, making the two circuits interchangeable.

■ INTEGRATED MATH APPLICATION:

Give the two Boolean equations for the two logic-gate circuits shown in Figure 11-8(a) and (b).

A	B	Y
0	0	1
0	1	0
1	0	0
1	1	0

$Y = \overline{A} \cdot \overline{B}$

(a)

A	B	Y
0	0	1
0	1	0
1	0	0
1	1	0

$Y = \overline{A + B}$

(b)

FIGURE 11-8 **DeMorgan's First Theorem.**

■ *Solution:*

With the bubbled input AND gate shown in Figure 11-8(a), you can see that input A is first inverted by a NOT gate, giving $\overline{A}$. Input B is also complemented by a NOT gate, giving $\overline{B}$. These two complemented inputs are then ANDed, giving the following final function.

$$Y = \overline{A} \cdot \overline{B}$$

With the NOR gate shown in Figure 11-8(b), inputs A and B are first ORed and then the result is complemented, giving

$$Y = \overline{A + B}$$

DeMorgan's first theorem, therefore, is as follows:

> DeMorgan's first theorem: $\overline{A} \cdot \overline{B} = \overline{A + B}$

Studying the differences between the left equation and the right equation, you can see that there are actually two basic changes: the line is either broken or solid, and the logical sign is different. Therefore, if we wanted to convert the left equation into the right equation, all we would have to do is follow this simple procedure:

$$Y = \overline{A} \cdot \overline{B} \rightarrow Y = \overline{A + B}$$

Mend
the
line, and
change
the sign.

To convert the right equation into the left equation, all we would have to do is follow this simple procedure:

Break
the
line, and
change
the sign.

$$Y = \overline{A} \cdot \overline{B} \leftarrow Y = \overline{A + B}$$

A	B	$Y = \overline{A \cdot B}$
0	0	$Y = \overline{0 \cdot 0} = \overline{0} = 1$
0	1	$Y = \overline{0 \cdot 1} = \overline{0} = 1$
1	0	$Y = \overline{1 \cdot 0} = \overline{0} = 1$
1	1	$Y = \overline{1 \cdot 1} = \overline{1} = 0$

(a)   **(b)**

FIGURE 11-9   **The Boolean Expression for a NAND Gate.**

This ability to interchange an AND function with an OR function and to interchange a bubbled input with a bubbled output will come in handy when we are trying to simplify logic circuits, as you will see in this chapter. For now, remember the rules on how to "De-Morganize" an equation.

> *Mend the line, and change the sign.*
> *Break the line, and change the sign.*

11-1-5   *The NAND Expression*

The Boolean expression for the two-input NAND gate shown in Figure 11-9(a) is

$$Y = \overline{A \cdot B} \text{ or } Y = \overline{AB} \text{ (pronounced "Y equals not A and B")}$$

This expression states that the two inputs A and B are first ANDed (indicated by the $A \cdot B$ part of the equation), and then the result is complemented (indicated by the bar over the AND expression).

     The truth table shown in Figure 11-9(b) tests this Boolean equation for all possible input combinations.

■   **INTEGRATED MATH APPLICATION:**

DeMorganize the following equation, and then sketch the logically equivalent circuits with their truth tables and Boolean expressions.

$$Y = \overline{A} + \overline{B}$$

■   *Solution:*

To apply DeMorgan's theorem to the equation $Y = \overline{A} + \overline{B}$, simply apply the rule "mend the line, and change the sign," as follows:

$$\overline{A} + \overline{B} = \overline{A + B}$$

The circuits for these two equations can be seen in Figure 11-10(a) and (b), and if you study the truth tables you can see that the outputs of both logic gates are identical for all input combinations.

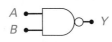

A	B	Y
0	0	1
0	1	1
1	0	1
1	1	0

$Y = \overline{A} + \overline{B}$

(a)

A	B	Y
0	0	1
0	1	1
1	0	1
1	1	0

$Y = \overline{A \cdot B}$

(b)

FIGURE 11-10   DeMorgan's Second Theorem.

This observation—that a bubbled input OR gate is interchangeable with a bubbled output AND gate, or NAND gate—was first made by DeMorgan and is referred to as DeMorgan's second theorem.

DeMorgan's second theorem: $\overline{A} + \overline{B} = \overline{A \cdot B}$

11-1-6   *The XOR Expression*

In Boolean algebra, the $\oplus$ symbol is used to describe the exclusive-OR action. This means that the Boolean expression for the two-input XOR gate shown in Figure 11-11(a) is

$Y = A \oplus B$ (pronounced "Y equals A exclusive-OR B")

The truth table in Figure 11-11(b) shows how this Boolean expression is applied to all possible input combinations.

■   INTEGRATED MATH APPLICATION:

Give the Boolean expression and truth table for the circuit shown in Figure 11-12(a).

■   *Solution:*

Looking at the circuit in Figure 11-12(a), you can see that the upper AND gate has an inverted A input ($\overline{A}$) and a B input, giving a result of $\overline{A} \cdot B$. This $\overline{A} \cdot B$ output is then ORed with

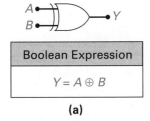

Boolean Expression
$Y = A \oplus B$

(a)

A	B	$Y = A \oplus B$
0	0	$Y = 0 \oplus 0 = 0$
0	1	$Y = 0 \oplus 1 = 1$
1	0	$Y = 1 \oplus 0 = 1$
1	1	$Y = 1 \oplus 1 = 0$

(b)

FIGURE 11-11   The Boolean Expression for an XOR Gate.

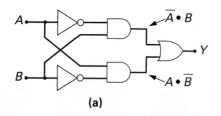

(a)

A B	$Y = \overline{A}B + A\overline{B}$
0 0	$Y = \overline{(0 \bullet 0)} + (0 \bullet \overline{0}) = (1 \bullet 0) + (0 \bullet 1) = 0 + 0 = \mathbf{0}$
0 1	$Y = \overline{(0 \bullet 1)} + (0 \bullet \overline{1}) = (1 \bullet 1) + (0 \bullet 0) = 1 + 0 = \mathbf{1}$
1 0	$Y = \overline{(1 \bullet 0)} + (1 \bullet \overline{0}) = (0 \bullet 0) + (1 \bullet 1) = 0 + 1 = \mathbf{1}$
1 1	$Y = \overline{(1 \bullet 1)} + (1 \bullet \overline{1}) = (0 \bullet 1) + (1 \bullet 0) = 0 + 0 = \mathbf{0}$

(b)

FIGURE 11-12   **(NOT *A* AND *B*) or (*A* AND NOT *B*).**

the input from the lower AND gate, which has an inverted *B* input and *A* input giving a result of $A \cdot \overline{B}$. The final equation therefore is

$$Y = (\overline{A} \cdot B) + (A \cdot \overline{B}) \quad \text{or} \quad Y = \overline{A}\,B + A\,\overline{B}$$

The truth table for this logic circuit that combines five logic gates is shown in Figure 11-12(b). Looking at the input combinations and the output, you can see that the output is 1 only when an odd number of 1s appears at the input. This circuit is therefore acting as an XOR gate, and so

$$\overline{A}\,B + A\,\overline{B} = A \oplus B$$

11-1-7   *The XNOR Expression*

The Boolean expression for the two-input XNOR gate shown in Figure 11-13(a) will be

$$Y = \overline{A \oplus B} \text{ (pronounced "Y equals not A exclusive-or B"}$$

From this equation, you can see that the two inputs *A* and *B* are first XORed and then the result is complemented.

The truth table in Figure 11-13(b) tests this Boolean expression for all possible input combinations.

Boolean Expression

$Y = \overline{A \oplus B}$

(a)

A	B	$Y = \overline{A \oplus B}$
0	0	$Y = \overline{0 \oplus 0} = \overline{0} = 1$
0	1	$Y = \overline{0 \oplus 1} = \overline{1} = 0$
1	0	$Y = \overline{1 \oplus 0} = \overline{1} = 0$
1	1	$Y = \overline{1 \oplus 1} = \overline{0} = 1$

(b)

FIGURE 11-13   **The Boolean Expression for an XNOR Gate.**

Give the Boolean expression and truth table for the circuit shown in Figure 11-14(a).

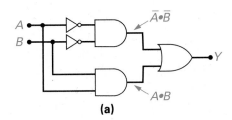

(a)

A	B	$Y = \bar{A}\bar{B} + AB$
0	0	$Y = (\overline{0 \bullet 0}) + (0 \bullet 0) = (1 \bullet 1) + (0 \bullet 0) = 1 + 0 = \mathbf{1}$
0	1	$Y = (\overline{0 \bullet 1}) + (0 \bullet 1) = (1 \bullet 0) + (0 \bullet 1) = 0 + 0 = \mathbf{0}$
1	0	$Y = (\overline{1 \bullet 0}) + (1 \bullet 0) = (0 \bullet 1) + (1 \bullet 0) = 0 + 0 = \mathbf{0}$
1	1	$Y = (\overline{1 \bullet 1}) + (1 \bullet 1) = (0 \bullet 0) + (1 \bullet 1) = 0 + 1 = \mathbf{1}$

(b)

FIGURE 11-14 (**NOT** *A* **AND NOT** *B*) **or** (*A* **AND** *B*).

■ *Solution:*

Studying the circuit in Figure 11-14(a), you can see that the upper AND gate has an inverted *A* input and an inverted *B* input, giving $\bar{A} \cdot \bar{B}$ at its output. The lower AND gate simply ANDs the *A* and *B* input, giving $A \cdot B$ at its output. The two AND gate outputs are then ORed, so the final equation for this circuit will be

$$Y = \bar{A} \cdot \bar{B} + A \cdot B \quad \text{or} \quad Y = \bar{A}\,\bar{B} + A\,B$$

The truth table for this circuit is shown in Figure 11-14(b). Comparing this truth table with an XNOR gate's truth table, you can see that the two are equivalent, since a 1 output is present only when an even number of 1s are applied to the input; therefore,

$$\bar{A}\,\bar{B} + A\,B = \overline{A \oplus B}$$

SELF-TEST EVALUATION POINT FOR SECTION 11-1

Now that you have completed this section, you should be able to:

■ *Objective 1.* *Give the Boolean expressions for each of the seven basic logic gates.*

■ *Objective 2.* *Describe how to write an equivalent Boolean expression for a logic circuit and how to sketch an equivalent logic circuit for a Boolean expression.*

Use the following questions to test your understanding of Section 11-1:

1. Which logic gate would be needed to perform each of the following Boolean expressions?
   a. $A\,B$ c. $\overline{N + M}$ e. $\overline{S + T}$
   b. $A \oplus B$ d. $\overline{X \cdot Y}$ f. $G\,H$

2. State DeMorgan's first theorem using Boolean expressions.

3. Apply DeMorgan's theorem to the following:
   a. $\overline{A} + \overline{B}$ b. $Y = \overline{(A + B) \cdot C}$

4. Sketch the logic gates or circuits for the following expressions.
   a. ABC c. $\overline{A + B}$ e. $A \cdot (L \oplus D)$
   b. $\overline{L \oplus D}$ d. $(\overline{A \cdot B}) + \overline{C}$ f. $\overline{(A + B) \cdot C}$

In this section we will discuss some of the rules and laws that apply to Boolean algebra. Many of these rules and laws are the same as ordinary algebra and, as you will see in this section, are quite obvious.

11-2-1 The Commutative Law

Commutative

Combining elements in such a manner that the result is independent of the order in which the elements are taken.

The word **commutative** is defined as "combining elements in such a manner that the result is independent of the order in which the elements are taken." This means that the order in which the inputs to a logic gate are ORed or ANDed, for example, is not important since the result will be the same, as shown in Figure 11-15.

In Figure 11-15(a), you can see that ORing A and B will achieve the same result as reversing the order of the inputs and ORing B and A. As stated in the previous section, the OR function is described in Boolean algebra as logical addition, so the *commutative law of addition* can be algebraically written as

$$\text{Commutative law of addition: } A + B = B + A$$

Figure 11-15(b) shows how this law relates to an AND gate, which in Boolean algebra is described as logical multiplication. The *commutative law of multiplication* can be algebraically written as

$$\text{Commutative law of multiplication: } A \cdot B = B \cdot A$$

11-2-2 The Associative Law

Associative

Combining elements such that when the order of the elements is preserved, the result is independent of the grouping.

The word **associative** is defined as "combining elements such that when the order of the elements is preserved, the result is independent of the grouping." To explain this in simple terms, Figure 11-16 shows that how you group the inputs in an ORing process or ANDing process has no effect on the output.

In Figure 11-16(a), you can see that ORing B and C and then ORing the result with A will achieve the same result as ORing A and B and then ORing the result with C. This *associative law of addition* can be algebraically written as

$$\text{Associative law of addition: } A + (B + C) = (A + B) + C$$

$Y = A + B$

Achieves the Same Output As

$Y = B + A$

(a)

$Y = A \bullet B$

Achieves the Same Output As

$Y = B \bullet A$

(b)

FIGURE 11-15 **The Commutative Law (a) Logical Addition (b) Logical Multiplication.**

$Y = A + (B + C)$

$Y = A \bullet (B \bullet C)$

Achieves the Same Output As   Achieves the Same Output As

$Y = (A + B) + C$

$Y = (A \bullet B) \bullet C$

(a)   **(b)**

FIGURE 11-16   **The Associative Law (a) Logical Addition (b) Logical Multiplication.**

Figure 11-16(b) shows how this law relates to an AND gate, which in Boolean algebra is described as logical multiplication. The *associative law of multiplication* can be algebraically written as

Associative law of multiplication: $A \cdot (B \cdot C) = (A \cdot B) \cdot C$

11-2-3   *The Distributive Law*

By definition, the word **distributive** means "producing the same element when operating on a whole, as when operating on each part, and collecting the results." It can be algebraically stated as

Distributive law: $A \cdot (B + C) = (A \cdot B) + (A \cdot C)$

Figure 11-17 illustrates this law by showing that ORing two or more inputs and then ANDing the result, as shown in Figure 11-17(a), achieves the same output as ANDing the single variable (A) with each of the other inputs (B and C) and then ORing the results, as shown in Figure 11-17(b). To help reinforce this algebraic law, let us use an example involving actual values for A, B, and C.

■ **INTEGRATED MATH APPLICATION:**

Prove the distributive law by inserting the values $A = 2$, $B = 3$ and $C = 4$.

■ *Solution:*

$$A \cdot (B + C) = (A \cdot B) + (A \cdot C)$$
$$2 \times (3 + 4) = (2 \times 3) + (2 \times 4)$$
$$2 \times 7 = 6 + 8$$
$$14 = 14$$

$A \bullet (B + C)$ = $(A \bullet B) + (A \bullet C)$

(a)   **(b)**

FIGURE 11-17   **The Distributive Law.**

In some instances, we may wish to reverse the process performed by the distributive law to extract the common factor. For example, consider the following equation:

$$Y = \overline{A} B + AB$$

Since "· B" (AND B) seems to be a common factor in this equation, by *factoring,* we obtain

$$Y = \overline{A} B + AB \qquad \text{Original expression}$$
$$Y = (\overline{A} + A) \cdot B \qquad \text{Factoring AND } B$$

11-2-4 *Boolean Algebra Rules*

Now that we have covered the three laws relating to Boolean algebra, let us now concentrate on these Boolean algebra rules, beginning with those relating to OR operations.

OR Gate Rules

The first rule can be seen in Figure 11-18(a). This shows what happens when one input to an OR gate is always 0 and the other input (A) is a variable. If $A = 0$, the output equals 0, and if $A = 1$, the output equals 1. Therefore, a variable input ORed with 0 will always equal the variable input. Stated algebraically:

$$A + 0 = A$$

FIGURE 11-18   **Boolean Rules for OR Gates.**

Another OR gate Boolean rule is shown in Figure 11-18(b), and it states that

$$A + A = A$$

As you can see in Figure 11-18(b), when a variable is ORed with itself, the output will always equal the logic level of the variable input.

Figure 11-18(c) shows the next OR gate Boolean rule. This rule states that any 1 input to an OR gate will result in a 1 output, regardless of the other input. Stated algebraically:

$$A + 1 = 1$$

The final Boolean rule for OR gates is shown in Figure 11-18(d). This rule states that when any variable (A) is ORed with its complement ($\overline{A}$) the result will be a 1.

$$A + \overline{A} = 1$$

AND Gate Rules

Like the OR gate, the AND gate has four Boolean rules. The first is illustrated in Figure 11-19(a) and states that if a variable input A is ANDed with a 0, the output will be 0 regardless of the other input.

$$A \cdot 0 = 0$$

(a)

(b)

(c)

(d)

FIGURE 11-19 Boolean Rules for AND Gates.

The second Boolean rule for AND gates is shown in Figure 11-19(b). In this illustration you can see that if a variable is ANDed with a 1, the output will equal the variable.

$$A \cdot 1 = A$$

Another AND gate Boolean rule is shown in Figure 11-19(c). In this instance, any variable that is ANDed with itself will always give an output that is equal to the variable.

$$A \cdot A = A$$

The last of the AND gate Boolean rules is shown in Figure 11-19(d). In this illustration you can see that if a variable (A) is ANDed with its complement ($\overline{A}$), the output will always equal 0.

$$A \cdot \overline{A} = 0$$

Double-Inversion Rule

The *double-inversion rule* is illustrated in Figure 11-20(a) and states that if a variable is inverted twice, then the variable will be back to its original state. To state this algebraically, we use the double bar, as follows:

$$\overline{\overline{A}} = A$$

In Figure 11-20(b), you can see that if the NOR gate were replaced with an OR gate, the INVERTER would not be needed, since the double inversion returns the logic level to its original state.

DeMorgan's Theorems

DeMorgan's first and second theorems were discussed earlier in the previous section and are repeated here, since they also apply as Boolean rules.

DeMorgan's first theorem: $\overline{A} \cdot \overline{B} = \overline{A + B}$

DeMorgan's second theorem: $\overline{A} + \overline{B} = \overline{A \cdot B}$

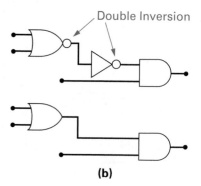

FIGURE 11-20   Double-Inversion Rule.

CHAPTER 11 / BOOLEAN EXPRESSIONS AND ALGEBRA

Apply DeMorgan's theorem and the double-inversion rule to find equivalent logic circuits of the following:

$$\text{a. } \overline{\overline{A} + B} \qquad \text{b. } \overline{A \cdot \overline{B}}$$

■ *Solution:*

With both equations we simply "break the line and change the sign" and then cancel any double inversions to find the equivalent logic circuits, as follows:

$$\text{a. } \overline{\overline{A} + B} = \overline{\overline{A}} \cdot \overline{B} = A \cdot \overline{B}$$

In the first example we found that a NOR gate with a bubbled A input is equivalent to an AND gate with a bubbled B input.

$$\text{b. } \overline{A \cdot \overline{B}} = \overline{A} + \overline{\overline{B}} = \overline{A} + B$$

In the second example we found that a NAND gate with a bubbled B input is equivalent to an OR gate with a bubbled A input.

The Duality Theorem

The *duality theorem* is very useful, since it allows us to change Boolean equations into their *dual* and to produce new Boolean equations. The two steps to follow for making the change are very simple.

a. First, change each OR symbol to an AND symbol and each AND symbol to an OR symbol.

b. Second, change each 0 to a 1 and each 1 to a 0.

To see if this works, let us apply the duality theorem to a simple example.

■ INTEGRATED MATH APPLICATION:

Apply the duality theorem to $A + 1 = 1$ and then state whether the resulting rule is true.

■ *Solution:*

Changing the OR symbol to an AND symbol and changing the 1s to 0s, we arrive at the following:

Original rule: $\qquad\qquad A + 1 = 1$
Dual rule: $\qquad\qquad\quad A \cdot 0 = 0$

OR symbol is changed to AND symbol. ——————— 1s are changed to 0s.

The dual Boolean rule states that any variable ANDed with a 0 will always yield a 0 output, which is true.

■ INTEGRATED MATH APPLICATION:

Determine the dual Boolean equation for the following:

$$A \cdot (B + C) = (A \cdot B) + (A \cdot C)$$

Once you have determined the new Boolean equation, sketch circuits and give the truth tables for both the left and right parts of the equation.

$$A + (B \bullet C) \qquad\qquad = \qquad\qquad (A + B) \bullet (A + C)$$

$$Y = A + (BC) \qquad = \qquad Y = (A+B)(A+C)$$

A	B	C	$Y = A + (B \bullet C)$
0	0	0	$Y = 0 + (0 \bullet 0) = 0$
0	0	1	$Y = 0 + (0 \bullet 1) = 0$
0	1	0	$Y = 0 + (1 \bullet 0) = 0$
0	1	1	$Y = 0 + (1 \bullet 1) = 1$
1	0	0	$Y = 1 + (0 \bullet 0) = 1$
1	0	1	$Y = 1 + (0 \bullet 1) = 1$
1	1	0	$Y = 1 + (1 \bullet 0) = 1$
1	1	1	$Y = 1 + (1 \bullet 1) = 1$

=

A	B	C	$Y = (A + B) \bullet (A + C)$
0	0	0	$Y = (0 + 0) \bullet (0 + 0) = 0$
0	0	1	$Y = (0 + 0) \bullet (0 + 1) = 0$
0	1	0	$Y = (0 + 1) \bullet (0 + 0) = 0$
0	1	1	$Y = (0 + 1) \bullet (0 + 1) = 1$
1	0	0	$Y = (1 + 0) \bullet (1 + 0) = 1$
1	0	1	$Y = (1 + 0) \bullet (1 + 1) = 1$
1	1	0	$Y = (1 + 1) \bullet (1 + 0) = 1$
1	1	1	$Y = (1 + 1) \bullet (1 + 1) = 1$

FIGURE 11-21 *A* **OR** (*B* **AND** *C*) **Equals** (*A* **OR** *B*) **AND** (*A* **OR** *C*).

■   *Solution:*

Applying the duality theorem to $A \cdot (B + C) = (A \cdot B) + (A \cdot C)$ means that we will have to change all the AND symbols to OR symbols, and vice versa.

Original equation: $A \cdot (B + C) = (A \cdot B) + (A \cdot C)$
Dual equation: $A + (B \cdot C) = (A + B) \cdot (A + C)$

Figure 11-22 shows the dual equation and its circuits and truth tables. As you can see by comparing the truth tables, the left and right parts of the equation are equivalent.

As a summary, here is a list of the Boolean laws and rules described in this section, with their dual equations.

		ORIGINAL EQUATION	DUAL EQUATION
1.	Commutative law	$A + B = B + A$	$A \cdot B = B \cdot A$
2.	Associative law	$A \cdot (B \cdot C) = (A \cdot B) \cdot C$	$A + (B + C) = (A + B) + C$
3.	Distributive law	$A + (B \cdot C) = (A + B) \cdot (A + C)$	$A \cdot (B + C) = (A \cdot B) + (A \cdot C)$
4.	OR-AND rules	$A + 0 = A$	$A \cdot 1 = A$
		$A + A = A$	$A \cdot A = A$
		$A + 1 = 1$	$A \cdot 0 = 0$
		$A + \overline{A} = 1$	$A \cdot \overline{A} = 0$
5.	DeMorgan's laws	$\overline{A \cdot B} = \overline{A} + \overline{B}$	$\overline{A + B} = \overline{A} \cdot \overline{B}$

SELF-TEST EVALUATION POINT FOR SECTION 11-2

Now that you have completed this section, you should be able to:

■   ***Objective 3.***   *List and explain the laws and rules that apply to Boolean algebra.*

Use the following questions to test your understanding of Section 11-2:

1. Give the answers to the following Boolean algebra rules.
   a. $A + 1 = ?$ **c.** $A \cdot 0 = ?$
   b. $A + \overline{A} = ?$ **d.** $A \cdot A = ?$
2. What is the dual equation of $A + (B \cdot C) = (A + B) \cdot (A + C)$?
3. Apply the associative law to the equation $A \cdot (B \cdot C)$.
4. Apply the distributive law to the equation $(A \cdot B) + (A \cdot C) + (A \cdot D)$.

Now that you understand the Boolean rules and laws for logic gates, let us put this knowledge to some practical use. If you need a logic-gate circuit to perform a certain function, the best way to begin is with a truth table that details what input combinations should drive the output HIGH and what input combinations should drive the output LOW. As an example, let us assume that we need a logic circuit that will follow the truth table shown in Figure 11-22(a). This means that the output (Y) should be 1 only when

$$A = 0, B = 0, C = 1$$
$$A = 0, B = 1, C = 1$$
$$A = 1, B = 0, C = 1$$
$$A = 1, B = 1, C = 1$$

In the truth table in Figure 11-22(a), you can see the **fundamental products** listed for each of these HIGH outputs. A fundamental product is a Boolean expression that describes what the inputs will need to be in order to generate a HIGH output. For example, the first fundamental product ($Y = \overline{A} \cdot \overline{B} \cdot C$) states that A must be 0, B must be 0, and C must be 1 for the output Y to be 1 ($Y = \overline{A} \cdot \overline{B} \cdot C = \overline{0} \cdot \overline{0} \cdot 1 = 1 \cdot 1 \cdot 1 = 1$). The second fundamental product ($Y = \overline{A} \cdot B \cdot C$) states that A must be 0, B must be 1, and C must be 1 for the output Y to be 1 ($Y = \overline{A} \cdot B \cdot C = \overline{0} \cdot 1 \cdot 1 = 1 \cdot 1 \cdot 1 = 1$). The third fundamental product ($Y = A \cdot \overline{B} \cdot C$) states that A must be 1, B must be 0, and C must be 1 for Y to be 1 ($Y = A \cdot \overline{B} \cdot C = 1 \cdot \overline{0} \cdot 1 = 1 \cdot 1 \cdot 1 = 1$). The fourth and final fundamental product ($Y = A \cdot B \cdot C$) states that A must be 1, B must be 1, and C must be 1 for Y to be 1 ($Y = A \cdot B \cdot C = 1 \cdot 1 \cdot 1 = 1$). The fundamental products for each of the HIGH outputs are listed in the truth table in Figure 11-22(a). By ORing all these fundamental products, we can derive a Boolean equation as follows:

$$Y = \overline{A}\,\overline{B}\,C + \overline{A}\,B\,C + A\,\overline{B}\,C + A\,B\,C$$

From this Boolean equation we can create a logic network that is the circuit equivalent of the truth table. To complete this step, we will need to study in detail the different parts of the Boolean equation. First, the Boolean equation states that the outputs of four three-input AND gates are connected to the input of a four-input OR gate. The first AND gate is connected to inputs $\overline{A}\,\overline{B}$ and C, the second to inputs $\overline{A}\,B$ and C, the third to inputs $A\,\overline{B}$ and C, and finally the fourth to inputs $A\,B$ and C. The resulting logic circuit equivalent for the truth table in Figure 11-22(a) is shown in Figure 11-22(b).

The Boolean equation for the logic circuit in Figure 11-22 is in the **sum-of-products (SOP) form.** To explain what this term means, we must recall once again that in Boolean

Fundamental Products

The truth table input combinations that are essential, since they produce a HIGH output.

Sum-of-Products (SOP) Form

A Boolean expression that describes the ORing of two or more AND functions.

A	B	C	Y	Fundamental Product
0	0	0	0	
0	0	1	1	$\overline{A}\overline{B}C$
0	1	0	0	
0	1	1	1	$\overline{A}BC$
1	0	0	0	
1	0	1	1	$A\overline{B}C$
1	1	0	0	
1	1	1	1	ABC

(a)

$$Y = \overline{A}\overline{B}C + \overline{A}BC + A\overline{B}C + ABC$$

(b)

FIGURE 11-22   Sum-of-Products (SOP) Form.

algebra, an AND gate's output is the *logical product* of its inputs. On the other hand, the OR gate's output is the *logical sum* of its inputs. A sum-of-products expression, therefore, describes the ORing together of (sum of) two or more AND functions (products). Here are some examples of sum-of-product expressions:

$$Y = AB + \overline{A}B$$
$$Y = A\overline{B} + C\overline{D}E$$
$$Y = \overline{A}BCD + AB\overline{CD} + A\overline{B}C\overline{D}$$

Product-of-Sums (POS) Form

A Boolean expression that describes the ANDing of two or more OR functions.

The other basic form for Boolean expressions is the **product-of-sums (POS) form.** This form describes the ANDing of (product of) two or more OR functions (sums). Here are some examples of product-of-sums expressions:

$$Y = (A + B + \overline{C}) \cdot (C + D)$$
$$Y = (\overline{A} + \overline{B}) \cdot (A + B)$$
$$Y = (A + B + C + \overline{D}) \cdot (\overline{A} + B + \overline{C} + D) \cdot (A + \overline{B} + \overline{C} + \overline{D})$$

■ INTEGRATED MATH APPLICATION:

Is the circuit shown in Figure 11-23 an example of sum-of-products or product-of-sums, and what is this circuit's Boolean expression?

■ *Solution:*

The circuit in Figure 11-23 is a product-of-sums equivalent circuit, and its Boolean expression is

$$Y = (A + B) \cdot (C + D)$$

FIGURE 11-23
SOP or POS?

■ INTEGRATED MATH APPLICATION:

Determine the logic circuit needed for the truth table shown in Figure 11-24(a).

■ *Solution:*

The first step is to write the fundamental product for each HIGH output in the truth table, as seen in Figure 11-24(a). By ORing (+) these fundamental products, we can obtain the following sum-of-products Boolean equation:

$$Y = \overline{ABC}D + \overline{AB}\overline{C}\,\overline{D} + A\overline{B}\,\overline{C}\,D + AB\overline{C}D$$

The next step is to develop the circuit equivalent of this equation. The equation states that the outputs of four four-input AND gates are connected to a four-input OR gate. The first AND gate has inputs $\overline{A}\,\overline{B}\,\overline{C}\,D$, the second has $\overline{A}B\overline{C}\,\overline{D}$, the third has $A\overline{B}\,\overline{C}\,D$, and the fourth has $AB\overline{C}D$. The circuit equivalent for this Boolean expression can be seen in Figure 11-24(b). This circuit will operate in the manner detailed by the truth table in Figure 11-24(a) and generate a 1 output only when

$$A = 0, B = 0, C = 0, D = 1$$
$$A = 0, B = 1, C = 0, D = 0$$
$$A = 1, B = 0, C = 0, D = 1$$
$$A = 1, B = 1, C = 0, D = 1$$

In all other instances, the output of this circuit will be 0.

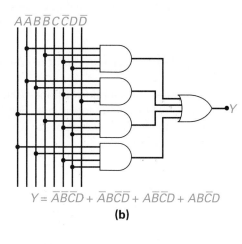

$$Y = \overline{A}\,\overline{B}CD + \overline{A}B\overline{C}D + A\overline{B}CD + AB\overline{C}D$$

(b)

(a)

FIGURE 11-24   **Sum-of-Products Example.**

■ **INTEGRATED MATH APPLICATION:**

Apply the distributive law to the following Boolean expression in order to convert it into a sum-of-products form.

$$Y = A\overline{B} + C(D\overline{E} + F\overline{G})$$

■ *Solution:*

$$Y = A\overline{B} + C(D\overline{E} + F\overline{G}) \qquad \text{Original expression}$$
$$Y = A\overline{B} + CD\overline{E} + CF\overline{G} \qquad \text{Distributive law applied}$$

SELF-TEST EVALUATION POINT FOR SECTION 11-3

Now that you have completed this section, you should be able to:

■ *Objective 4.*   *Describe the logic circuit design process from truth table to gate circuit.*

■ *Objective 5.*   *Explain the difference between a sum-of-products equation and a product-of-sums equation.*

Use the following questions to test your understanding of Section 11-3:

1. A two-input AND gate will produce at its output the logical _____ of A, and B, while an OR gate will produce at its output the logical _____ of A and B.

2. What are the fundamental products of each of the following input words?
    a. $A, B, C, D = 1011$ **b.** $A, B, C, D = 0110$

3. The Boolean expression $Y = AB + \overline{A}B$ is an example of a(n) _____ (SOP/POS) equation.

4. Describe the steps involved to develop a logic circuit from a truth table.

11-4   GATE CIRCUIT SIMPLIFICATION

In the last section we saw how we could convert a desired set of conditions in a truth table to a sum-of-products equation, and then to an equivalent logic circuit. In this section we will see how we can simplify the Boolean equation using Boolean laws and rules and therefore simplify the final logic circuit. As to why we would want to simplify a logic circuit, the answer is: A simplified circuit performs the same function, but is smaller, easier to construct, consumes less power, and costs less.

11-4-1 *Boolean Algebra Simplification*

Generally, the simplicity of one circuit compared with another circuit is judged by counting the number of logic-gate inputs. For example, the logic circuit in Figure 11-25(a) has a total of six inputs (two on each of the AND gates and two on the OR gate). The Boolean equation for this circuit is

$$Y = \overline{A}B + AB$$

Since "· B" (AND B) seems to be a common factor in this equation, by factoring we obtain the following:

$$Y = \overline{A}B + AB \qquad \text{Original expression}$$
$$Y = (\overline{A} + A) \cdot B \qquad \text{Factoring AND } B$$

This equivalent logic circuit is shown in Figure 11-25(b), and if you count the number of logic-gate inputs in this circuit (two for the OR and two for the AND, a total of four), you can see that the circuit has been simplified from a six-input circuit to a four-input circuit.

   This, however, is not the end of our simplification process, since another rule can be applied to further simplify the circuit. Remembering that any variable ORed with its complement results in a 1, we can replace $\overline{A} + A$ with a 1, as follows:

$$Y = (\overline{A} + A) \cdot B$$
$$Y = (1) \cdot B$$
$$Y = 1 \cdot B$$

The resulting equation $Y = 1 \cdot B$ can be further simplified, since any variable ANDed with a 1 will always cause the output to equal the variable. Therefore, $Y = 1 \cdot B$ can be replaced with B, as follows:

$$Y = 1 \cdot B$$
$$Y = B$$

   This means that the entire circuit in Figure 11-25(a) can be replaced with a single wire connected from the B input to the output, as seen in Figure 11-25(c). This fact is confirmed in the truth table in Figure 11-25(d), which shows that output Y exactly follows the input B.

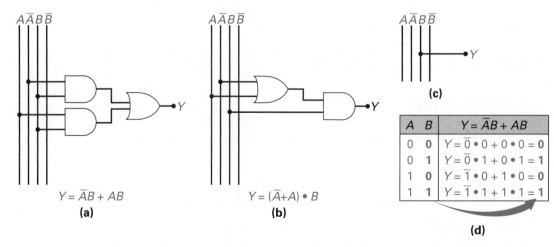

FIGURE 11-25 **Simplifying Gate Circuits.**

CHAPTER 11 / BOOLEAN EXPRESSIONS AND ALGEBRA

■ INTEGRATED MATH APPLICATION:

Simplify the Boolean equation shown in Figure 11-26(a).

■ *Solution:*

The Boolean equation for this circuit is

$$Y = A\overline{B}\,\overline{C} + \overline{A}B\overline{C} + A\overline{B}C + \overline{A}\,\overline{B}\,\overline{C}$$

The first step in the simplification process is to try and spot common factors in each of the ANDed terms, and then to rearrange these terms as shown in Figure 11-26(b). The next step is to factor out the common expressions, which are $A\overline{B}$ and $\overline{A}\,\overline{C}$, resulting in

$Y = A\overline{B}\,\overline{C} + \overline{A}B\overline{C} + A\overline{B}C + \overline{A}\,\overline{B}\,\overline{C}$	Original equation
$Y = A\overline{B}\,\overline{C} + A\overline{B}C + \overline{A}B\overline{C} + \overline{A}\,\overline{B}\,\overline{C}$	Rearranged equation
$Y = A\overline{B}(\overline{C} + C) + \overline{A}\,\overline{C}(B + \overline{B})$	Factored equation

Studying the last equation you can see that we can apply the OR rule $(\overline{A} + A) = 1$ and simplify the terms within the parentheses to obtain the following:

$$Y = A\overline{B}(\overline{C} + C) + \overline{A}\,\overline{C}(B + \overline{B})$$
$$Y = A\overline{B} \cdot (1) + \overline{A}\,\overline{C} \cdot (1)$$

Now, we can apply the $A \cdot 1 = A$ rule to further simplify the terms, since $A\overline{B} \cdot 1 = A\overline{B}$ and $\overline{A}\,\overline{C} \cdot 1 = \overline{A}\,\overline{C}$; therefore:

$$Y = A\overline{B} + \overline{A}\,\overline{C}$$

This equivalent logic circuit is illustrated in Figure 11-26(c) and has only 6 logic gate inputs compared with the original circuit shown in Figure 11-26(a), which has 16 logic gate inputs. The truth table in Figure 11-26(d) compares the original equation with the simplified equation and shows how the same result is obtained.

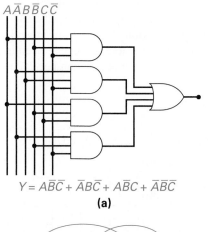

$$Y = A\overline{B}\overline{C} + \overline{A}B\overline{C} + A\overline{B}C + \overline{A}\,\overline{B}\,\overline{C}$$

(a)

$$Y = \overbrace{A\overline{B}\overline{C}} + \overbrace{\overline{A}B\overline{C}} + \overbrace{A\overline{B}C} + \overbrace{\overline{A}\,\overline{B}\overline{C}}$$

$$Y = A\overline{B}\overline{C} + A\overline{B}C + \overline{A}B\overline{C} + \overline{A}\,\overline{B}\overline{C}$$

(b)

(c)

A	B	C	$Y = A\overline{B}\overline{C} + \overline{A}B\overline{C} + A\overline{B}C + \overline{A}\,\overline{B}\overline{C}$	$Y = A\overline{B} + \overline{A}\overline{C}$
0	0	0	$Y = 0\overline{0}\,\overline{0} + \overline{0}0\overline{0} + 0\overline{0}0 + \overline{0}\,\overline{0}\,\overline{0} = 1$	$Y = 0\overline{0} + \overline{0}\,\overline{0} = 1$
0	0	1	$Y = 0\overline{0}\,\overline{1} + \overline{0}0\overline{1} + 0\overline{0}1 + \overline{0}\,\overline{0}\,\overline{1} = 0$	$Y = 0\overline{0} + \overline{0}\,\overline{1} = 0$
0	1	0	$Y = 0\overline{1}\,\overline{0} + \overline{0}1\overline{0} + 0\overline{1}0 + \overline{0}\,\overline{1}\,\overline{0} = 1$	$Y = 0\overline{1} + \overline{0}\,\overline{0} = 1$
0	1	1	$Y = 0\overline{1}\,\overline{1} + \overline{0}1\overline{1} + 0\overline{1}1 + \overline{0}\,\overline{1}\,\overline{1} = 0$	$Y = 0\overline{1} + \overline{0}\,\overline{1} = 0$
1	0	0	$Y = 1\overline{0}\,\overline{0} + \overline{1}0\overline{0} + 1\overline{0}0 + \overline{1}\,\overline{0}\,\overline{0} = 1$	$Y = 1\overline{0} + \overline{1}\,\overline{0} = 1$
1	0	1	$Y = 1\overline{0}\,\overline{1} + \overline{1}0\overline{1} + 1\overline{0}1 + \overline{1}\,\overline{0}\,\overline{1} = 1$	$Y = 1\overline{0} + \overline{1}\,\overline{1} = 1$
1	1	0	$Y = 1\overline{1}\,\overline{0} + \overline{1}1\overline{0} + 1\overline{1}0 + \overline{1}\,\overline{1}\,\overline{0} = 0$	$Y = 1\overline{1} + \overline{1}\,\overline{0} = 0$
1	1	1	$Y = 1\overline{1}\,\overline{1} + \overline{1}1\overline{1} + 1\overline{1}1 + \overline{1}\,\overline{1}\,\overline{1} = 0$	$Y = 1\overline{1} + \overline{1}\,\overline{1} = 0$

(d)

FIGURE 11-26 **Boolean Equation Simplification.**

FIGURE 11-27 **The Karnaugh Map—A Rearranged Truth Table.**

A	B	Y
0	0	$\overline{A}\overline{B}$
0	1	$\overline{A}B$
1	0	$A\overline{B}$
1	1	AB

	$\overline{B}$	B
$\overline{A}$	$\overline{A}\overline{B}$	$\overline{A}B$
A	$A\overline{B}$	AB

11-4-2 *Karnaugh Map Simplification*

In most instances, engineers and technicians will simplify logic circuits using a **Karnaugh map,** or *K-map*. The Karnaugh map, named after its inventor, is quite simply a rearranged truth table, as shown in Figure 11-27. With this map the essential gating requirements can be more easily recognized and reduced to their simplest form.

The total number of boxes or cells in a K-map depends on the number of input variables. For example, in Figure 11-27, only two inputs (*A* and *B*) and their complements ($\overline{A}$ and $\overline{B}$) are present, and therefore the K-map contains (like a two-variable truth table) only four combinations (00, 01, 10, and 11).

Each cell of this two-variable K-map represents one of the four input combinations. In practice, the input labels are placed outside the cells, as shown in Figure 11-28(a), and apply to either a column or row of cells. For example, the row label $\overline{A}$ applies to the two upper cells, and the row label *A* applies to the two lower cells. Running along the top of the K-map, the label $\overline{B}$ applies to the two left cells, and the label *B* applies to the two right cells. As an example, the upper right cell represents the input combination $\overline{A} B$. Figure 11-28(b) and (c) shows the formats for a three-variable ($2^3 = 8$ cells) and four-variable ($2^4 = 16$ cells) K-map.

Now that we have an understanding of the K-map, let us see how it can be used to simplify a logic circuit. As an example, imagine that we need to create an equivalent logic circuit for the truth table given in Figure 11-29(a). The first step is to develop a sum-of-products Boolean expression. This is achieved by writing the fundamental product for each HIGH output in the truth table and then ORing all the fundamental products, as shown in Figure 11-29(b). The equivalent logic circuit for this equation is shown in Figure 11-29(c). The next step is to plot this Boolean expression on a two-variable K-map, as seen in Figure 11-29(d). When plotting a sum-of-products expression on a K-map, remember that each cell

FIGURE 11-28 **Karnaugh Map (a) Two-Variable (b) Three-Variable (c) Four-Variable.**

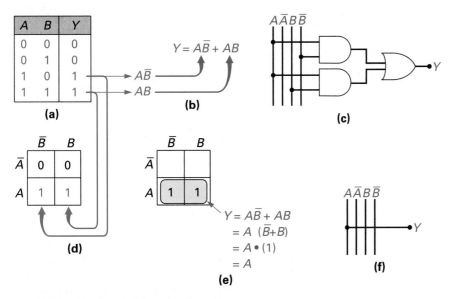

A	B	Y
0	0	0
0	1	0
1	0	1
1	1	1

(a)

$\rightarrow A\bar{B}$
$\rightarrow AB$

$Y = A\bar{B} + AB$

(b)

$A\,\bar{A}\,B\,\bar{B}$

(c)

	$\bar{B}$	B
$\bar{A}$	0	0
A	1	1

(d)

	$\bar{B}$	B
$\bar{A}$		
A	1	1

$Y = A\bar{B} + AB$
$= A\,(\bar{B}+B)$
$= A \bullet (1)$
$= A$

(e)

$A\,\bar{A}\,B\,\bar{B}$

(f)

FIGURE 11-29   **Karnaugh Map Simplification.**

corresponds to each of the input combinations in the truth table. A HIGH output in the truth table should appear as a 1 in its equivalent cell in the K-map, and a LOW output in the truth table should appear as a 0 in its equivalent cell. A 1, therefore, will appear in the lower left cell (corresponding to $A\bar{B}$) and in the lower right cell (corresponding to AB). The other input combinations ($\bar{A}\,\bar{B}$ and $\bar{A}\,B$) both yield a 0 output, and therefore a 0 should be placed in these two upper cells.

Reducing Boolean equations is largely achieved by applying the rule of complements, which states that $A + \bar{A} = 1$. Now that the SOP equation has been plotted on the K-map shown in Figure 11-29(d), the next step is to group terms and then factor out the common variables. If you study the K-map in Figure 11-29(d), you will see that adjacent cells differ by only one input variable. This means that if you move either horizontally or vertically from one cell to an adjacent cell, only one variable will change. Grouping adjacent cells containing a 1, as seen in Figure 11-29(e), allows cells to be compared and simplified (using the rule of complements) to create one-product terms. In this example, cells $A\bar{B}$ and AB contain B and $\bar{B}$, so these opposites or complements cancel, leaving A, as follows:

$$Y = A\bar{B} + AB \qquad \text{Grouped pair}$$
$$Y = A \cdot (\bar{B} + B) \qquad \text{Factoring } A \text{ AND}$$
$$Y = A \cdot 1$$
$$Y = A$$

This procedure can be confirmed by studying the original truth table in Figure 11-29(a) in which you can see that output Y exactly follows input A. The equivalent circuit is shown in Figure 11-29(f).

■ INTEGRATED MATH APPLICATION:

Determine the simplest logic circuit for the truth table shown in Figure 11-30(a). Illustrate each step in the process.

■ Solution:

Since this example has three input variables, the first step is to draw a three-variable K-map, as shown in Figure 11-30(b). The next step is to look for the HIGH outputs in the truth table in Figure 11-30(a) and to plot these 1s in their equivalent cells in the K-map, as shown in Figure 11-30(b). After inserting 0s in the remaining cells, group the 1s into pairs,

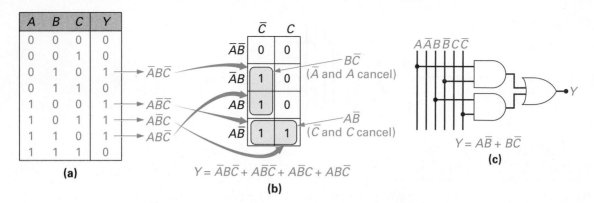

FIGURE 11-30   Three-Variable K-Map Simplification.

as shown in Figure 11-30(b), and then study the row and column variable labels associated with that grouped pair to see which variable will drop out due to the rule of complements. In the upper group, $\bar{A}$ and A will cancel, leaving $B\bar{C}$, and in the lower group $\bar{C}$ and C will cancel, leaving $A\bar{B}$. These reduced products will form an equivalent Boolean equation and logic circuit, as seen in Figure 11-30(c). In this example, the original 16-input equation in Figure 11-30(b) has been simplified to the equivalent 6-input logic gate circuit shown in Figure 11-30(c).

The 1s in a K-map can be grouped in pairs (groups of two), quads (groups of four), octets (groups of eight), and all higher powers of 2. Figure 11-31 shows some examples of grouping and how the K-map has been used to reduce large Boolean equations. Notice that the larger groups will yield smaller terms and therefore gate circuits with fewer inputs. For this reason, you should begin by looking for the largest possible groups, and then step down in group size if none are found (for example, begin looking for octets, then quads, and finally pairs). Also notice that you can capture 1s on either side of a map by wrapping a group around behind the map.

■   INTEGRATED MATH APPLICATION:

What would be the sum-of-products Boolean expression for the truth table in Figure 11-32(a)? After determining the Boolean expression, plot it on a K-map to see if it can be simplified.

■   *Solution:*

The first step in developing the sum-of-products Boolean expression is to write down the fundamental products for each HIGH output in the truth table, as seen in Figure 11-32(a). From this, we can derive an SOP equation:

$$Y = \bar{A}\,\bar{B}\,\bar{C}D + \bar{A}\,\bar{B}CD + \bar{A}BC\bar{D} + \bar{A}BCD + A\bar{B}CD + ABCD$$

The next step is to draw a four-variable K-map, as seen in Figure 11-32(b), and then plot the HIGH outputs of the truth table in their equivalent cells in the K-map. Looking at Figure 11-32(b), you can see that the 1s can be grouped into two quads. With the square-shaped quad, row labels B and $\bar{B}$ will cancel, and column labels C and $\bar{C}$ will cancel, leaving $\bar{A}D$. With the rectangular-shaped quad, all its row labels will cancel, leaving CD. The simplified Boolean expression, therefore, is

$$Y = \bar{A}D + CD$$

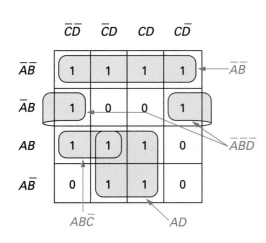

$\overline{AB}$ $\overline{AB}$

AB

$A\overline{B}$

$AB\overline{C}$ AD

Before : $Y = \overline{A}\,\overline{B}\,\overline{C}\,\overline{D} + \overline{A}\,\overline{B}\,\overline{C}D + \overline{A}\,\overline{B}CD + \overline{A}\,\overline{B}C\overline{D} +$
$\overline{A}B\overline{C}\,\overline{D} + \overline{A}BCD + AB\overline{C}\,\overline{D} + AB\overline{C}D +$
$ABCD + A\overline{B}\,\overline{C}D + A\overline{B}CD$

After : $Y = \quad AB\overline{C} + AD + \overline{A}B\overline{D} + \overline{A}\,\overline{B}$

(a)

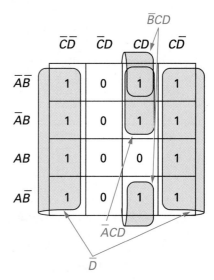

$\overline{BCD}$

$\overline{AB}$

$\overline{AB}$

AB

$A\overline{B}$

$\overline{A}CD$

$\overline{D}$

Before : $Y = \overline{A}\,\overline{B}\,\overline{C}\,\overline{D} + \overline{A}\,\overline{B}C\overline{D} + \overline{A}\,\overline{B}CD +$
$\overline{A}B\overline{C}\,\overline{D} + \overline{A}BC\overline{D} + \overline{A}BCD +$
$AB\overline{C}\,\overline{D} + ABC\overline{D} + A\overline{B}\,\overline{C}\,\overline{D} +$
$A\overline{B}CD + A\overline{B}C\overline{D}$

After : $Y = \quad \overline{A}CD + \overline{B}CD + \overline{D}$

(b)

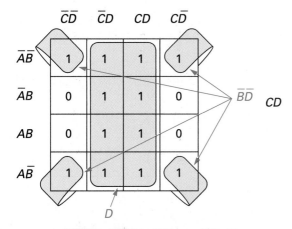

$\overline{CD}$ $\overline{C}D$ CD $C\overline{D}$

$\overline{AB}$

$\overline{AB}$

AB

$A\overline{B}$

$\overline{BD}$ CD

D

Before : $Y = \overline{A}\,\overline{B}\,\overline{C}\,\overline{D} + \overline{A}\,\overline{B}\,\overline{C}D + \overline{A}\,\overline{B}CD + \overline{A}\,\overline{B}C\overline{D} +$
$\overline{A}B\overline{C}D + \overline{A}BCD + AB\overline{C}D + ABCD +$
$A\overline{B}\,\overline{C}\,\overline{D} + A\overline{B}\,\overline{C}D + A\overline{B}CD + A\overline{B}C\overline{D}$

After : $Y = \quad \overline{B}\overline{D} + D$

(c)

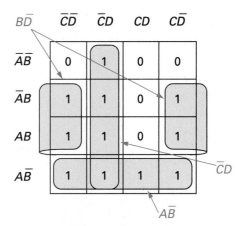

$B\overline{D}$ $\overline{CD}$ $\overline{C}D$ CD $C\overline{D}$

$\overline{AB}$

$\overline{AB}$

AB

$A\overline{B}$

$\overline{CD}$

$A\overline{B}$

Before : $Y = \overline{A}\,\overline{B}\,\overline{C}D + \overline{A}B\overline{C}\,\overline{D} + \overline{A}B\overline{C}D + \overline{A}BC\overline{D} +$
$AB\overline{C}\,\overline{D} + AB\overline{C}D + ABC\overline{D} + A\overline{B}\,\overline{C}\,\overline{D} +$
$A\overline{B}\,\overline{C}D + A\overline{B}CD + A\overline{B}C\overline{D}$

After : $Y = \quad A\overline{B} + B\overline{D} + \overline{C}D$

(d)

FIGURE 11-31   **K-Map Grouping Examples.**

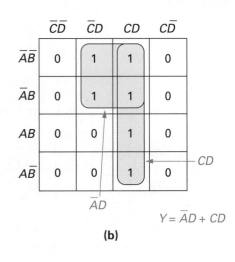

$Y = \bar{A}D + CD$

(b)

(a)

FIGURE 11-32   Example of Four-Variable K-Map.

Referring to the truth table in Figure 11-33(a), you can see that input words 0000 through 1001 have either a 0 or 1 at the output Y. Input words 1010 through 1111, on the other hand, have an "X" written in the output column to indicate that the output Y for these input combinations is unimportant (output can be either a 0 or 1). These Xs in the output are appropriately called *don't-care conditions,* and can be treated as either a 0 or 1. In K-maps, these Xs can be grouped with 1s to create larger groups, and therefore a simpler logic circuit, as we will see in the next example.

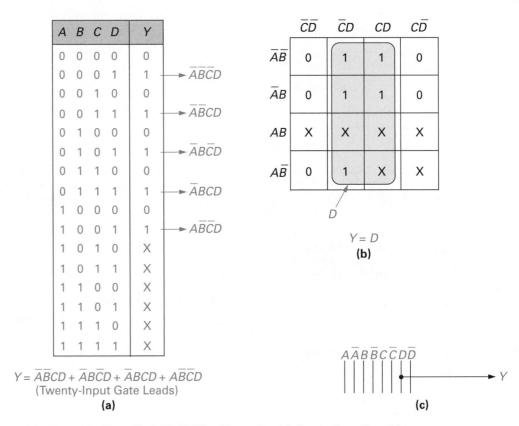

$Y = \bar{A}\bar{B}\bar{C}D + \bar{A}\bar{B}CD + \bar{A}BCD + A\bar{B}\bar{C}D$
(Twenty-Input Gate Leads)
(a)

$Y = D$
(b)

(c)

FIGURE 11-33   Four-Variable K-Map Example with Don't-Care Conditions.

■ INTEGRATED MATH APPLICATION:

Sketch an equivalent logic circuit for the truth table given in Figure 11-33(a) using the fewest number of inputs.

■ *Solution:*

Figure 11-33(b) shows a four-variable K-map with the 0s, 1s, and Xs (don't-care conditions) inserted into their appropriate cells. Since a larger group will always yield a smaller term, visualize the Xs as 1s, and then try for as big a group as possible. In this example, the Xs have allowed us to group an octet, resulting in the single product *D*. As a result, the 20-input Boolean equation in Figure 11-33(a) can be replaced with a single line connected to *D,* as shown in Figure 11-33(c).

SELF-TEST EVALUATION POINT FOR SECTION 11-4

Now that you have completed this section, you should be able to:

■ **Objective 6.** *Describe how Boolean algebra and Karnaugh maps can be used to simplify logic circuits.*

Use the following questions to test your understanding of Section 11-4.

1. A two-variable K-map will have _____ cells.
2. Simplify the equation $Y = \overline{A} B + A B$.
3. When grouping 1s in a K-map, the _____ (smaller/larger) the groups, the smaller the terms.
4. Can the don't-care conditions in a K-map be grouped with 0s or with 1s, or does it matter?

REVIEW QUESTIONS

Multiple Choice Questions

1. In Boolean, the OR function is described as logical _____ and the AND function is described as logical _____.

   a. Addition, division
   b. Multiplication, addition
   c. Addition, multiplication
   d. Multiplication, subtraction

2. The OR gate is said to produce at its output the logical _____ of its inputs, while the AND gate is said to produce at its output the logical _____ of its inputs.

   a. Sum, product
   b. Product, quotient
   c. Sum, subtrahend
   d. Product, sum

3. If inputs *A* and *B* are ANDed together and then the result is ORed with *C,* the Boolean expression equation describing this operation would be _____.

   a. $Y = AB + BC$
   b. $Y = (A + B)C$
   c. $Y = AB + C$
   d. $Y = AB + AC$

4. The Boolean expression $Y = (A \cdot B) + (A \cdot C)$ describes a logic circuit that has _____.

   a. Six inputs
   b. Two OR and one AND gate
   c. Two AND and one OR gate
   d. Four inputs

5. The Boolean expression $Y = \overline{A + B}$ describes a two-input _____.

   a. NAND gate   c. AND gate
   b. NOR gate   d. OR gate

6. DeMorgan's first theorem states that _____.

   a. $\overline{A \cdot B} = \overline{A} + \overline{B}$ c. $\overline{A} + \overline{B} = \overline{A \cdot B}$
   b. $\overline{A + B} = \overline{A} + \overline{B}$ d. $\overline{A \cdot B} = A \cdot B$

7. DeMorgan's second theorem states that _____.

   a. $\overline{A \cdot B} = \overline{A} + \overline{B}$ c. $\overline{A} + \overline{B} = \overline{A \cdot B}$
   b. $\overline{A + B} = \overline{A} + \overline{B}$ d. $\overline{A \cdot B} = A \cdot B$

8. In Boolean algebra which symbol is used to describe the exclusive OR action?

   a. $+$ c. $\times$
   b. $\cdot$ d. $\oplus$

9. Which of the following algebraically states the commutative law?

   a. $A + B = B + A$
   b. $A \cdot (B \cdot C) = (A \cdot B) \cdot C$
   c. $A + (B \cdot C) = (A + B) \cdot (A + C)$
   d. None of the above

10. $A + (B \cdot C) = (A + B) \cdot (A + C)$ is the algebraic definition of the _____.

   a. Associative law   c. Commutative law
   b. Distributive law   d. Complements law

11. The four Boolean rules for an OR gate are $A + 0 = ?, A + A = ?, A + 1 = ?, A + \overline{A} = ?$.

    a. $A, 0, 1, 1$ **c.** $A, A, 1, 1$
    b. $A, A, 0, 0$ **d.** $A, A, 0, 1$

12. The four Boolean rules for an AND gate are $A \cdot 1 = ?, A \cdot A = ?, A \cdot 0 = ?, A \cdot \overline{A} = ?$.

    a. $A, 0, 1, 1$ **c.** $A, A, 1, 1$
    b. $A, A, 0, 0$ **d.** $A, A, 0, 1$

13. If you applied DeMorgan's theorem and the double-inversion rule to the equation $Y = \overline{A} + BC$, what would be the result?

    a. $A\overline{B} + C$ **c.** $\overline{A}\,B + C$
    b. $A + \overline{BC}$ **d.** $AB + \overline{C}$

14. What would be the dual equation of $A + (B \cdot C) = (A + B) \cdot (A + C)$?

    a. $A \cdot (B + C) = (A \cdot B) \cdot (A \cdot C)$
    b. $\overline{A} + B = AB + \overline{C}$
    c. $A \cdot (B \cdot C) = (A \cdot B) + (A \cdot C)$
    d. $A + (B \cdot C) = (A \cdot B) \cdot (A + C)$

15. $Y = AB + BC + AC$ is an example of a _____.

    a. Product-of-sums expression
    b. Sum-of-products expression
    c. Quotient-of-sums expression
    d. Sum-of-quotients expression

16. What would be the fundamental product of the input word $ABCD = 1101$?

    a. $ABCD$ **c.** $AB\overline{C}D$
    b. $\overline{AB}CD$ **d.** $ABC\overline{D}$

17. How many cells would a three-variable Karnaugh map have?

    a. 4   **c.** 8
    b. 6   **d.** 16

18. What would be the Boolean expression for a four-variable logic circuit that NANDs inputs A and B and also NANDs inputs C and D, and then NANDs the results?

    a. $\overline{\overline{AB} + \overline{CD}}$ **c.** $\overline{(\overline{AB}) + (\overline{CD})}$
    b. $(\overline{AB}) \cdot (\overline{CD})$ **d.** $\overline{(A + B) \cdot (C + D)}$

19. If you applied DeMorgan's theorem and the double-inversion rule to the answer in question 18, you would obtain which of the following equations?

    a. $AB + CD$ **c.** $(A + B) \cdot (C + D)$
    b. $A + B + C + D$ **d.** $(A + B) + (CD)$

20. In reference to questions 18 and 19, you can see that the DeMorgan equivalent of a NAND-NAND circuit is a(n) _____.

    a. OR-AND   **c.** AND-OR
    b. AND-NOR   **d.** NOR-AND

Communication Skill Questions

21. List the Boolean expression for the seven basic logic gates. (11-1)

22. Describe briefly, with an example, the process of converting a desired set of output conditions into an equivalent logic circuit. (11-2)

23. What is the difference between an SOP and a POS equation? (11-3)

24. What is a Karnaugh map? (11-4-2)

25. Briefly describe how K-maps can be used to simplify a logic circuit. (11-4-2)

26. How is the simplicity of a logic circuit generally judged? (11-4)

27. Why is it necessary to simplify logic circuits? (11-4)

28. What is meant by the terms *logical product* and *logical sum*? (11-1)

29. How can DeMorgan's theorems be stated using Boolean algebra? (11-2)

30. Briefly describe why larger groups of 1s on a K-map yield smaller terms, and why smaller terms are desired. (11-4-2)

Practice Problems

31. Determine the Boolean expressions for the logic circuits shown in Figure 11-34.

32. Which of the expressions in question 31 would be considered POS, and which SOP?

33. Sketch the equivalent logic circuit for the following Boolean equations.

    a. $Y = (AB) + (\overline{AB}) + (\overline{A + B}) + (A \oplus B)$
    b. $Y = \overline{(\overline{A} + B)} \oplus ABC$

34. Determine the Boolean expression for the Y_2 output in Table 11-1.

35. Use a K-map to simplify the Y_2 expression from question 34.

36. How much was the Y_2 logic circuit simplified in question 35?

37. Referring to Table 11-1, (a) determine the Boolean expression for the Y_1 output. (b) Simplify the expression using a K-map. (c) Describe how much the logic circuit was simplified.

38. Determine the Boolean expression for the Y_3 output in Table 11-1, simplify the expression using a K-map, and then describe how much the logic circuit was simplified.

39. Sketch the simplified logic circuits for the Y_1, Y_2, and Y_3 outputs listed in Table 11-1.

40. If the don't-care conditions in the Y_3 output in Table 11-1 were 0s, what would the simplified expression be? Is the expression more or less simplified with don't-care conditions?

(a)

(b)

(c)

(d)

FIGURE 11-34   From Logic Circuit to Boolean Equation.

TABLE 11-1   The Y_1, Y_2, and Y_3
Truth Table

A	B	C	D	Y_1	Y_2	Y_3
0	0	0	0	1	1	1
0	0	0	1	1	1	0
0	0	1	0	1	0	1
0	0	1	1	1	0	0
0	1	0	0	0	0	1
0	1	0	1	0	0	1
0	1	1	0	0	1	1
0	1	1	1	0	1	1
1	0	0	0	0	1	0
1	0	0	1	1	1	0
1	0	1	0	0	0	0
1	0	1	1	1	0	0
1	1	0	0	1	0	1
1	1	0	1	1	0	x
1	1	1	0	1	0	x
1	1	1	1	1	0	x

Web Site Questions

Go to the web site http://www.prenhall.com/cook, select the textbook *Introductory Computer Mathematics, 2e,* select this chapter, and then follow the instructions when answering the multiple-choice practice problems.

Binary Arithmetic

Woolen Mill Makes Minis

At the age of 24 in 1950, Kenneth Olsen went to work at MIT's Digital Computer Laboratory as a research assistant. For the next seven years, Olsen worked on the SAGE (Semi Automatic Ground Environment) project, which was a new computer system designed to store the constantly changing data obtained from radars tracking long-range enemy bombers capable of penetrating American air space. IBM won the contract to build the SAGE network and Olsen traveled to the plant in New York to supervise production. After completing the SAGE project, Olsen was ready to move on to bigger and better things; however, he didn't want to go back to MIT and was not happy with the rigid environment at a large corporation like IBM.

In September 1957, Olsen and fellow MIT colleague Harlan Anderson leased an old brick woolen mill in Maynard, Massachusetts, and after some cleaning and painting they began work in their new company, which they named Digital Equipment Corporation or DEC. After three years, the company came out with their first computer, the PDP 1 (programmed data processor), and its ease of use, low cost, and small size made it an almost instant success. In 1965, the desktop PDP 8 was launched, and since in that era miniskirts were popular, it was probably inevitable that the small machine was nicknamed a minicomputer. In 1969, the PDP 11 was unveiled, and like its predecessors, this number-crunching data processor went on to be accepted in many applications, such as tracking the millions of telephone calls received on the 911 emergency line, directing welding machines in automobile plants, recording experimental results in laboratories, and processing all types of data in offices, banks, and department stores. Just 20 years after introducing low-cost computing in the 1960s, DEC was second only to IBM as a manufacturer of computers in the United States.

Outline and Objectives

Introduction

At the heart of every digital electronic computer system is a circuit called a microprocessor unit (MPU). To perform its function, the MPU contains several internal circuits, one of which is called the arithmetic–logic unit, or ALU. This unit is the number-crunching circuit of every digital system and, as its name implies, it can be controlled to perform either arithmetic operations (such as addition and subtraction) or logic operations (such as AND and OR).

In order to understand the operation of an ALU, we must begin by discussing how binary numbers can represent any value, whether it is large, small, positive, or negative. We must also discuss how these numbers can be manipulated to perform arithmetic operations such as addition, subtraction, multiplication, and division.

12-1 BINARY ARITHMETIC

The addition, subtraction, multiplication, and division of binary numbers is performed in exactly the same way as the addition, subtraction, multiplication, and division of decimal numbers. The only difference is that the decimal number system has 10 digits, whereas the binary number system has 2.

As we discuss each of these arithmetic operations, we will use decimal arithmetic as a guide to binary arithmetic so that we can compare the known to the unknown.

12-1-1 Binary Addition

Figure 12-1(a) illustrates how the decimal values 29,164 and 63,729 are added together. First the units 4 and 9 are added, producing a sum of 13, or 3 carry 1. The carry is then added to the 6 and 2 in the tens column, producing a sum of 9 with no carry. This process is then repeated for each column, moving right to left, until the total sum of the addend, augend, and their carries has been obtained.

In the addition of two decimal values, many combinations are available, since decimal has 10 different digits. With binary, however, only 2 digits are available, and therefore only four basic combinations are possible. These four rules of binary addition are shown in Figure 12-1(b). To help explain these rules, let us apply them to the example in Figure 12-1(c), in which we will add 11010 (decimal 26) to 11100 (decimal 28). As in decimal addition, we will begin at the right in the ones column and proceed to the left. In the ones column $0 + 0 = 0$ with a carry of 0 (Rule 1). In the twos column $1 + 0$ and the 0 carry from the ones column $= 1$ with no carry (Rule 2). In the fours column we have $0 + 1$ plus a carry of 0 from the twos column. This results in a fours column total of 1, with a carry of 0 (Rule 2). In the eights column we have $1 + 1$ plus a 0 carry from the fours column, producing a total of 10_2 ($1 + 1 =$ decimal 2, which is 10 in binary). In the eights column this 10 (decimal 2) is written down as a total of 0 with a carry of 1 (Rule 3). In the sixteens column, we have $1 + 1$, plus a carry of 1 from the eights column, producing a total of 11_2 ($1 + 1 + 1 =$ decimal 3, which is 11 in binary). In the sixteens column, this 11 (decimal 3) is written down as a total of 1 with a carry of 1 (Rule 4). Since there is only a carry of 1 in the thirty-twos column, the total in this column will be 1.

Converting the binary addend and augend to their decimal equivalents as seen in Figure 12-1(c), we see that 11010 (decimal 26) plus 11100 (decimal 28) will yield a sum or total of 110110 (decimal 54).

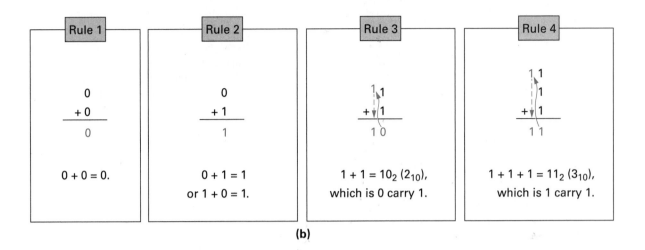

	(10,000)	(1,000)	(100)	(10)	(1)	
Carry	0	1	0	0	1	
Addend:		2	9,	1	6	4
Augend:	+	6	3,	7	2	9
Sum:		9	2,	8	9	3

$4 + 9 = 13$ or 3 1 carry

(a)

Rule 1	Rule 2	Rule 3	Rule 4
0 + 0 ― 0	0 + 1 ― 1	1 + 1 ― 1 0	1 1 1 + 1 ― 1 1
$0 + 0 = 0$.	$0 + 1 = 1$ or $1 + 0 = 1$.	$1 + 1 = 10_2 (2_{10})$, which is 0 carry 1.	$1 + 1 + 1 = 11_2 (3_{10})$, which is 1 carry 1.

(b)

	(32)	(16)	(8)	(4)	(2)	(1)			(10)	(1)
Carry:	1	1	0	0	0				1	
Addend:		1	1	0	1	0_2			2	6_{10}
Augend:	+	1	1	1	0	0_2	=	+	2	8_{10}
Sum:	1	1	0	1	1	0_2			5	4_{10}

(c)

FIGURE 12-1   **Binary Addition.**

■   **INTEGRATED MATH APPLICATION:**

Find the sum of the following binary numbers.

 a. $1011 + 1101 = ?$ b. $1000110 + 1100111 = ?$

■   *Solution:*

a. *Carry:*   1 1 1 1
   Augend: 1 0 1 1_2 (11_{10})
   Addend: +   1 1 0 1_2 (13_{10})
   Sum: 1 1 0 0 0_2 (24_{10})

b. *Carry:*   1 0 0 0 1 1 0
   Augend: 1 0 0 0 1 1 0_2 (70_{10})
   Addend: +   1 1 0 0 1 1 1_2 (103_{10})
   Sum: 1 0 1 0 1 1 0 1_2 (173_{10})

12-1-2   *Binary Subtraction*

Figure 12-2(a) reviews the decimal subtraction procedure by subtracting 4615 from 7003. Starting in the units column, you can see that since we cannot take 5 away from 3, we will have to go to a higher-order minuend unit and borrow. Since the minuend contains no tens or hundreds, we will have to go to the thousands column. From this point a chain of borrowing occurs as a thousand is borrowed and placed in the hundreds column (leaving 6 thousands), one of the hundreds is borrowed and placed in the tens column (leaving 9 hundreds), and one of the tens is borrowed and placed in the units column (leaving 9 tens). After borrowing 10, the minuend units digit has a value of 13, and if we now perform the subtraction, the result or difference of $13 - 5 = 8$. Since all the other minuend digits are now greater than their respective subtrahend digits, there will be no need for any further borrowing to obtain the difference.

In the subtraction of one decimal value from another, many combinations are available since decimal has 10 different digits. With binary, however, only 2 digits are available, and therefore only four basic combinations are possible. These four rules of binary subtraction are shown in Figure 12-2(b). To help explain these rules, let us apply them to the example in Fig-

(a)

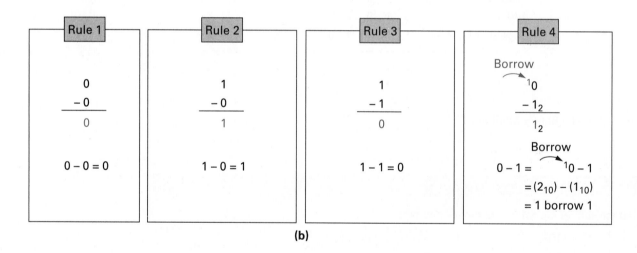

(b)

(c)

FIGURE 12-2   Binary Subtraction.

ure 12-2(c) in which we will subtract 11100 (decimal 28) from 1010110 (decimal 86). As in decimal subtraction, we will begin at the right in the ones column and proceed to the left. In the ones column $0 - 0 = 0$ (Rule 1). In the twos column $1 - 0 = 1$ (Rule 2), and in the fours column $1 - 1 = 0$ (Rule 3). In the eights column we cannot subtract 1 from nothing or 0, so therefore we must borrow 1 from the sixteens column, making the minuend in the eights column 10 (decimal 2). The difference can now be calculated in the eights column since $10 - 1 = 1$ (decimal $2 - 1 = 1$, Rule 4). Due to the previous borrow, the minuend in the sixteens column is now 0 and therefore we need to borrow a 1 from the thirty-twos column. Since the thirty-twos column is also 0, we will need to borrow from the sixty-fours column. Borrowing 1 from the sixty-fours column will leave a minuend of 0 and make the minuend in the thirty-twos column equal to 10 (decimal 2). Borrowing 1 from the 10 (borrowing decimal 1 from 2) in the thirty-twos column will leave a minuend of 1 and make the minuend in the sixteens column equal to 10 (decimal 2). We can now subtract the subtrahend of 1 from the minuend of 10 (decimal 2) in the sixteens column to obtain a difference of 1. Due to the previous borrow, the minuend in the thirty-twos column is equal to 1, and $1 - 0 = 1$ (Rule 2). Finally, since the minuend and subtrahend are both 0 in the sixty-fours column, the subtraction is complete.

Converting the binary minuend and subtrahend to their decimal equivalents, as seen in Figure 12-2(c), we see that 1010110 (decimal 86) minus 11100 (decimal 28) will result in a difference of 111010 (decimal 58).

■ INTEGRATED MATH APPLICATION:

Find the difference between the following binary numbers.

a. $100111 - 11101 = ?$ b. $101110111 - 1011100 = ?$

■ *Solution:*

a. Minuend: $^0\cancel{1}$ $^1\cancel{0}$ ^{10} 1 1 1_2 (39_{10})
   Subtrahend: $-$ 1 1 1 0 1_2 (29_{10})
   Difference: 0 0 1 0 1 0_2 (10_{10})

b. Minuend: 1 0 1 $^0\cancel{1}$ $^1\cancel{1}$ ^{10} 1 1 1_2 (375_{10})
   Subtrahend: $-$ 1 0 1 1 1 0 0_2 (92_{10})
   Difference: 1 0 0 0 1 1 0 1 1_2 (283_{10})

12-1-3   *Binary Multiplication*

Figure 12-3(a) reviews the decimal multiplication procedure by multiplying 463 by 23. To perform this operation, we begin by multiplying each digit of the multiplicand by the units digit of the multiplier to obtain the first partial product. Second, we multiply each digit of the multiplicand by the tens digit of the multiplier to obtain the second partial product. Finally, we add all the partial products to obtain the final product.

In the multiplication of one decimal value by another, many combinations are available, since decimal has 10 different digits. With binary multiplication, however, only 2 digits are available, and therefore only four basic combinations are possible. These four rules of binary multiplication are shown in Figure 12-3(b). To help explain these rules, let us apply them to the example in Figure 12-3(c) in which we will multiply 1011 (decimal 11) by 101 (decimal 5). As in decimal multiplication, we begin by multiplying each bit of the multiplicand by the ones column multiplier bit to obtain the first partial product. Second, we multiply each bit of the multiplicand by the twos column multiplier bit to obtain the second partial product. Remember that the LSB of the partial product must always be directly under its respective multiplier bit, and therefore the LSB of the second partial product should be under the twos column multiplier bit. In the next step, we add together the first and second partial products to obtain a sum of the partial products. With decimal multiplication we usually calculate all the partial products and then add them together to obtain the final product.

<table>
<tr><th></th><th>(10,000)</th><th>(1,000)</th><th>(100)</th><th>(10)</th><th>(1)</th></tr>
<tr><td>Multiplicand:</td><td></td><td></td><td>4</td><td>6</td><td>3</td></tr>
<tr><td>Multiplier:</td><td></td><td>×</td><td></td><td>2</td><td>3</td></tr>
<tr><td>First Partial Product:</td><td></td><td>1,</td><td>3</td><td>8</td><td>9</td></tr>
<tr><td>Second Partial Product:</td><td></td><td>9,</td><td>2</td><td>6</td><td></td></tr>
<tr><td>Final Product:</td><td>1</td><td>0,</td><td>6</td><td>4</td><td>9</td></tr>
</table>

(a)

Rule 1	Rule 2	Rule 3	Rule 4
0	1	0	1
× 0	× 0	× 1	× 1
0	0	0	1
$0 \times 0 = 0$	$0 \times 1 = 0$	$1 \times 0 = 0$	$1 \times 1 = 1$

(b)

<table>
<tr><th></th><th>(8)</th><th>(4)</th><th>(2)</th><th>(1)</th><th></th><th></th></tr>
<tr><td>Multiplicand:</td><td>1</td><td>0</td><td>1</td><td>1_2</td><td>=</td><td>11_{10}</td></tr>
<tr><td>Multiplier:</td><td>×</td><td>1</td><td>0</td><td>1_2</td><td></td><td>× 5_{10}</td></tr>
<tr><td>First Partial Product:</td><td>1</td><td>0</td><td>1</td><td>1</td><td></td><td>55_{10}</td></tr>
<tr><td>Second Partial Product:</td><td>0</td><td>0</td><td>0</td><td>0 ←</td><td></td><td></td></tr>
<tr><td>Sum of Partial Products:</td><td>0 1</td><td>0</td><td>1</td><td>1</td><td></td><td></td></tr>
<tr><td>Third Partial Product:</td><td>1 0</td><td>1</td><td>1 ←</td><td></td><td></td><td></td></tr>
<tr><td>Final Product:</td><td>1 1</td><td>0</td><td>1</td><td>1</td><td>1_2</td><td></td></tr>
</table>

(c)

FIGURE 12-3   **Binary Multiplication.**

This procedure can also be used with binary multiplication; however, it is better to add only two partial products at a time, since the many carries in binary become hard to keep track of and easily lead to errors. In the next step, we multiply each bit of the multiplicand by the fours column multiplier bit to obtain the third partial product. Remember that the LSB of this third partial product must be placed directly under its respective multiplier bit. The final step is to add the third partial product to the previous partial sum to obtain the final product or result.

If you study the partial products obtained in this example, you will notice that they were either exactly equal to the multiplicand (when the multiplier was 1) or all 0s (when the multiplier was 0). You can use this shortcut in future examples; however, be sure always to place the LSB of the partial products directly below their respective multiplier bits.

■ INTEGRATED MATH APPLICATION:

Multiply 1101_2 by 1011_2.

■ *Solution:*

Multiplicand:				1	1	0	1_2	13_{10}
Multiplier:				$\times$ 1	0	1	1_2	$\times$ 11_{10}
First Partial Product:				1	1	0	1	143_{10}
Second Partial Product:			1	1	0	1	$\leftarrow$	
Sum of Partial Product:		1	0	0	1	1	1	
Third Partial Product:		0	0	0	0	$\leftarrow$	$\leftarrow$	
Sum of Partial Product:		1	0	0	1	1	1	
Fourth Partial Product:	1	1	0	1	$\leftarrow$	$\leftarrow$	$\leftarrow$	
Final Product:	1 0 0 0	1	1	1	1_2			

12-1-4   *Binary Division*

Figure 12-4(a) reviews the decimal division procedure by dividing 830 by 23. We begin this long-division procedure by determining how many times the divisor (23) can be subtracted from the first digit of the dividend (8). Since the dividend is smaller, the quotient is 0. Next,

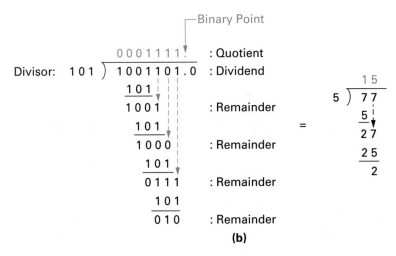

FIGURE 12-4   **Binary Division.**

we see how many times the divisor (23) can be subtracted from the first two digits of the dividend (83). Since 23 can be subtracted three times from 83, a quotient of 3 results, and the product of 3×23, or 69, is subtracted from the first two digits of the dividend, giving a remainder of 14. Then the next digit of the dividend (0) is brought down to the remainder (14), and we see how many times the divisor (23) can be subtracted from the new remainder (140). In this example, 23 can be subtracted six times from 140, a quotient of 6 results, and the product of 6×23, or 138, is subtracted from the remainder, giving a new remainder of 2. Therefore, $830 \div 23 = 36$, remainder 2.

Figure 12-4(b) illustrates an example of binary division, which is generally much simpler than decimal division. To perform this operation, we must first determine how many times the divisor (101_2) can be subtracted from the first bit of the dividend (1_2). Since the dividend is smaller than the divisor, a quotient of 0_2 is placed above the first digit of the dividend. Next, we see how many times the divisor (101_2) can be subtracted from the first two bits of the dividend (10_2), and this again results in a quotient of 0_2. Similarly, the divisor (101_2) cannot be subtracted from the first three bits of the dividend, so the quotient is once again 0_2. Finally, we find a dividend (1001_2) that is greater than the divisor (101_2), and therefore a 1_2 is placed in the quotient, and the divisor is subtracted from the first four bits of the dividend, resulting in a remainder of 100_2. Then the next bit of the dividend (1_2) is brought down to the remainder (100_2), and we see how many times the divisor (101_2) can be subtracted from the new remainder (1001_2). Since 101_2 is smaller than 1001_2, a quotient of 1_2 results. The divisor is subtracted from the remainder, resulting in a new remainder of 100_2. Then, the next bit of the dividend (0_2) is brought down to the remainder (100_2), and we see how many times the divisor (101_2) can be subtracted from the new remainder (1000_2). Since 101_2 is smaller than 1000_2, a quotient of 1 results. The divisor is subtracted from the remainder, resulting in a new remainder of 11_2. Then, the next bit of the dividend (1_2) is brought down to the remainder (11_2), and we see how many times the divisor (101_2) can be subtracted from the new remainder (111_2). Since 101_2 is smaller than 111_2, a quotient of 1 results. The divisor is subtracted from the remainder, resulting in a final remainder of 10. Therefore, $1001101_2 \div 101_2 = 1111_2$, remainder 10 ($77 \div 5 = 15$, remainder 2).

■ INTEGRATED MATH APPLICATION:

Divide 1110101_2 by 110_2.

■ *Solution:*

```
                0 0 1 0 0 1 1₂          Quotient          019₁₀
Divisor:  110)1 1 1 0 1 0 1₂           Dividend         6₁₀)117₁₀
              1 1 0 ↓ ↓ ↓ ↓
              0 0 1 0 1 0 ↓            Remainder
                    1 1 0 ↓                                 6
                    1 0 0 1            Remainder            57
                      1 1 0                                 54
                      0 0 1 1₂         Remainder            3₁₀
```

Now that you have completed this section, you should be able to:

■ *Objective 1.* *Describe how to complete the four basic arithmetic operations—binary addition, binary subtraction, binary multiplication, and binary division.*

Use the following questions to test your understanding of Section 12-1:

1. Perform the following binary arithmetic operations.
   a. $1011 + 11101 = ?$ **c.** $101 \times 10 = ?$
   b. $1010 - 1011 = ?$ **d.** $10111 \div 10 = ?$

CHAPTER 12 / BINARY ARITHMETIC

12-2 REPRESENTING POSITIVE AND NEGATIVE NUMBERS

Many digital systems, such as calculators and computers, are used to perform mathematical functions on a wide range of numbers. Some of these numbers are positive and negative signed numbers; therefore, a binary code is needed to represent these numbers.

12-2-1 *The Sign and Magnitude Number System*

The **sign and magnitude number system** is a binary code system used to represent positive and negative numbers. A sign–magnitude number contains a *sign bit* (0 for positive, 1 for negative) followed by the *magnitude bits*.

As an example, Figure 12-5(a) shows some 4-bit sign–magnitude positive numbers. All these sign–magnitude binary numbers—0001, 0010, 0011, and 0100—have an MSB, or sign bit, of 0, and therefore they are all positive numbers. The remaining 3 bits in these 4-bit words indicate the magnitude of the number based on the standard 421 binary column weight ($001_2 = 1_{10}$, $010_2 = 2_{10}$, $011_2 = 3_{10}$, $100_2 = 4_{10}$).

As another example, Figure 12-5(b) shows some 4-bit sign–magnitude negative numbers. All of these sign–magnitude binary numbers—1100, 1101, 1110, and 1111—have an MSB, or sign bit, of 1, and therefore they are all negative numbers. The remaining 3 bits in these 4-bit words indicate the magnitude of the number based on the standard 421 binary column weight ($100_2 = 4_{10}$, $101_2 = 5_{10}$, $110_2 = 6_{10}$, $111_2 = 7_{10}$).

If you need to represent larger decimal numbers, you simply use more bits, as shown in Figure 12-5(c) and (d). The principle still remains the same in that the MSB indicates the sign of the number, and the remaining bits indicate the magnitude of the number.

The sign–magnitude number system is an ideal example of how binary numbers could be coded to represent positive and negative numbers. This code system, however, requires complex digital hardware for addition and subtraction, and is therefore seldom used.

Sign Bit	Magnitude Bits
0 = + 1 = −	= Binary Equivalent

4-Bit Sign-Magnitude Numbers

0	0	0	1_2	=	$+1_{10}$
0	0	1	0_2	=	$+2_{10}$
0	0	1	1_2	=	$+3_{10}$
0	1	0	0_2	=	$+4_{10}$

Sign Bit   Magnitude Bits
(a)

1	1	0	0_2	=	-4_{10}
1	1	0	1_2	=	-5_{10}
1	1	1	0_2	=	-6_{10}
1	1	1	1_2	=	-7_{10}

Sign Bit   Magnitude Bits
(b)

8-Bit Sign-Magnitude Numbers

0	0	0	0	1	1	1	1_2	=	$+15_{10}$
0	1	0	1	0	0	0	0_2	=	$+80_{10}$
1	0	0	0	1	0	1	0_2	=	-10_{10}
1	1	1	1	1	1	1	1_2	=	-127_{10}

(c)

16-Bit Sign-Magnitude Numbers

0	0	0	0	0	0	0	0	0	0	0	0	0	0	1_2		=	$+1_{10}$
0	0	0	0	0	0	0	0	1	0	1	0	0	1	0	0_2	=	$+164_{10}$
1	0	0	0	0	0	0	0	0	0	1	1	0	0	1	0_2	=	-50_{10}
1	0	0	0	1	1	1	1	0	0	0	0	0	0	0	0_2	=	-3840_{10}

(d)

FIGURE 12-5   **The Sign–Magnitude Number System.**

What are the decimal equivalents of the following sign–magnitude number system codes?

a. 0111   b. 1010   c. 0001 1111   d. 1000 0001

■ *Solution:*

a. +7   b. −2   c. +31   d. −1

What are the 8-bit sign–magnitude number system codes for the following decimal values?

a. +8   b. −65   c. +127   d. −127

■ *Solution:*

a. 0000 1000   b. 1100 0001   c. 0111 1111   d. 1111 1111

12-2-2   *The One's Complement (1's Complement) Number System*

One's Complement Number System

A binary code used to represent positive and negative numbers, in which negative values are determined by inverting all the binary digits of an equivalent positive value.

The **one's complement number system** became popular in early digital systems. It also used an MSB sign bit that was either a 0 (indicating a positive number) or a 1 (indicating a negative number).

Figure 12-6(a) shows some examples of 8-bit one's complement positive numbers. As you can see, one's complement positive numbers are no different from sign–magnitude positive numbers. The only difference between the one's complement number system and the sign–magnitude number system is the way in which negative numbers are represented. To understand the code used to represent negative numbers, we must first understand the meaning of the term *one's complement*. To one's complement a binary number means to invert or change all the 1s in the binary word to 0s, and all the 0s in the binary word to 1s. Performing the one's complement operation on a positive number will change the number from a one's complement positive number to a one's complement negative number. For example,

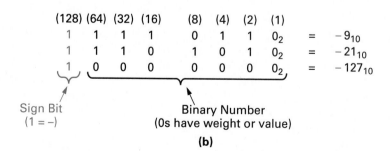

FIGURE 12-6   **The One's Complement Number System.**

by one's complementing the number 00000101, or +5, we will get 11111010, which is −5. In Figure 12-6(b), you can see the negative one's complement codes of the positive one's complement codes shown in Figure 12-6(a).

The next question you may have is, How can we determine the decimal value of a negative one's complement code? The answer is to do the complete opposite, or complement, of the positive one's complement code. This means count the 1s in a positive one's complement code, and count the 0s in a negative one's complement code. For example, with a positive one's complement code (MSB = 0), the value is determined by simply adding together all the column weights that contain a 1 (00000110 has a 1 in the fours column and twos column, so this number is 4 + 2, or +6). On the other hand, for a negative one's complement code (MSB = 1), the value is determined by doing the opposite and adding together all the column weights that contain a 0 (11110101 has a 0 in the eights column and twos column, so this number is 8 + 2, or −10).

■ **INTEGRATED MATH APPLICATION:**

What are the decimal equivalents of the following one's complement number system codes?

 a. 0010 0001   b. 1101 1101

■ *Solution:*

 a. MSB = 0 (number is positive, so add together all the column weights that contain a 1). The number 0010 0001 has a 1 in the thirty-twos column and a one in the units column, so 0010 0001 = +33.

 b. MSB = 1 (number is negative, so add together all the column weights that contain a 0). The number 1101 1101 has a 0 in the thirty-twos column and a 0 in the twos column, so 1101 1101 = −34.

Since the term *one's complement* can be used to describe both an operation (change all of the 1s to 0s, and all the 0s to 1s), and a positive and negative number system, some confusion can occur. To avoid this, remember that if you are asked to one's complement a number, it means to change all the 1s in the binary word to 0s and all 0s to 1s. If, on the other hand, you are asked to find the decimal equivalent of a one's complement number, you should decode the sign bit and binary value to determine the decimal equivalent value.

■ **INTEGRATED MATH APPLICATION:**

One's complement the following numbers.

 a. 0101 1111   b. 1110 0101

■ *Solution:*

To one's complement a number means to change all the 1s to 0s and all the 0s to 1s.

 a. 1010 0000   b. 0001 1010

■ **INTEGRATED MATH APPLICATION:**

Find the decimal equivalent of the following one's complement numbers.

 a. 0101 1111   b. 1110 0101

■ *Solution:*

 a. MSB = 0 (number is positive, so add together all the column weights that contain a 1). The number 0101 1111 = +95.

 b. MSB = 1 (number is negative, so add together all the column weights that contain a 0). The number 1110 0101 = −26.

The **two's complement number system** is used almost exclusively in digital systems to represent positive and negative numbers.

Positive two's complement numbers are no different from positive sign–magnitude and one's complement numbers in that an MSB sign bit of 0 indicates a positive number, and the remaining bits of the word indicate the value of the number. Figure 12-7(a) shows two examples of positive two's complement numbers.

Negative two's complement numbers are determined by two's complementing a positive two's complement number code. To two's complement a number first one's complement it (change all the 1s to 0s and 0s to 1s), and then add 1. For example, if we were to two's complement the number 0000 0101 (+5), we would obtain the two's complement number code for −5. The procedure to follow to two's complement 0000 0101 (+5) would be as follows:

0000 0101	Original number (+5)
1111 1010	One's complement (inverted)
+ 1	Plus 1
1111 1011	Two's complement code for −5

The new two's complement code obtained after this two's complement operation is the negative equivalent of the original number. In our example, therefore:

Original number:	0000 0101 = +5
After two's complement operation:	1111 1011 = −5

Any negative equivalent two's complement code therefore can be obtained by simply two's complementing the same positive two's complement number. For example, to find the negative two's complement code for −125, we would perform the following operation.

Two's	0111 1101	Original number (+125)
complement ⟶	1000 0010	One's complement (inverted)
	+ 1	Plus 1
operation	1000 0011	Two's complement code for −125

(a)

(b)

FIGURE 12-7   **The Two's Complement Number System.**

■ INTEGRATED MATH APPLICATION:

Find the negative two's complement codes for

    a. -26 b. -67

■ *Solution:*

a. 0001 1010   Two's complement code for $+26$
      1110 0101   One's complement (inverted)
      $+$ 1   Plus 1
      1110 0110   Two's complement code for -26

b. 0100 0011   Two's complement code for $+67$
      1011 1100   One's complement (inverted)
      $+$ 1   Plus 1
      1011 1101   Two's complement code for -67

Performing the two's complement operation on a number twice will simply take the number back to its original value and polarity. For example, let's take the two's complement code for $+6$ (0000 0110) and two's complement it twice.

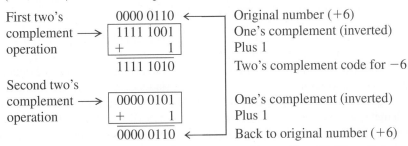

The decimal equivalent of a positive two's complement number (MSB = 0) is easy to determine, since the binary value following the sign bit applies the same as any binary number, as shown in Figure 12-7(a). On the other hand, the decimal equivalent of a negative two's complement number (MSB = 1) is not determined in the same way. To determine the value of a negative two's complement number, refer to the remaining bits following the sign bit, add the weight of all columns containing a 0, and then add 1 to the result. For example, the negative two's complement number in Figure 12-7(b) has 0s in the eights and twos columns, and therefore its value will equal the sum of 8 and 2, plus the additional 1 (due to two's complement). As a result:

$$\overset{(8)\,(4)\,(2)\,(1)}{1\ 1\ 1\ 1\ 0\ 1\ 0\ 1} = (8+2) + 1 = -11$$

■ INTEGRATED MATH APPLICATION:

Determine the decimal equivalent of the following negative two's complement numbers.

    a. 1110 0001   c. 1000 0000
    b. 1011 1011   d. 1111 1111

■ *Solution:*

a. $\overset{(\pm)\,(64)\,(32)\,(16)\,(8)\,(4)\,(2)\,(1)}{1\ \ \ 1\ \ \ 1\ \ \ 0\ \ \ 0\ \ 0\ \ 0\ \ 1_2} = (16 + 8 + 4 + 2) + 1 = -31_{10}$

b. $\overset{(\pm)\,(64)\,(32)\,(16)\,(8)\,(4)\,(2)\,(1)}{1\ \ \ 0\ \ \ 1\ \ \ 1\ \ \ 1\ \ 0\ \ 1\ \ 1_2} = (64 + 4) + 1 = -69_{10}$

c. $\overset{(\pm)\,(64)\,(32)\,(16)\,(8)\,(4)\,(2)\,(1)}{1\ \ \ 0\ \ \ 0\ \ \ 0\ \ \ 0\ \ 0\ \ 0\ \ 0_2} = (64 + 32 + 16 + 8 + 4 + 2 + 1) + 1 = -128_{10}$

d. $\overset{(\pm)\,(64)\,(32)\,(16)\,(8)\,(4)\,(2)\,(1)}{1\ \ \ 1\ \ \ 1\ \ \ 1\ \ \ 1\ \ 1\ \ 1\ \ 1_2} = (0) + 1 = -1_{10}$

4-Bit Word	Standard Binary Value	2's Complement Value
0 0 0 0	0	0
0 0 0 1	1	+1
0 0 1 0	2	+2
0 0 1 1	3	+3 8 Positive
0 1 0 0	4	+4 Numbers
0 1 0 1	5	+5
0 1 1 0	6	+6
0 1 1 1	7	+7
1 0 0 0	8	−8
1 0 0 1	9	−7
1 0 1 0	10	−6
1 0 1 1	11	−5 8 Negative
1 1 0 0	12	−4 Numbers
1 1 0 1	13	−3
1 1 1 0	14	−2
1 1 1 1	15	−1

$2^4 =$ 16 Words

FIGURE 12-8   **Four-Bit Two's Complement Codes.**

As with the sign–magnitude and one's complement number systems, if you need to represent larger decimal numbers with a two's complement code, you will have to use a word containing more bits. For example, a total of 16 different combinations is available for a 4-bit word ($2^4 = 16$), as shown in Figure 12-8. With standard binary, these 16 different combinations give a count from 0 to 15, as shown in the center column in Figure 12-8. As for two's complement, we split these 16 different codes in half, and end up with 8 positive two's complement codes and 8 negative two's complement codes, as shown in the right column in Figure 12-8. Since decimal zero uses one of the positive two's complement codes, the 16 different combinations give a range of -8_{10} to $+7_{10}$ (the maximum positive value, +7, is always one less than the maximum negative value, −8, since decimal zero uses one of the positive codes).

An 8-bit word will give 256 different codes ($2^8 = 256$), as shown in Figure 12-9. Dividing this total in half will give 128 positive two's complement codes and 128 negative two's complement codes. Since decimal zero uses one of the positive two's complement codes, the 256 different combinations give a range of -128_{10} to $+127_{10}$.

■   INTEGRATED MATH APPLICATION:

What would be the two's complement range for a 16-bit word?

■   *Solution:*

$2^{16} = 65,536$; half of 65,536 = 32,768; therefore, the range will be $-32,768$ to $+32,767$.

SELF-TEST EVALUATION POINT FOR SECTION 12-2

Now that you have completed this section, you should be able to:

■   *Objective 2.   Explain how the sign–magnitude, one's complement, and two's complement number systems are used to represent positive and negative numbers.*

Use the following questions to test your understanding of Section 12-2:

1. Two's complement the following binary numbers.
   a. 1011   **b.** 0110 1011   **c.** 0011   **d.** 1000 0111
2. What is the decimal equivalent of the following two's complement numbers?
   a. 1011   **b.** 0110   **c.** 1111 1011   **d.** 0000 0110

8-Bit Word								Standard Binary Value	2's Complement Value
0	0	0	0	0	0	0	0	0	0
0	0	0	0	0	0	0	1	1	+1
0	0	0	0	0	0	1	0	2	+2
0	0	0	0	0	0	1	1	3	+3
0	0	0	0	0	1	0	0	4	+4
0	0	0	0	0	1	0	1	5	+5
0	0	0	0	0	1	1	0	6	+6
0	0	0	0	0	1	1	1	7	+7
								.	.
0	1	1	1	1	0	1	1	123	+123
0	1	1	1	1	1	0	0	124	+124
0	1	1	1	1	1	0	1	125	+125
0	1	1	1	1	1	1	0	126	+126
0	1	1	1	1	1	1	1	127	+127
1	0	0	0	0	0	0	0	128	−128
1	0	0	0	0	0	0	1	129	−127
1	0	0	0	0	0	1	0	130	−126
1	0	0	0	0	0	1	1	131	−125
1	0	0	0	0	1	0	0	132	−124
1	0	0	0	0	1	0	1	133	−123
								.	.
1	1	1	1	1	0	1	1	251	−5
1	1	1	1	1	1	0	0	252	−4
1	1	1	1	1	1	0	1	253	−3
1	1	1	1	1	1	1	0	254	−2
1	1	1	1	1	1	1	1	255	−1

$2^8 =$ 256 Words

128 Positive Numbers

128 Negative Numbers

FIGURE 12-9 Eight-Bit Two's Complement Codes.

12-3 TWO'S COMPLEMENT ARITHMETIC

To perform its function, a computer's MPU contains several internal circuits, and one of these circuits is called the **arithmetic–logic unit, or ALU.** This unit is the number-crunching circuit of every digital system, and as its name implies, it can be controlled to perform either arithmetic operations (such as addition or subtraction) or logic operations (such as AND and OR). The arithmetic–logic unit has two parallel inputs, one parallel output, and a set of function select control lines, as shown in Figure 12-10.

The ALU has to be able to perform the four basic arithmetic operations, namely, addition, subtraction, multiplication, and division. One would expect, therefore, that an ALU would contain a separate circuit for each of these operations. This is not so, since all four operations can be performed using a binary-adder circuit. To explain this point, because the binary-adder circuit's function is to perform addition, and since multiplication is simply repeated addition, the binary-adder circuit can also be used to perform any multiplication operations. For example, 3×4 means that the number 4 is added three times ($4 + 4 + 4 = 12$, therefore $3 \times 4 = 12$). But what about subtraction and division? The answer to this problem is the two's complement number system. As you will see in this section, by representing all values as two's complement numbers, we can perform a subtraction operation by

Arithmetic–Logic Unit (ALU)

The section of a computer that performs all arithmetic and logic operations.

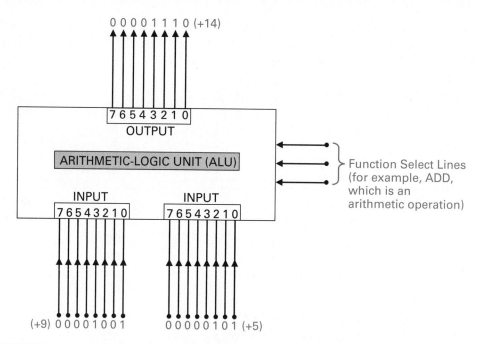

FIGURE 12-10 Block Diagram of the Arithmetic-Logic Unit (ALU).

slightly modifying the second number and then adding the two numbers together. This means that the binary-adder circuit can also be used to perform subtraction operations, and since division is simply repeated subtraction, the subtraction operation of the binary-adder circuit can also be used to perform any division operation. For example, $12 \div 4$ means that you take the number 12 and see how many times you can subtract 4 from it ($12 - 4 - 4 - 4 = 0$; therefore, there are three 4s in 12, $12 \div 4 = 3$).

By using two's complement numbers, therefore, we can use a binary-adder circuit within an ALU to add, subtract, multiply, and divide. Having one circuit able to perform all four mathematical operations means that the circuit will be small in size and fast to operate, will consume very little power, and will be cheap to manufacture. In this section we will see how we can perform all four mathematical operations by always adding two's complement numbers.

12-3-1 *Adding Positive Numbers*

To begin, let us prove that the addition of two positive two's-complement numbers will render the correct result.

■ **INTEGRATED MATH APPLICATION:**

Add +9 (0000 1001) and +5 (0000 0101).

■ *Solution:*

$$
\begin{array}{r}
0000\ 1001 \\
+\ 0000\ 0101 \\
\hline
\text{MSB of 0 = positive number} \longrightarrow \quad 0000\ 1110
\end{array}
$$

When both numbers to be added are positive two's complement numbers (both have an MSB of 0), it makes sense that the sum will have an MSB of 0, indicating a positive result. This is always true, unless the range of the two's complement word is exceeded. For example, the maximum positive number that can be represented by an 8-bit word is $+127$ (0111 1111). If the sum of the two positive two's complement numbers to be added exceeds the upper positive range limit, there will be a *two's complement overflow* into the sign bit. For instance:

$$
\begin{array}{r}
\text{(\pm) (64) (32) (16) (8) (4) (2) (1)} \\
0 \quad 1 \quad 1 \quad 0 \quad 0 \quad 1 \quad 0 \quad 0 \qquad (+100) \\
+ \quad 0 \quad 0 \quad 0 \quad 1 \quad 1 \quad 1 \quad 1 \quad 0 \qquad + \quad (+30) \\
\hline
1 \quad 0 \quad 0 \quad 0 \quad 0 \quad 0 \quad 1 \quad 0 \qquad (-126)
\end{array}
$$

MSB of 1 = negative number $\longrightarrow$

If two positive numbers are applied to an arithmetic circuit and it is told to add the two, the result should also be a positive number. Most arithmetic circuits are able to detect this problem if it occurs by simply monitoring the MSB of the two input words and the MSB of the output word, as shown in Figure 12-11. If the two input sign bits are the same, but the output sign bit is different, then a two's complement overflow has occurred.

12-3-2   *Adding Positive and Negative Numbers*

By using the two's complement number system, we can also add numbers with unlike signs and obtain the correct result. As an example, let us add $+9$ (0000 1001) and -5 (1111 1011).

$$
\begin{array}{r}
0000 \ 1001 \qquad (+9) \\
+ \ 1111 \ 1011 \qquad + \ (-5) \\
\hline
1 \quad 0000 \ 0100 \qquad (+4)
\end{array}
$$

Ignore the final carry $\longrightarrow$

A final carry will always be generated whenever the sum is a positive number; however, if this carry is ignored because it is beyond the 8-bit two's complement word, the answer will be correct.

FIGURE 12-11   Two's Complement Overflow Detection.

Add $+10$ (0000 1010) and -12 (1111 0100).

■ *Solution:*

$$
\begin{array}{ll}
0000\ 1010 & (+10) \\
\underline{+\ 1111\ 0100} & \underline{+\ (-12)} \\
1111\ 1110 & (-2)
\end{array}
$$

12-3-3 *Adding Negative Numbers*

As a final addition test for the two's complement number system, let us add two negative numbers to see if we can obtain the correct negative sum.

■ **INTEGRATED MATH APPLICATION:**

Add -5 (1111 1011) and -4 (1111 1100).

■ *Solution:*

$$
\begin{array}{lll}
& 1111\ 1011 & (-5) \\
& \underline{+\ 1111\ 1100} & \underline{+\ (-4)} \\
\text{Ignore the final carry} \longrightarrow 1 & 1111\ 0111 & (-9)
\end{array}
$$

The sum in this example is the correct two's complement code for -9.

Two's complement overflow can also occur in this condition if the sum of the two negative numbers exceeds the negative range of the word. For example, with an 8-bit two's complement word, the largest negative number is -128, and if we add -100 (1001 1100) and -30 (1110 0010), we will get an error, as follows:

$$
\begin{array}{lll}
& 1001\ 1100 & (-100) \\
& \underline{+\ 1110\ 0010} & \underline{+\ (-30)} \\
\text{Ignore the final carry} \longrightarrow 1 & 0111\ 1110 & (+126)
\end{array}
$$

Once again, the two's complement overflow detect circuit, shown in Figure 12-11, will indicate such an error, since two negative numbers were applied to the input and a positive number appeared at the output.

In all the addition examples mentioned so far (adding positive numbers, adding positive and negative numbers, and adding negative numbers) you can see that an ALU added the bits of the input bytes without any regard for whether they were signed or unsigned numbers. This brings up a very important point, which is often misunderstood. The ALU adds the bits of the input words using the four basic rules of binary addition, and it does not know if these input and output bit patterns are coded two's complement words or standard unsigned binary words. It is therefore up to us to maintain a consistency, which means that if we are dealing with coded two's complement words at the input of an ALU, we must interpret the output of an ALU as a two-coded two's complement word. On the other hand, if we are dealing with standard unsigned binary words at the input of an ALU, we must interpret the output of an ALU as a standard unsigned binary word.

12-3-4 *Subtraction through Addition*

The key advantage of the two's complement number system is that it enables us to perform a subtraction operation using the adder circuit within the ALU. To explain how this is done, let us take an example and assume that we want to subtract 3 (0000 0011) from 8 (0000 1000), as follows:

Minuend:	0000 1000	(+8)
Subtrahend:	+ 0000 0011	− (+3)
Difference:	0000 0101	+5

If we subtract the subtrahend of +3 from the minuend of +8, we will obtain the correct result of +5 (0000 0101), however, we must find the difference to this problem by using addition and not subtraction. The solution to this is achieved by *two's complementing the subtrahend, and then adding the result to the minuend.* The following display shows how this is done.

As you can see from this operation, adding the −3 to +8 achieves exactly the same result as subtracting +3 from +8. We can therefore perform a subtraction by two's complementing the subtrahend of a subtraction problem, and then adding the result to the minuend.

Digital electronic calculators and computer systems perform the previously described steps automatically. That is, they first instruct the ALU to one's complement the subtrahend input (function select control inputs = invert, which is a logic operation), then add 1 to the result (function select control inputs = +1, which is an arithmetic operation), and finally add the result of the two previous operations (which is the two's complemented subtrahend) to the minuend (function select = ADD, arithmetic operation). Although this step-by-step procedure or **algorithm** seems complicated, it is a lot simpler than having to include a separate digital electronic circuit to perform all subtraction operations. Having one circuit able to perform both addition and subtraction means that the circuit will be small in size and fast to operate, will consume very little power, and will be cheap to manufacture.

Flowcharts will be discussed in more detail in the next chapter, but for now Figure 12-12 shows a **flowchart** summarizing the subtraction through addition procedure. Flowcharts provide a graphic way of describing a step-by-step procedure (algorithm) or a program of instructions.

Algorithm
A set of rules for the solution of a problem.

Flowchart
A graphic representation of a step-by-step procedure or program of instructions.

12-3-5 *Multiplication through Repeated Addition*

Most digital electronic calculators and computers do not include a logic circuit that can multiply, since the adder circuit can also be used to perform this arithmetic operation. For example, to multiply 6 × 3, you simply add 6 three times.

Multiplicand:	6		6
Multiplier:	× 3	=	+ 6
Product:	18		+ 6
			18

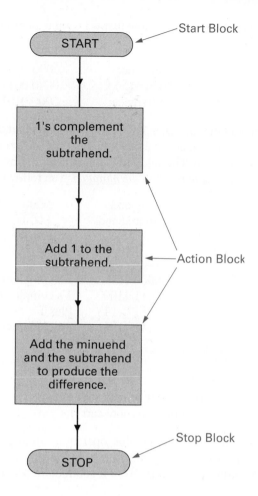

FIGURE 12-12　Flowchart for Subtraction through Addition.

Start Block

START

1's complement the subtrahend.

Add 1 to the subtrahend.

Action Block

Add the minuend and the subtrahend to produce the difference.

Stop Block

STOP

As with subtraction through addition, an algorithm must be followed when a digital system needs to perform a multiplication operation. Figure 12-13 shows the flowchart for multiplication through repeated addition. Let us apply this procedure to our previous example of 6 × 3 and see how the multiplicand 6 is added three times to obtain a final product of 18. The steps are as follows:

Start: multiplicand = 6, multiplier = 3, product = 0
After first pass: multiplicand = 6, multiplier = 2, product = 6
After second pass: multiplicand = 6, multiplier = 1, product = 12
After third pass: multiplicand = 6, multiplier = 0, product = 18
Stop.

In this example we cycled through the flowchart three times (until the multiplier was 0) and then finally obtained a product of 18.

Although this algorithm seems complicated, it is a lot simpler than having to include a separate digital circuit for multiplication. Having one circuit able to perform addition, subtraction, and multiplication means that the circuit will be small in size and fast to operate, will consume very little power, and will be cheap to manufacture.

12-3-6　Division through Repeated Subtraction

Up to this point we have seen how we can use the two's complement number system and the adder circuit within an ALU for addition, subtraction, and multiplication. Just as multiplication can be achieved through repeated addition, division can be achieved through repeated

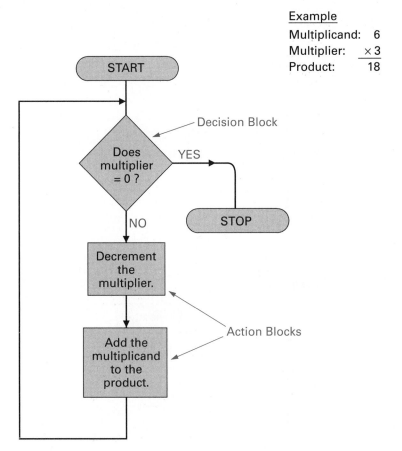

Example

Multiplicand: 6

Multiplier: × 3

Product: 18

FIGURE 12-13   Flowchart for Multiplication through Repeated Addition.

subtraction. For example, to divide 31 by 9 we simply see how many times 9 can be subtracted from 31, as follows:

$$
\begin{array}{r}
3 \\
\text{Divisor: } 9\overline{)31} \\
27 \\
4
\end{array}
\quad
\begin{array}{l}
\text{Quotient} \\
\text{Dividend} \\
\\
\text{Remainder}
\end{array}
\quad = \quad
\begin{array}{r}
31 \\
-9 \\
\hline
22 \\
-9 \\
\hline
13 \\
-9 \\
\hline
4
\end{array}
\quad
\begin{array}{l}
\text{First subtraction} \\
\\
\text{Second subtraction} \\
\\
\text{Third subtraction}
\end{array}
$$

Since we were able to subtract 9 from 31 three times with a remainder of 4, then $31 \div 9 = 3$, remainder 4.

As with multiplication, an algorithm must be followed when a digital system needs to perform a division. Figure 12-14 shows the flowchart for division through repeated subtraction. Let us apply this procedure to our previous example of $31 \div 3$ and see how the quotient is incremented to 3, leaving a remainder of 4. The steps would be as follows:

Start: dividend = 31, divisor = 9, quotient = 0, remainder = 31

After first pass: dividend = 22, divisor = 9, quotient = 1, remainder = 22

After second pass: dividend = 13, divisor = 9, quotient = 2, remainder = 13

After third pass: dividend = 4, divisor = 9, quotient = 3, remainder = 4

Stop.

SECTION 12-3 / TWO'S COMPLEMENT ARITHMETIC   **299**

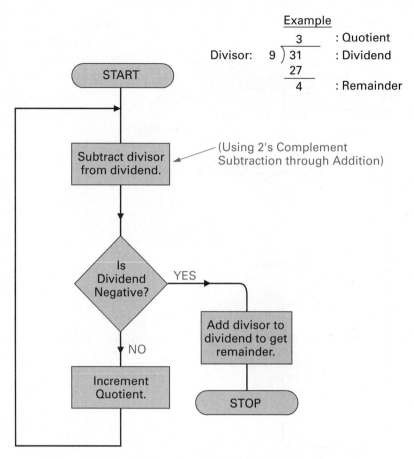

FIGURE 12-14 Flowchart for Division through Repeated Subtraction.

When this program of instructions cycles or loops back for the third pass, the divisor of 9 is subtracted from the remaining dividend of 4, resulting in −5. When the question, Is the dividend negative? is asked, the answer is yes, so the branch on the right is followed. This branch of the flowchart will add the divisor of 9 to the dividend of −5 to restore it to the correct remainder of 4 before the program stops. To summarize, these are the steps followed by this program:

$$
\begin{array}{rl}
(+31) & \\
-(+\ 9) & \\
\hline
(+22) & \text{Quotient} = 1 \\
-(+\ 9) & \\
\hline
(+13) & \text{Quotient} = 2 \\
-(+\ 9) & \\
\hline
(+\ 4) & \text{Quotient} = 3 \\
-(+\ 9) & \\
\hline
(-\ 5) & \\
-(+\ 9) & \text{Add divisor to dividend} \\
\hline
(+\ 4) & \text{Remainder} = 4
\end{array}
$$

When a digital system is instructed to perform a division, it will use the ALU for all arithmetic and logic operations. Since division is achieved through repeated subtraction, and subtraction is achieved through addition, the ALU's adder circuit can also be used for division.

Although this algorithm seems complicated, it is a lot simpler than having to include a separate digital circuit for division. As previously mentioned, having one circuit able to perform addition, subtraction, multiplication, and division means that the circuit will be small in size and fast to operate, will consume very little power, and will be cheap to manufacture.

Now that you have completed this section, you should be able to:

■ **Objective 3.** *Describe how an adder circuit can add, subtract, multiply, and divide two's complement numbers.*

Use the following questions to test your understanding of Section 12-3.

1. The two's complement number system makes it possible for us to perform addition, multiplication and division all with a/an _____ circuit.

12-4 REPRESENTING LARGE AND SMALL NUMBERS

Although two's complementing enables us to represent positive and negative numbers, the range is limited to the number of bits in the word. For example, the range of an 8-bit two's complement word is from -128 to $+127$, and the range of a 16-bit two's complement word is from $-32,768$ to $+32,767$. To extend this range, we can keep increasing the number of bits; however, these large words become difficult to work with. Another problem arises when we wish to represent fractional numbers such as 1.57.

These problems are overcome by using the **floating-point number system,** which uses scientific notation to represent a wider range of values. Figure 12-15 explains the 2-byte (16-bit) floating-point number system. The first byte is called the **mantissa,** and it is always an unsigned binary number that is interpreted as a value between 0 and 1. The second byte is called the **exponent,** and it is always a two's complement number and therefore represents a range from 10^{-128} to 10^{+127}. As an example, let us see how the floating point number system can be used to represent a large-value number such as 27,000,000,000,000 and a small-value number such as 0.0000000015.

> **Floating-Point Number System**
> A number system in which a value is represented by a mantissa and an exponent.
>
> **Mantissa**
> The fractional part of a logarithm.
>
> **Exponent**
> A small number placed to the right and above a value to indicate the number of times that symbol is a factor.

■ **INTEGRATED MATH APPLICATION:**

Give the 2-byte floating-point binary values for 27,000,000,000,000.

■ *Solution:*

The value 27,000,000,000,000 is written in scientific notation as 0.27×10^{14} and, when coded as a floating-point number, appears as follows:

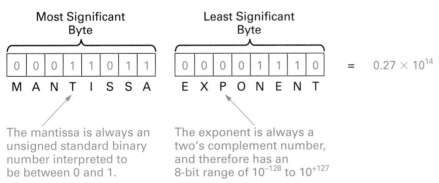

FIGURE 12-15 **The Floating-Point Number System.**

In this example, the mantissa byte (standard binary number) is equal to 27 (0001 1011) and is therefore interpreted as 0.27, and the exponent byte (two's complement number) is equal to +14 (0000 1110) and is therefore interpreted as 10^{14}.

■ INTEGRATED MATH APPLICATION:

Give the 2-byte floating-point binary values for 0.0000000015.

■ *Solution:*

The value 0.0000000015 is written in scientific notation as $0.15 \times 3\ 10^{-28}$ and, when coded as a floating-point number, appears as follows:

$$0.0000000015$$
$$= \quad 0.15 \times 10^{-8}$$

$$= \quad \boxed{0000\ 1111} \qquad \boxed{1111\ 1000}$$

Mantissa (unsigned binary) = 15, which is interpreted as 0.15.

Exponent (two's complement = −8, which is interpreted as 10^{-8}.

In this example, the mantissa byte (standard binary number) is equal to 15 (0000 1111) and is therefore interpreted as 0.15, and the exponent byte (two's complement number) is equal to −8 (1111 1000) and is therefore interpreted as 10^{-8}.

As with all binary number coding methods, we must know the number representation system being used if we are to accurately decipher the bit pattern. For instance, the 16-bit word in Figure 12-15 would be very different if it were interpreted as an unsigned binary number, a two's complement number, and as a floating-point number.

SELF-TEST EVALUATION POINT FOR SECTION 12-4

Now that you have completed this section, you should be able to:

■ *Objective 4.* *Describe how the floating-point number system can be used to represent a large range of values.*

Use the following questions to test your understanding of Section 12-4:

1. Give the 2-byte floating-point binary values for each.
   a. 65,000,000
   b. 0.18×10^{-12}

REVIEW QUESTIONS

Multiple Choice Questions

1. The sum of 101_2 and 011_2 is
   a. 0111_2.
   b. 8_{10}.
   c. 1001_2.
   d. 6_{10}.

2. Which digital logic unit is used to perform arithmetic operations and logic functions?
   a. MPU
   b. ALU
   c. CPU
   d. ICU

3. With binary addition, $1 + 1 = ?$
   a. 10_2
   b. 0 carry 1
   c. 2_{10}
   d. All the above

4. With binary subtraction, $0 - 1 = ?$
   a. 10_2
   b. 1 borrow 0
   c. 1 borrow 1
   d. All the above

5. What would be the result of $1011101_2 \times 01_2$?
   a. 1011101_2
   b. 1110101_2
   c. 000111_2
   d. 1011100_2

6. The four rules of binary multiplication are $0 \times 0 =$ _____, $1 \times 0 =$ _____, $0 \times 1 =$ _____, and $1 \times 1 =$ _____.
   a. 0, 1, 0, 1
   b. 0, 0, 1, 1
   c. 0, 0, 0, 0
   d. 0, 0, 0, 1

7. What would be the result of $1001_2 \div 0011_2$?

   a. 3_{10} **c.** 0011_2
   b. 3, remainder zero   **d.** All the above

8. The one's complement of $0101\ 0001_2$ would be _____.

   a. $0101\ 0010_2$ **c.** $1010\ 1111_2$
   b. $1010\ 1110_2$ **d.** All the above

9. What would be the decimal equivalent of 1001_2 if it were a two's complement number?

   a. $+9$ **c.** $+7$
   b. -9 **d.** -7

Communication Skill Questions

11. List the rules for each of the following: (12-1)

   a. Binary addition   **c.** Binary multiplication
   b. Binary subtraction   **d.** Binary division

12. Using an example, briefly describe how to perform the following operations. (12-1)

   a. Binary addition   **c.** Binary multiplication
   b. Binary subtraction   **d.** Binary division

13. Briefly describe how the following number systems are used to represent positive and negative binary numbers. (12-2)

   a. Sign–magnitude   **c.** Two's complement
   b. One's complement

14. Why is the two's complement number system most frequently used to represent positive and negative numbers? (12-2-3)

Practice Problems

19. Add the following 8-bit numbers.

   a. 0110 1000 and 0001 1111
   b. 0111 1111 and 0100 0011
   c. 0001 1010 and 0011 1010

20. Subtract the following 8-bit numbers.

   a. 0001 1010 from 1010 1010
   b. 0001 1111 from 0011 0000
   c. 0010 0000 from 1000 0000

21. Multiply the following binary numbers.

   a. 0101 1011 by 011   **c.** 1110 by 010
   b. 0001 1111 by 1011

22. Divide the following binary numbers.

   a. 0110 into 0111 0110   **c.** 1001 into 1011 1110
   b. 101 into 1010 1010

23. Convert each of the following decimal numbers to an 8-bit sign–magnitude number.

   a. $+34$ **c.** $+83$
   b. -83

10. Which number system makes it possible for us to add, subtract, multiply, and divide binary numbers all with an adder circuit?

   a. Sign–magnitude   **c.** Two's complement
   b. One's complement   **d.** Floating-point

15. How would you determine the maximum positive and maximum negative value of a 10-bit two's complement number? (12-2-3)

16. Describe how to two's complement a binary number. (12-2-3)

17. Using the two's complement number system, briefly describe how to perform each of the following operations. (12-3)

   a. Add two positive numbers
   b. Add two negative numbers
   c. Add a positive and negative number
   d. Subtract through addition
   e. Multiply through repeated addition
   f. Divide through repeated subtraction

18. What is the floating-point number system? (12-4)

24. One's complement the following numbers.

   a. 0110 1001   **c.** 1111 0000
   b. 1010 1010

25. Convert all the following two's complement numbers to their decimal equivalents.

   a. 0001 0000   **c.** 1111 0101
   b. 1011 1110

26. Convert each of the following decimal numbers to an 8-bit two's complement number.

   a. $+119$ **c.** -34
   b. -119

27. Two's complement the following numbers.

   a. 0101 1010   **c.** 1111 1101
   b. 1000 0001

28. What would be the two's complement range of a 6-bit binary number?

Web Site Questions

Go to the web site http://www.prenhall.com/cook, select the textbook *Introductory Computer Mathematics, 2e,* select this chapter, and then follow the instructions when answering the multiple-choice practice problems.

Introduction to Computer Programming

Making an Impact

John Von Neuman, a mathematics professor at the Institute of Advanced Studies, delighted in amazing his students by performing complex computations in his head faster than they could with pencil, paper, and reference books. He possessed a photographic memory, and at his frequently held lavish parties in his home in Princeton, New Jersey, he gladly occupied center stage to recall from memory entire pages of books read years previously, the lineage of European royal families, and a store of controversial limericks. His memory, however, failed him in his search for basic items in a house he had lived in for 17 years. On many occasions when traveling, he would become so completely absorbed in mathematics that he would have to call his office to find out where he was going and why.

Born in Hungary, he was quick to demonstrate his genius. At the age of six he would joke with his father in classical Greek. At the age of eight, he had mastered calculus, and in his mid-20s he was teaching and making distinct contributions to the science of quantum mechanics, which is the cornerstone of nuclear physics.

Next to fine clothes, expensive restaurants, and automobiles, which he had to replace annually due to smash-ups, was his love of his work. His interest in computers began when he became involved in the top-secret Manhattan Project at Los Alamos, New Mexico, where he proved mathematically the implosive method of detonating an atom bomb. Working with the then-available computers, he became aware that they could become much more than a high-speed calculator. He believed that they could be an all-purpose scientific research tool, and he published these ideas in a paper. This was the first document to outline the logical organization of the electronic digital computer and was widely circulated to all scientists throughout the world. In fact, even to this day, scientists still refer to computers as "Von Neuman machines."

Von Neuman collaborated on a number of computers of advanced design for military applications, such as the development of the hydrogen bomb and ballistic missiles.

In 1957, at the age of 54, he lay in a hospital dying of bone cancer. Under the stress of excruciating pain, his brilliant mind began to break down. Since Von Neuman had been privy to so much highly classified information, the Pentagon had him surrounded with only medical orderlies specially cleared for security for fear he might, in pain or sleep, give out military secrets.

13

Outline and Objectives

Note: Material in this chapter relating to the SAM microcomputer (5036A) courtesy of Hewlett-Packard Company.

Introduction

The microprocessor is a large, complex integrated circuit containing all the computation and control circuitry for a small computer. The introduction of the microprocessor caused a dramatic change in the design of digital electronic systems. Before microprocessors, random or hard-wired digital circuits were designed using individual logic blocks, such as gates, flip-flops, registers, and counters. These building blocks were interconnected to achieve the desired end, as required by the application. Using random logic, each application required a unique design, and there was little similarity between one system and another. This approach was very similar to analog circuit design, in that the structure of the circuit was governed by the function that needed to be performed. Once constructed, the function of the circuit was difficult to change.

The microprocessor, on the other hand, provides a general-purpose control system that can be adapted to a wide variety of applications with only slight circuit modification. The individuality of a microprocessor-based system is provided by a list of instructions (called the program) that controls the system's operation. A microprocessor-based system therefore has two main elements—the actual components, or hardware, and the programs, or software.

13-1 COMPUTER HARDWARE

The earliest electronic computers were constructed using thousands of vacuum tubes. These machines were extremely large and unreliable and were mostly a laboratory curiosity. Figure 13-1 compares the electronic numeric integrator and computer, or ENIAC, which was unveiled in 1946, with today's personal computer. The ENIAC weighed 38 tons, measured 18 feet wide and 88 feet long, and used 17,486 vacuum tubes. These vacuum tubes produced a great deal of heat and developed frequent faults requiring constant maintenance. Today's personal computer, on the other hand, makes use of semiconductor integrated circuits and is

FIGURE 13-1 Microcomputer Systems—Past and Present. (Photo (b) Courtesy of Apple Computer, Inc.)

far more powerful, versatile, portable, and reliable than the ENIAC. Another advantage is cost: In 1946 the ENIAC calculator cost $400,000 to produce, whereas a present-day high-end personal computer can be purchased for about $1,000.

Figure 13-2 shows the block diagram of a basic microcomputer system. The microprocessor (also called central processor unit, or CPU) is the "brains" of the system. It contains all the logic circuitry needed to recognize and execute the program of instructions stored in memory. The input port connects the processor to the keyboard so that data can be read from this input device. The output port connects the processor to the display so that we can write data to this output device. Combined in this way, the microprocessor unit, memory unit, and input/output unit form a **microcomputer.**

The blocks within the microcomputer are interconnected by three buses. The microprocessor uses the address bus to select which memory location, input port, or output port it wishes to put information into or take information out of. Once the microprocessor has selected the location using the address bus, data or information is transferred via the data bus. In most cases, data will travel either from the processor to memory, from memory to the processor, from the input port to the processor, or from the processor to the output port. The control bus is a group of control signal lines that are used by the processor to coordinate the transfer of data within the microcomputer.

A list of instructions is needed to direct a microcomputer system so that it will perform a desired task. For example, if we wanted the system in Figure 13-2 to display the number of any key pressed on the keyboard, the program would be as follows:

1. Read the data from the keyboard.
2. Write the data to the display.
3. Repeat (go to step 1).

For a microprocessor to perform a task from a list of instructions, the instructions must first be converted into codes that the microprocessor can understand. These instruction codes are stored, or programmed, into the system's memory. When the program is run, the microprocessor begins by reading the first coded instruction from memory, decoding its meaning, and then performing the indicated operation. The processor then reads the instruction from the next location in memory, decodes the meaning of this next instruction, and performs the indicated operation. This process is then repeated, one memory location after another, until all the instructions within the program have been fetched, decoded and then executed.

Microcomputer

Complete system, including CPU, memory, and I/O interfaces.

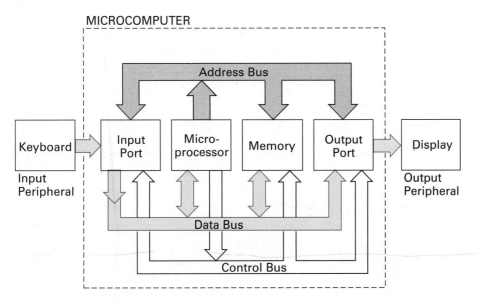

FIGURE 13-2   **Basic Microcomputer System.**

MICROCOMPUTER

Input → Analog-to-Digital Converter

Function Selector Switches

Input Ports

Micro-processor

Memory

Output Port

Display

Address Bus

Data Bus

Control Bus

Flow Chart

FIGURE 13-3   A Microprocessor-Based Digital Voltmeter. (Photo courtesy of Fluke Corporation. Reproduced with permission.)

Peripheral

Any interface device connected to a computer. Also, a mass storage or communications device connected to a computer.

The input/output devices connected to the input and output ports (the keyboard and display, for example) are called the **peripherals.** Peripherals are the system's interface with the user, or the system's interface with other equipment such as printers or data storage devices. As an example, Figure 13-3 shows how a microcomputer could be made to operate as a microprocessor-based digital voltmeter. Its input peripherals are an analog-to-digital converter (ADC), and a range and function switches. The output peripheral is a digital display.

As you come across different microcomputer applications, you will see that the basic microcomputer system is always the same. The only differences between one system and another are in the program and peripherals. Whether the system is a personal computer or a digital voltmeter, the only differences are in the program of instructions that controls the microcomputer and the peripherals connected to the microcomputer.

SELF-TEST EVALUATION POINT FOR SECTION 13-1

Now that you have completed this section, you should be able to:

■ **Objective 1.** Define the term microprocessor.

■ **Objective 2.** Describe the difference between a hard-wired digital system and a microprocessor-based digital system.

■ **Objective 3.** List and describe the function of the three basic blocks that make up a microcomputer.

■ **Objective 4.** Define the difference between computer hardware and software.

Use the following questions to test your understanding of Section 13-1:

1. What three blocks are interconnected to form a microcomputer, and what is the function of each block?

2. What three buses are needed to interconnect the three blocks within a microcomputer, and what is the function of each bus?

3. What is the difference between hardware and software?

13-2   COMPUTER SOFTWARE

Microcomputer programs are first written in a way that is convenient for the person writing the program, or *programmer.* Once written, the program must then be converted and stored as a code that can be understood by the microprocessor. When writing a computer program, the programmer must tell the computer what to do down to the most minute detail. Computers can act with tremendous precision and speed for long periods of time, but they must be told exactly what to do. A computer can respond to a change in conditions, but only if it contains a program that says "if this condition occurs, do this."

13-2-1 *The Microcomputer as a Logic Device*

To understand how a program controls a microcomputer, let us see how we could make a microcomputer function as a simple two-input AND gate. As can be seen in Figure 13-4, the two AND gate inputs will be applied to the input port, and the single AND gate output will appear at the output port. To make the microcomputer operate as an AND gate, a program of instructions will first have to be loaded into the system's memory. This list of instructions will be as follows:

1. Read the input port.
2. Go to step 5 if all inputs are HIGH; otherwise, continue.
3. Set output LOW.
4. Go to step 1.
5. Set output HIGH.
6. Go to step 1.

Studying the steps in this program, you can see that first the input port is read. Then, the inputs are examined to see if they are all HIGH, since that is the function of an AND gate. If the inputs are all HIGH, the output is set HIGH. If the inputs are not all HIGH, the output is set LOW. Once the procedure has been completed, the program jumps back to step 1 and repeats indefinitely, so that the output will continuously follow any changes at the input.

You may be wondering why we would use this complex system to perform a simple AND gate function. If this were the only function that we wanted the microcomputer to perform, then it would definitely be easier to use a simple AND logic gate; however, the microcomputer provides tremendous flexibility. It allows the function of the gate to be arbitrarily redefined just by changing the program. You could easily add more inputs, and the gate function could be extremely complex. For example, we could have the microcomputer function as an eight-input electronic lock. The output of this lock would go HIGH only if the inputs were turned ON in a specific order. Using traditional logic, this would require a complex circuit; however, this function could be easily implemented using the same microprocessor system used for the simple AND gate. Of course, a new program would be needed, more complex than the simple AND gate program, but the hardware would not change. Additionally, the combination of this electronic lock could easily be changed by simply modifying the program.

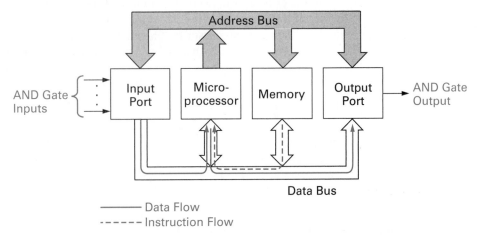

FIGURE 13-4 **A Microprocessor-Based AND Gate.**

13-2-2   *Flowcharts*

To develop a program of instructions to solve a problem, a programmer would generally follow three steps:

1. Define the problem.
2. Flowchart the solution.
3. Write the program.

To understand these steps, let us begin by examining the elements of the flowchart in more detail.

Flowcharts provide a graphic way of describing the operation of a program. They are composed of different types of blocks interconnected with lines, as shown in Figure 13-5. A rectangular block describes each action that the program takes. A diamond-shaped block is used for each decision, such as testing the value of a variable. An oval block is used at the beginning of the flowchart with the name of the program placed inside it. This oval block can also be used to mark the end of the flowchart.

As an example, Figure 13-6 shows a flowchart for the AND gate program discussed previously. For each line of the program there is a block, except for the two "go to" instructions. These are represented as lines indicating how the program will flow from one block to another. The flowchart contains the same information as the program list but in a graphic form. When you first set out to write a program, a flowchart is a good way to organize your thoughts and document what the program must do. By going through the flowchart manually, you can check the logic of the flowchart and then write the actual program from it. Flowcharts are also useful for understanding a program that has been written in the past.

13-2-3   *Programming Languages*

Writing programs in English is convenient, since it is the language most people understand. Unfortunately, English is meaningless to a microprocessor. The language understood by the microprocessor is called **machine language,** or machine code. Because microprocessors deal directly with digital signals, machine language instructions are binary codes, such as 00111100 and 11100111.

Machine language is not easy for people to use, since 00111100 has no obvious meaning. It can be made easier to work with by using the hexadecimal representation of 0011 1100, which is 3C; however, this still does not provide the user with any clue as to the meaning of this instruction. To counteract this problem, microprocessor manufacturers replace each instruction code with a short name called a **mnemonic,** or memory aid. As an example, Intel's 8085 microprocessor uses the mnemonic "INR A" for the code 3C, since this code instructs the 8085 microprocessor to "increment its internal A register." The mnemonics are much easier to remember than the machine codes. By assigning

FIGURE 13-5   Flowchart Symbols.

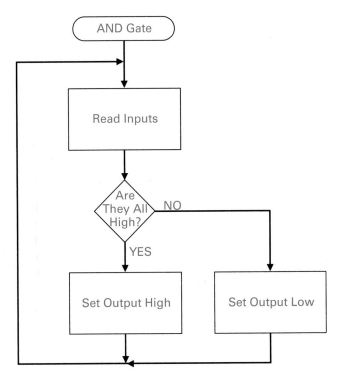

FIGURE 13-6 **AND Gate Flowchart.**

a mnemonic to each instruction code, you can write programs using the mnemonics instead of machine codes. Once the program is written, the mnemonics can easily be converted to their equivalent machine codes. Programs written using mnemonics are called **assembly language** programs.

The machine language is generally determined by the design of the microprocessor chip and cannot be modified. The assembly language mnemonics are made up by the microprocessor's manufacturer as a convenience for programmers and are not set by the microprocessor's design. For example, you could write INC A instead of INR A, as long as both were translated to the machine code 3C. A microprocessor is designed to recognize a specific list or group of codes called the **instruction set,** and each microprocessor type has its own set of instructions.

Although assembly language is a vast improvement over machine language, it is still difficult to use for complex programs. To make programming easier, **high-level languages** have been developed. These are similar to English and are generally independent of any particular microprocessor. For example, a typical instruction might be "LET COUNT = 10" or "PRINT COUNT." These instructions give a more complicated command than those that the microprocessor can understand. Therefore, microcomputers using high-level languages also contain long, complex programs (permanently stored in their memory) that translate the high-level language program into a machine language program. A single high-level instruction may translate into dozens of machine language instructions. Such translator programs are called **compilers.**

13-2-4   *A Programming Example*

To show the difference between machine language, assembly language, and a high-level language, let us use a simple programming example. Figure 13-7(a) shows the flowchart for a program that counts to 10. There are no input or output operations in this program, since

Assembly Language

A program written as a series of statements using mnemonic symbols that suggest the definition of the instruction. It is then translated into machine language by an assembler program.

Instruction Set

Total group of instructions that can be executed by a given microprocessor.

High-level Language

A language closer to the needs of the problem to be handled than to the language of the machine on which it is to be implemented.

Compiler

Translation program that converts high-level instructions into a set of binary instructions (machine code) for execution.

Line No.	Instruction	Description
1	LET COUNT = 0	Set Count to 0
2	LET COUNT = COUNT + 1	Increment Count
3	IF COUNT = 10 THEN 1	Go to 1 if Count = 10
4	GO TO 2	Otherwise go to 2

(b)

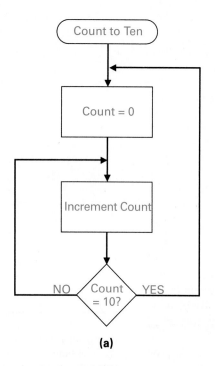

(a)

Label	Instruction	Comments
START:	MVI A,0	;Set A register to 0
LOOP:	INR A	;Increment A register
	CPI 10	;Compare A register to 10
	JZ START	;Go to beginning if A = 10
	JMP LOOP	;Repeat

(c)

Memory Address (Hex)	Memory Contents (Hex)	(Binary)
07F0	3E	00111110
07F1	00	00000000
07F2	3C	00111100
07F3	FE	11111110
07F4	0A	00001010
07F5	CA	11001010
07F6	F0	11110000
07F7	07	00000111
07F8	C3	11000011
07F9	F2	11110010
07FA	07	00000111

(d)

BASIC Language		8085 Assembly Language			8085 Machine Language		
Line No.	Instruction	Label	Instruction		Address	Contents	
1	LET COUNT = 0	START:	MVI	A,0	07F0	3E	Opcode
					07F1	00	Data
2	LET COUNT = COUNT + 1	LOOP:	INR	A	07F2	3C	Opcode
3	IF COUNT = 10 THEN 1		CPI	10_{10}	07F3	FE	Opcode
					07F4	0A	Data
			JZ	START	07F5	CA	Opcode
					07F6	F0	Address
					07F7	07	
4	GO TO 2		JMP	LOOP	07F8	C3	Opcode
					07F9	F2	Address
					07FA	07	

(e)

FIGURE 13-7   A Count-to-10 Programming Example. (a) Flowchart. (b) High-Level Language Program. (c) Assembly Language Program. (d) Machine Language Program. (e) Summary of Program Languages.

all we will be doing is having the contents of a designated memory location count from zero to 10, and then repeat.

The translation from the flowchart to a high-level language is fairly simple, as seen in Figure 13-7(b). The high-level language used in this example is called **BASIC,** which stands for Beginner's All-purpose Symbolic Instruction Code and has the advantage of being simple and similar to English. Following the program listing shown in Figure 13-7(b), you can see that the first two lines of the program correspond exactly to the first two action blocks of the count-to-10 flowchart. In the first line, the memory location called COUNT is set to 0. In the second line, LET COUNT = COUNT + 1 is simply a way of saying "increment the count." Lines 3 and 4 perform the function of the decision block in the count-to-10 flow chart. Line 3 specifies that if COUNT = 10, then the next instruction executed should be line 1. If the count is not equal to 10, the COUNT = 10 instruction has no effect, and the program continues with line 4, which says "go to line 2." To test whether this BASIC program will count to 10 and then repeat, try following the program step by step to see if it works, starting with a value of zero.

Assembly language is not one specific language but a class of languages. Each microprocessor has its own machine language and therefore its own assembly language, which is defined by the manufacturer. Figure 13-7(c) shows the assembly language listing for the count-to-10 program. This program is certainly more cryptic than the BASIC language program, but it performs exactly the same function.

The three columns in Figure 13-7(c) are for labels, instructions, and comments. The label provides the same function as the line number in the BASIC program. Instead of numbering every line, you simply make up a name (called a label) for each line to which you need to refer. A colon (:) is used to identify the label. A line needs a label only if there is another instruction in the program that refers to that line. The comments are an aid to understanding the program, and a semicolon (;) is used to identify the beginning of a comment. High-level language programs do not always need many comments because the instructions themselves are more descriptive. For assembly language programs, however, comments are an invaluable aid for people other than the programmer, or when the programmer returns to a program after some time.

The first instruction is MVI A, 0, which means "move immediately to the accumulator the data zero." The **accumulator** is also called the A register and is a storage location inside the microprocessor. This assembly language instruction is equivalent to the BASIC instruction LET COUNT = 0, except instead of making up a name for the variable (COUNT), we used a preassigned name (A) for a register inside the microprocessor. This MVI A, 0 instruction will therefore load the data 0 into the microprocessor's internal A register. The next instruction, INR A, means "increment the value in the accumulator," and this assembly language instruction is equivalent to the BASIC instruction LET COUNT = COUNT + 1. The next three instructions implement the decision function. The instruction CPI 10 means "compare the value in the accumulator with the value 10." The result of this comparison will determine whether a special flip-flop within the microprocessor called the *zero flag* is SET or RESET. If the value in the accumulator is not equal to 10, the zero flag is RESET LOW, whereas if the value in the accumulator is equal to 10, the zero flag is SET HIGH. The next instruction, JZ START, means "jump if the zero flag is SET to the line with the label START." This instruction tests the zero flag, and if it is SET (accumulator = 10), it will cause the program to jump to the line with the label START. Together, these two instructions (CPI 10 and JZ START) perform the function of the BASIC statement IF COUNT = 0 THEN 1. The last instruction, JMP LOOP, means "jump to the line with the label LOOP." This instruction simply causes the program to jump to the line with the label LOOP and is equivalent to the BASIC statement GO TO 2.

Figure 13-7(d) shows the machine language listing for the count-to-10 program. Although this language looks the most alien to us, its sequence of 1s and 0s is the only language that the microprocessor understands. In this example, each memory location holds eight bits of data. To program the microcomputer, we will have to store these 8-bit codes in the microcomputer's memory. Each instruction begins with an **opcode,** or operation code, that specifies the operation to be performed, and since all 8085 opcodes are eight bits, each

BASIC

An easy-to-learn and easy-to-use language available on most microcomputer systems.

Accumulator

One or more registers associated with the arithmetic and logic unit (ALU), which temporarily store sums and other arithmetical and logical results of the ALU.

Opcode

The first part of a machine language instruction that specifies the operation to be performed. The other parts specify the data, address, or port. For the 8085, the first byte of each instruction is the opcode.

opcode will occupy one memory location. An 8085 opcode may be followed by one byte of data, two bytes of data, or no bytes, depending upon the instruction type used.

Stepping through the machine language program shown in Figure 13-7(d), you can see that the first byte ($3E_{16}$) at address $07F0_{16}$ is the opcode for the instruction MVI A. The MVI A instruction is made up of two bytes. The first byte is the opcode specifying that you want to move some data into the accumulator, and the second byte stored in the very next memory location (address $07F1_{16}$) contains the data 00_{16} to be stored in the accumulator. The third memory location (address 07F2) contains the opcode for the second instruction, INR A. This opcode (3C) tells the microprocessor to increment the accumulator. The INR A instruction has no additional data, and therefore the instruction occupies only one memory location. The next code, FE, is the opcode for the compare instruction, CPI. Like the MVI A, 0 instruction, the memory location following the opcode contains the data required by the instruction. Since the machine language program is shown in hexadecimal notation, the data 10 (decimal) appears as 0A (hex). This instruction compares the accumulator with the value 10 and sets the zero flag if they are equal, as described earlier. The next instruction, JZ, has the opcode CA and appears at address 07F5. This opcode tells the microprocessor to jump if the zero flag is set. The next two memory locations contain the address to jump to. Because addresses in an 8085 system are 16 bits long, it takes two memory locations (8 bits each) to store an address. The two parts of the address are stored in an order that is the reverse of what you might expect. The least significant half of the address is stored first and then the most significant half of the address is stored next. The address 07F0 is therefore stored as F0 07. The assembly language instruction JZ START means that the processor should jump to the instruction labeled START. The machine code must therefore use the actual address that corresponds to the label START, which in this case is address 07F0. The last instruction, JMP LOOP, is coded in the same way as the previous jump instruction. The only difference is that this jump instruction is independent of any flag condition, and therefore no flags will be tested to determine whether this jump should occur. When the program reaches the C3 jump opcode, it will always jump to the address specified in the two bytes following the opcode, which in this example is address 07F2. The machine language program, therefore, contains a series of bytes, some of which are opcodes, some are data, and some are addresses.

You must know the size and format of an instruction if you want to be able to interpret the operation of the program correctly. To compare the differences and show the equivalents, Figure 13-7(e) combines the high-level language, assembly language, and machine language listings for the count-to-10 program.

SELF-TEST EVALUATION POINT FOR SECTION 13-2

Now that you have completed this section, you should be able to:

■ **Objective 5.** *Describe how flowcharts can help a programmer write programs.*

■ **Objective 6.** *Describe the meaning, structure, and differences among programs that are written in machine language, assembly language, and a high-level language.*

Use the following questions to test your understanding of Section 13-2:

1. Briefly describe the three following languages:
   a. machine language
   b. assembly language
   c. high-level language

2. What is a flowchart?
3. Define the following terms:
   a. instruction set
   b. peripheral
   c. programmer
   d. mnemonic
   e. compiler
   f. accumulator
   g. flag register
   h. opcode

REVIEW QUESTIONS

Multiple Choice Questions

1. Microprocessor-based systems are more flexible than hard-wired logic designs because _____.
   a. They are faster
   b. They use LSI devices
   c. Their operation is controlled by software
   d. The hardware is specialized

2. The peripherals of a microcomputer system are _____.
   a. The memory devices
   b. The microprocessor
   c. The software
   d. The I/O devices

3. The personality of a microprocessor-based system is determined primarily by _____.
   a. The microprocessor used
   b. The program and peripherals
   c. The number of data bus lines
   d. The type of memory ICs used

4. The language that the microprocessor understands directly is called _____.
   a. Assembly language
   b. High-level language
   c. English language
   d. Machine language

5. Programs written using mnemonics are called _____ programs.
   a. Assembly language
   b. High-level language
   c. English language
   d. Machine language

6. A program designed to translate a high-level language into machine language is called _____.
   a. An instruction set
   b. A compiler
   c. An opcode
   d. A mnemonic

7. Which unit within the microprocessor performs arithmetic and logic operations?
   a. Accumulator
   b. I/O
   c. ALU
   d. Memory

8. "INRA" is an example of _____.
   a. Assembly language
   b. BASIC
   c. Machine language
   d. C language

9. "LET COUNT = 0" is an example of _____.
   a. Assembly language
   b. BASIC
   c. Machine language
   d. C language

10. An _____ is the first part of a machine language instruction.
    a. Accumulator
    b. ALU
    c. Assembler
    d. Opcode

Communication Skill Questions

11. What is a microprocessor?   (Introduction)

12. What three blocks form a microcomputer?   (13-1)

13. What is the difference between programming a microcomputer and running a program in a microcomputer?   (13-1)

14. What is the difference between software and hardware?   (13-1)

15. What are peripherals?   (13-1)

16. What is a flowchart?   (13-2-2)

17. Briefly describe each of the following programming languages: (13-2-3)
    a. machine language
    b. assembly language
    c. high-level language

18. Define the following terms: (13-2)
    a. opcode   f. mnemonic
    b. instruction   g. program
    c. flag   h. peripheral
    d. accumulator   i. flowchart
    e. BASIC

19. List and describe the steps a programmer would follow to develop a program.   (13-2-4)

20. Why, do you think, are flowcharts useful?   (13-2-2)

Practice Problems

21. Develop a flowchart to control two traffic signals (A and B) at an intersection with the following specifications and sequence:

    Light A: Green for 3 minutes
              Yellow for 1 minute
              Red for as long as light B is green and yellow

    Light B: Green for 2 minutes
              Yellow for 1 minute
              Red for as long as light A is green and yellow

    Repeat cycle.

22. Modify the flowchart in question 21 so that light B will go into its cycle only if a car is detected by the B road sensors.

23. Develop a flowchart that monitors the number of stadium tickets sold and when that number equals 10,000, activates the "Stadium Full" sign.

24. Referring to the flowchart in Figure 13-8, step through the instructions to see if the example value will be converted from BCD (3 digits) to binary (8-bit).

25. Referring to the flowchart in Figure 13-9, step through the instructions to see if a value will be converted from binary (8-bit) to BCD (3 digit).

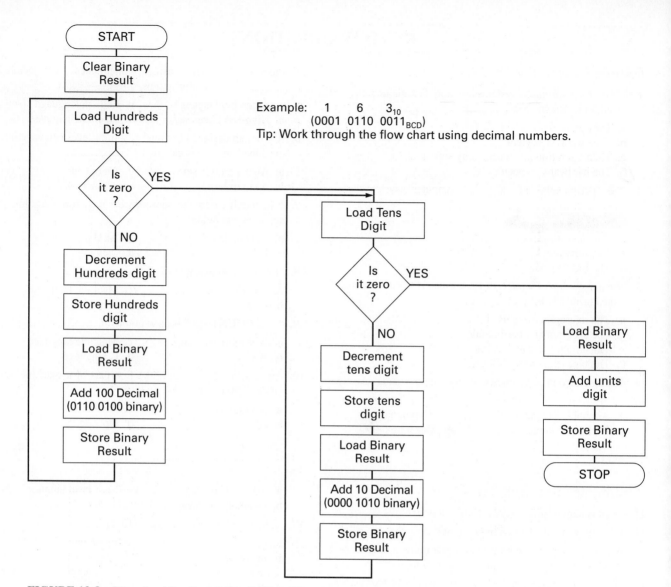

Example: 1   6   3₁₀

Example: 1   6 3_{10}
(0001 0110 0011_{BCD})
Tip: Work through the flow chart using decimal numbers.

FIGURE 13-8   **Flowchart for Converting BCD to Binary.**

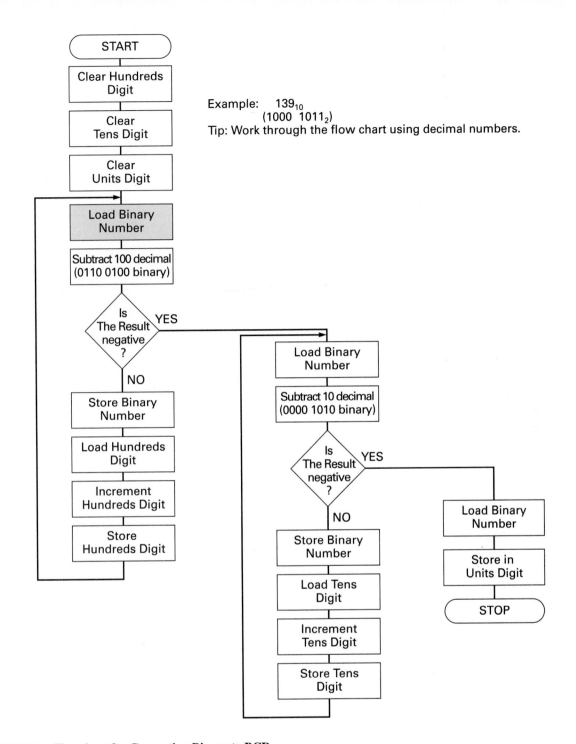

Example: 139_{10}
$(1000\ 1011_2)$
Tip: Work through the flow chart using decimal numbers.

FIGURE 13-9   **Flowchart for Converting Binary to BCD.**

Web Site Questions

Go to the web site http://prenhall.com/cook, select the textbook *Introductory Computer Mathematics, 2e,* select this chapter, and follow the instructions when answering the multiple-choice practice problems.

Answers to Self-Test Evaluation Points

STEP 1-1

1. (a) $\dfrac{6}{8}$ **(b)** $\dfrac{3}{6}$ **(c)** $\dfrac{5}{16}$

2. (a) $2\dfrac{1}{2}$ **(b)** $4\dfrac{1}{4}$ **(c)** $5\dfrac{3}{4}$ **(d)** $6\dfrac{7}{8}$ **(e)** $8\dfrac{3}{8}$

STEP 1-2

1. $\dfrac{4}{6} + \dfrac{1}{6} = \dfrac{4+1}{6} = \dfrac{5}{6}$

2. $\dfrac{3}{64} + \dfrac{2}{64} + \dfrac{25}{64} = \dfrac{3+2+25}{64} = \dfrac{30}{64} \div \dfrac{2}{2} = \dfrac{15}{32}$

3. $\dfrac{3}{9} + \dfrac{6}{18} = \dfrac{6+6}{18} = \dfrac{12}{18} \div \dfrac{6}{6} = \dfrac{2}{3}$

4. $\dfrac{1}{3} + \dfrac{1}{4} + \dfrac{3}{15} = \dfrac{}{?}$

$$
\begin{array}{c|ccc}
2 & 3 & 4 & 15 \\
2 & 3 & 2 & 15 \\
3 & 3 & 1 & 15 \\
5 & 1 & 1 & 5 \\
  & 1 & 1 & 1 \\
\end{array}
$$
$\longrightarrow 2 \times 2 \times 3 \times 5 = 60$

$\dfrac{1}{3} + \dfrac{1}{4} + \dfrac{3}{15} = \dfrac{20+15+12}{60} = \dfrac{47}{60}$

5. $\dfrac{3}{4} + \dfrac{4}{5} = \dfrac{15+16}{20} = \dfrac{31}{20} = 1\dfrac{11}{20}$

$$
\begin{array}{c|cc}
2 & 4 & 5 \\
2 & 2 & 5 \\
5 & 1 & 5 \\
  & 1 & 1 \\
\end{array}
$$
$\longrightarrow 2 \times 2 \times 5 = 20$

6. $\dfrac{9}{12} + \dfrac{4}{24} + \dfrac{3}{4} = \dfrac{18+4+18}{24} = \dfrac{40}{24} = 1\dfrac{16}{24} \div \dfrac{8}{8} = 1\dfrac{2}{3}$

7. $\dfrac{7}{9} + \dfrac{5}{9} + \dfrac{4}{18} = \dfrac{14+10+4}{18} = \dfrac{28}{18} = 1\dfrac{10}{18} \div \dfrac{2}{2} = 1\dfrac{5}{9}$

8. $\dfrac{15}{6} = 2\dfrac{3}{6} \div \dfrac{3}{3} = 2\dfrac{1}{2}$

9. $\dfrac{4}{24} = \dfrac{1}{6}$

10. $\dfrac{25}{100} \div \dfrac{25}{25} = \dfrac{1}{4}$

STEP 1-3

1. $\dfrac{4}{32} - \dfrac{2}{32} = \dfrac{4-2}{32} = \dfrac{2}{32} \div \dfrac{2}{2} = \dfrac{1}{16}$

2. $\dfrac{4}{5} - \dfrac{3}{10} = \dfrac{8-3}{10} = \dfrac{5}{10} \div \dfrac{5}{5} = \dfrac{1}{2}$

3. $\dfrac{6}{12} - \dfrac{4}{16} = \dfrac{24-12}{48} = \dfrac{12}{48} = \dfrac{1}{4}$

$$
\begin{array}{c|cc}
2 & 12 & 16 \\
2 & 6 & 8 \\
2 & 3 & 4 \\
2 & 3 & 2 \\
3 & 3 & 1 \\
  & 1 & 1 \\
\end{array}
$$
$\longrightarrow 2 \times 2 \times 2 \times 2 \times 3 = 48$

4. $15\dfrac{1}{3} - 2\dfrac{4}{12} = 13\dfrac{4-4}{12} = 13\dfrac{0}{12} = 13$

5. $5\dfrac{15}{36} - 2\dfrac{31}{36} = 3\dfrac{15-31}{36} \leftarrow$ need to borrow

$= 4\left(\dfrac{36}{36}\right) + \dfrac{15}{36} - 2\dfrac{31}{36}$

$= 4\dfrac{51}{36} - 2\dfrac{31}{36} = 2\dfrac{51-31}{36} = 2\dfrac{20}{36} = 2\dfrac{5}{9}$

STEP 1-4

1. $\dfrac{1}{3} \times \dfrac{2}{3} = \dfrac{1 \times 2}{3 \times 3} = \dfrac{2}{9}$

2. $\dfrac{4}{8} \times \dfrac{2}{9} = \dfrac{8 \div 8}{72 \div 8} = \dfrac{1}{9}$

3. $\dfrac{4}{17} \times \dfrac{3}{34} = \dfrac{12}{578} = \dfrac{6}{289}$

4. $\dfrac{1}{16} \times 5\dfrac{3}{7} = \dfrac{1}{16} \times \dfrac{38}{7} = \dfrac{38}{112} = \dfrac{19}{56}$

5. $2\dfrac{1}{7} \times 7\dfrac{2}{3} = \dfrac{15}{7} \times \dfrac{23}{3} = \dfrac{345}{21} = 16\dfrac{9}{21} = 16\dfrac{3}{7}$

STEP 1-5

1. $\dfrac{2}{3} \div \dfrac{1}{12} = \dfrac{2}{3} \times \dfrac{12}{1} = \dfrac{24}{3} = 8$

2. $3\dfrac{1}{4} \div \dfrac{3}{4} = \dfrac{13}{4} \div \dfrac{3}{4} = \dfrac{13}{4} \times \dfrac{4}{3} = \dfrac{52}{12} = 4\dfrac{4}{12} = 4\dfrac{1}{3}$

3. $8\dfrac{16}{20} \div 4\dfrac{4}{20} = \dfrac{176}{20} \div \dfrac{84}{20} = \dfrac{176}{20} \times \dfrac{20}{84}$

$= \dfrac{3520}{1680} = 2\dfrac{160}{1680} \div \dfrac{10}{10} = 2\dfrac{16}{168} \div \dfrac{4}{4} = 2\dfrac{4}{42} \div \dfrac{2}{2}$

$= 2\dfrac{2}{21}$

4. $4\dfrac{3}{8} \div 2\dfrac{5}{16} = \dfrac{35}{8} \div \dfrac{37}{16} = \dfrac{35}{8} \times \dfrac{16}{37}$

$= \dfrac{560}{296} = 1\dfrac{264}{296} \div \dfrac{4}{4} = 1\dfrac{66}{74} \div \dfrac{2}{2} = 1\dfrac{33}{37}$

STEP 1-6

1. $\dfrac{16}{\boxed{27}} \times \dfrac{\boxed{27}}{240} = \dfrac{\boxed{16}}{1} \times \dfrac{1}{\boxed{240}} = \dfrac{1}{1} \times \dfrac{1}{15} = \dfrac{1}{15}$

2. $\dfrac{\boxed{8}}{17} \times \dfrac{3}{\boxed{4}} = \dfrac{2}{17} \times \dfrac{3}{1} = \dfrac{6}{17}$

3. $\dfrac{15}{32} \div \dfrac{30}{4} = \dfrac{\textcircled{15}}{32} \times \dfrac{4}{\textcircled{30}} = \dfrac{1}{\textcircled{32}} \times \dfrac{\textcircled{4}}{2}$

$= \dfrac{1}{8} \times \dfrac{1}{2} = \dfrac{1}{16}$

4. $6\dfrac{2}{3} \times 2\dfrac{7}{10} = \dfrac{\textcircled{20}}{3} \times \dfrac{27}{\textcircled{10}} = \dfrac{2}{\textcircled{3}} \times \dfrac{\textcircled{27}}{1}$

$= \dfrac{2}{1} \times \dfrac{9}{1} = \dfrac{18}{1} = 18$

5. $6\dfrac{1}{2} \div \dfrac{9}{32} = \dfrac{13}{\textcircled{2}} \times \dfrac{\textcircled{32}}{9} = \dfrac{13}{1} \times \dfrac{16}{9} = \dfrac{208}{9} = 23\dfrac{1}{9}$

STEP 2-1
1. 10
2.
$$
\begin{aligned}
3 \times 10{,}000 &= 30{,}000 \\
5 \times 1{,}000 &= 5{,}000 \\
7 \times 100 &= 700 \\
2 \times 10 &= 20 \\
9 \times 1 &= 9 \\
\hline
&35{,}729
\end{aligned}
$$
3. See Section 2-1-3.
4.

499,997	
499,998	1
499,999	2
499,999	3
500,000	3
500,001	4
500,002	5

The units, tens, hundreds, thousands, and ten-thousands columns will reset and carry when the count is advanced by 3. The final count will be 500,002.

STEP 2-2
1.
$$
\begin{aligned}
1 \times 100 &= 100 \\
7 \times 10 &= 70 \\
8 \times 1 &= 8 \\
6 \times 0.1 &= 0.6 \\
4 \times 0.01 &= 0.04 \\
9 \times 0.001 &= 0.009 \\
\hline
&178.649
\end{aligned}
$$

2. (a) $1\dfrac{5}{2} = 1 + (5 \div 2) = 1 + 2.5 = 3.5$

(b) $192\dfrac{3}{4} = 192 + (3 \div 4) = 192 + 0.75 = 192.75$

(c) $67\dfrac{6}{9} = 67 + (6 \div 9) = 67 + 0.667 = 67.667$

3. (a) $\dfrac{1}{2500} = 0.0004$

(b) $\dfrac{1}{0.25} = 4$

(c) $\dfrac{1}{0.000025} = 40{,}000$

4. (a) $7.25 = 7\dfrac{4}{16}\left(\dfrac{1}{16} = 0.0625\right)$

(b) $156.90625 = 156\dfrac{29}{32}\left(\dfrac{1}{32} = 0.03125\right)$

STEP 2-3
1. Addition, subtraction
2. (a) Subtraction
(b) Addition
(c) Division
(d) Multiplication

3. (a)
$$
\begin{aligned}
&26.443 \\
&197.1 \\
&2.1103 \\
+\ &0.004 \\
\hline
&225.6573
\end{aligned}
$$

(b)
$$
\begin{aligned}
&19637.224 \\
-\ &866.43 \\
\hline
&18770.794
\end{aligned}
$$

(c)
$$
\begin{aligned}
&894.357 \\
\times\ &8.6 \\
\hline
&5366142 \\
&7154856 \\
\hline
&7691.4702
\end{aligned}
$$

(d) $0.015\,\overline{)1.3397}89.313333$

or $15\,\overline{)1339.7}$

4. $\dfrac{176 \div 8}{8 \div 8} = \dfrac{22}{1}$ or $22:1$ or 22.0 to 1 (ratio of twenty-two to one)

5. (a) 86.44
(b) 12,263,415.01
(c) 0.18
6. 86.43760   **(a)** 7
                (b) 6
12,263,415.00510   **(a)** 13
                    (b) 12
0.176600   **(a)** 6
           (b) 4
7. (a) 0.075
(b) 220
(c) 0.235
(d) 19.2
8. (a) 35
(b) 4004

STEP 3-1
1. (a) $+3 - (-4) = +7$
(b) $12 \times (+4) = +48$
(c) $-5 \div (-7) = 0.7142$
(d) $-0.63 \times (+6.4) = -4.032$
2. (a) $+6 + (-7)$
(b) $-0.75 \times (18)$ or -0.75×18
3. (a) $\boxed{7}\,\boxed{.}\,\boxed{5}\,\boxed{+/-}\,\boxed{+}\,\boxed{4}\,\boxed{.}\,\boxed{6}\,\boxed{+/-}\,\boxed{=}$
(b) $\boxed{5}\,\boxed{\times}\,\boxed{2}\,\boxed{+/-}\,\boxed{=}$
(c) $\boxed{3}\,\boxed{1}\,\boxed{6}\,\boxed{.}\,\boxed{6}\,\boxed{2}\,\boxed{9}\,\boxed{\div}\,\boxed{1}\,\boxed{.}\,\boxed{4}\,\boxed{4}\,\boxed{+/-}\,\boxed{=}$

STEP 3-2
1. $+15 + (+4) = +19$
2. $-3 + (+45) = +42$
3. $+114 + (-111) = +3$
4. $-357 + (-74) = -431$
5. $17 + (15) = +32$
6. $-8 + (-177) = -185$
7. $+4600 + (-3400) = +1200$
8. $-6.25 + (+0.34) = -5.91$

STEP 3-3
1. $+18 - (+7) = +11$
2. $-3.4 - (-5.7) = +2.3$
3. $19{,}665 - (-5{,}031) = +24{,}696$
4. $-8 - (+5) = -13$
5. $467 - 223 = +244$
6. $-331 - (-2.6) = -328.4$
7. $8 - (+25) = -17$
8. $-0.64 - (-0.04) = -0.6$

STEP 3-4
1. $4 \times (+3) = +12$
2. $+17 \times (-2) = -34$
3. $-8 \times (+16) = -128$
4. $-8 \times (-5) = +40$

5. $+12.6 \times (+15) = +189$
6. $-3.3 \times (+1.4) = -4.62$
7. $+0.3 \times (-4) = -1.2$
8. $-4.6 \times (-3.3) = +15.18$

STEP 3-5
1. $+16.7 \div (+2.3) = 16.7 \div 2.3 = 7.26$
2. $+18 \div (+6) = 3$
3. $-6 \div (+2) = -3$
4. $+18 \div (-4) = -4.5$
5. $+2 \div (-8) = -0.25$
6. $-8 \div (+5) = -1.6$
7. $-15 \div (-5) = 3$
8. $0.664 \div (-0.2) = -3.32$

STEP 3-6
1. $+6 + (+3) + (-7) + (-5) = -3$
2. $+9 - (+2) - (-13) - (-4) = +24$
3. $-6 \times (-4) \times (-5) = -120$
4. $+8 \div (+2) \div (-5) = -0.8$
5. $-4 \div (-2) \times (+8) = +16$
6. $-9 + (+5) - (-7) = +3$

STEP 4-1
1. (a) $16^4 = 16 \times 16 \times 16 \times 16 = 65,536$
   (b) $32^3 = 32 \times 32 \times 32 = 32,768$
   (c) $112^2 = 112 \times 112 = 12,544$
   (d) $15^6 = 15 \times 15 \times 15 \times 15 \times 15 \times 15 = 11,390,625$
   (e) $2^3 = 2 \times 2 \times 2 = 8$
   (f) $3^{12} = 3 \times 3 \times 3 \times 3 \times 3 \times 3 \times 3 \times 3 \times 3 \times 3 \times 3 \times 3$
    $= 531,441$
2. (a) $\sqrt[2]{144} = 12$
   (b) $\sqrt[3]{3375} = 15$
   (c) $\sqrt[2]{20} = 4.47$
   (d) $\sqrt[3]{9} = 2.08$
3. (a) $(9^2 + 14^2)^2 - \sqrt[3]{3 \times 7}$
    $= (81 + 196)^2 - \sqrt[3]{21}$
    $= 277^2 - 2.76$
    $= 76,729 - 2.76 = 76,726.24$
   (b) $\sqrt{3^2 \div 2^2} + \dfrac{151 - 9^2}{3.5^2}$
    $= \sqrt{9 \div 4} + \dfrac{151 - 81}{12.25}$
    $= \sqrt{2.25} + \dfrac{70}{12.25}$
    $= 1.5 + 5.71 = 7.21$

STEP 4-2
1. (a) 10^2
   (b) 10^0
   (c) 10^1
   (d) 10^6
   (e) 10^{-3}
   (f) 10^{-6}
2. (a) $6.3 \times 10^3 = 6.300. = 6300.0$ or 6300
   (b) $114,000 \times 10^{-3} = 114.000_o = 114.0$ or 114
   (c) $7,114,632 \times 10^{-6} = 7.114632_o = 7.114632$
   (d) $6624 \times 10^6 = 6624.000000. = 6,624,000,000.0$
3. (a) $\sqrt{3 \times 10^6} = \sqrt{3,000,000} = 1732.05$
   (b) $(2.6 \times 10^{-6}) - (9.7 \times 10^{-9}) = 0.0000025$ or 2.5×10^{-6}
   (c) $\dfrac{(4.7 \times 10^3)^2}{3.6 \times 10^6} = (4.7 \times 10^3)^2 \div (3.6 \times 10^6) = 6.14$

4. (a) $47,000 = 47\,000_o = 47 \times 10^3$
   (b) $0.00000025 = 0.000000250. = 250 \times 10^{-9}$
   (c) $250,000,000 = 250.000\,000_o = 250 \times 10^6$
   (d) $0.0042 = 0.004.2 = 4.2 \times 10^{-3}$

STEP 4-3
1. (a) 10^3
   (b) 10^{-2}
   (c) 10^{-3}
   (d) 10^6
   (e) 10^{-6}
2. (a) cm $\times 0.4 =$ inches, 15 cm $\times 0.4 = 6$ inches
   (b) kg $\times 2.2 =$ pounds, 23 kg $\times 2.2 = 50.6$ pounds
   (c) liters $\times 0.26 =$ gallons, 37 L $\times 0.26 = 9.62$ gallons
   (d) $\left(\dfrac{9}{5} \times {}^\circ\text{C}\right) + 32 = \left(\dfrac{9}{5} \times 23\right) + 32 = 73.4{}^\circ\text{F}$
3. (a) Miles/hour $\times 1.6 =$ kilometers/hour,
    55 mph $\times 1.6 = 88$ km/hour
   (b) gallons $\times 3.8 =$ liters, 16 gallons $\times 3.8 = 60.8$ liters
   (c) square yards $\times 0.8 =$ square meters, 3 yd^2 $\times 0.8 = 2.4$ m^2
   (d) $\dfrac{5}{9} \times ({}^\circ\text{F} - 32) = \dfrac{5}{9} \times (92 - 32) = 33.33{}^\circ\text{C}$
4. (a) meter (m)   (f) ampere (A)
   (b) gram (g)   (g) volt (V)
   (c) degree Celsius (°C) (h) liter (L)
   (d) second (s)   (i) joule (J)
   (e) watt (W)   (j) ohm (Ω)
5. (a) 25,000 volts $= 25.000. \times 10^3$ volts
    $= 25$ kilovolts
   (b) 0.014 watts $= 0.014. = 14 \times 10^{-3}$ watts
    $= 14$ milliwatts
   (c) 0.000016 microfarad $= 0.000016 \times 10^{-6}$ farad
    $= 0.000.016 \times 10^{-9}$ farad
    $= 0.016 \times 10^{-9}$ farad
    $= 0.016$ nanofarad

STEP 5-1
1. Yes
2. No
3. $\dfrac{144}{12} \times \square = \dfrac{36}{6} \times 2 \times \square = 60$
   $12 \times \square = 6 \times 2 \times \square = 60$
   $12 \times 5 = 12 \times 5 = 60$
   $\square = 5$
4. $\dfrac{(8 - 4) + 26}{5} = \dfrac{81 - 75}{2}$
   $\dfrac{4 + 26}{5} = \dfrac{6}{2}$
   $\dfrac{30}{5} = \dfrac{6}{2}$
   $6 = 3$ (equation is not equal)
5. Yes

STEP 5-2
1. (a) $x + x = 2x$
   (b) $x \times x = x^2$
   (c) $7a + 4a = 11a$
   (d) $2x - x = 1x$ or x
   (e) $\dfrac{x}{x} = 1$
   (f) $x - x = 0$
2. (a) $x + 14 = 30$
    $x + 14 - 14 = 30 - 14$ (-14 from both sides)
       $x = 30 - 14$
       $x = 16$

(b) $8 \times x = \dfrac{80 - 40}{10} \times 12$

$8 \times x = 4 \times 12$

$8 \times x = 48$

$\dfrac{8 \times x}{8} = \dfrac{48}{8}$ $\quad (\div 8)$

$x = \dfrac{48}{8}$

$x = 6$

(c) $y - 4 = 8$

$y - 4 + 4 = 8 + 4$ $\quad (+ 4)$

$y = 8 + 4$

$y = 12$

(d) $(x \times 3) - 2 = \dfrac{26}{2}$

$(x \times 3) - 2 + 2 = \dfrac{26}{2} + 2$ $\quad (+ 2)$

$x \times 3 = \dfrac{26}{2} + 2$

$x \times 3 = 15$

$\dfrac{x \times 3}{3} = \dfrac{15}{3}$ $\quad (\div 3)$

$x = \dfrac{15}{3}$

$x = 5$

(e) $x^2 + 5 = 14$

$x^2 + 5 - 5 = 14 - 5$ $\quad (- 5)$

$x^2 = 14 - 5$

$\sqrt{x^2} = \sqrt{14 - 5}$ $\quad (\sqrt{\ })$

$x = \sqrt{14 - 5}$

$x = \sqrt{9}$

$x = 3$

(f) $2(3 + 4x) = 2(x + 13)$

$6 + 8x = 2x + 26$ $\quad$ (remove parentheses)

$6 + 8x - 2x = 2x + 26 - 2x$ $\quad (- 2x)$

$6 + (8x - 2x) = 26$

$6 + 6x = 26$

$6 + 6x - 6 = 26 - 6$ $\quad (- 6)$

$6x = 26 - 6$

$6x = 20$

$\dfrac{6 \times x}{6} = \dfrac{20}{6}$ $\quad (\div 6)$

$x = \dfrac{20}{6}$

$x = 3.3333$

3. (a) $x + y = z, y = ?$

$x + y - x = z - x$ $\quad (- x)$

$y = z - x$

(b) $Q = C \times V, C = ?$

$\dfrac{Q}{V} = \dfrac{C \times V}{V}$ $\quad (\div V)$

$\dfrac{Q}{V} = C$

$C = \dfrac{Q}{V}$

(c) $X_L = 2 \times \pi \times f \times L, L = ?$

$\dfrac{X_L}{2 \times \pi \times f} = \dfrac{2 \times \pi \times f \times L}{2 \times \pi \times f}$ $\quad (\div 2 \times \pi \times f)$

$\dfrac{X_L}{2 \times \pi \times f} = L$

$L = \dfrac{X_L}{2 \times \pi \times f}$

(d) $V = I \times R, R = ?$

$\dfrac{V}{I} = \dfrac{I \times R}{I}$ $\quad (\div I)$

$\dfrac{V}{I} = R$

$R = \dfrac{V}{I}$

4. (a) $I^2 = 9$

$\sqrt{I^2} = \sqrt{9}$ $\quad (\sqrt{\ })$

$I = \sqrt{9}$

$I = 3$

(b) $\sqrt{Z} = 8$

$\sqrt{Z}^2 = 8^2$

$Z = 8^2$

$Z = 64$

STEP 5-3

1. $x = y \times z$ and $a = x \times y, y = 14, z = 5, a = ?$

$a = x \times y$

$a = y \times z \times y$ $\quad$ (substitute $y \times z$ for x)

$a = y^2 \times z$ $\quad (y \times y = y^2)$

$a = 14^2 \times 5$

$a = 980$

2. (a) For $I = \sqrt{\dfrac{P}{R}}$,

$P = V \times I$ to $P = I^2 \times R$

$\dfrac{P}{R} = \dfrac{I^2 \times R}{R}$ $\quad (\div R)$

$\dfrac{P}{R} = I^2$

$\dfrac{P}{R} = \sqrt{I^2}$ $\quad (\sqrt{\ })$

$\sqrt{\dfrac{P}{R}} = I$

$I = \sqrt{\dfrac{P}{R}}$

(b) For $I = \dfrac{P}{V}$,

$P = V \times I$

$\dfrac{P}{V} = \dfrac{V \times I}{V}$ $\quad (\div V)$

$\dfrac{P}{V} = I$

$I = \dfrac{P}{V}$

(c) For $V = \sqrt{P \times R}$

$P = V \times I$ to $P = \dfrac{V^2}{R}$

$P \times R = \dfrac{V^2 \times R}{R}$ $\quad (\times R)$

$P \times R = V^2$

$\sqrt{P \times R} = \sqrt{V^2}$ $\quad (\sqrt{\ })$

$\sqrt{P \times R} = V$

$V = \sqrt{P \times R}$

(d) For $R = \dfrac{V^2}{P}$,

$$P = V \times I \text{ to } P = \dfrac{V^2}{R}$$

$$P \times R = \dfrac{V^2}{R} \times R \qquad (\times R)$$

$$P \times R = V^2$$

$$\dfrac{P \times R}{P} = \dfrac{V^2}{P}$$

$$R = \dfrac{V^2}{P}$$

3. What percentage of 12 gives 2.5?

$$x \times 12 = 2.5$$

$$\dfrac{x \times 12}{12} = \dfrac{2.5}{12} \div 12$$

$$x = \dfrac{2.5}{12}$$

$$x = 0.2083 \quad (0.2083 \times 12 = 2.5)$$

$$x\% = 20.83\% \quad (20.83\% \times 12 = 2.5)$$

4. 60 is 22% of what number?

$$60 = 22\% \times x$$

$$60 = 0.22 \times x$$

$$\dfrac{60}{0.22} = \dfrac{0.22 \times x}{0.22} \qquad (\div 0.22)$$

$$\dfrac{60}{0.22} = x$$

$$x = 272.73 \quad (60 = 0.22 \times 272.73$$
$$\text{or } 60 = 22\% \times 272.73)$$

STEP 5-4

1. Parentheses, Exponents, Multiplication, Division, Addition, Subtraction.

2. Monomial: $3x$
Binomial: $3x + 4y$
Trinomial: $3x + 4y + 5z$
Polynomial: $2a + 2b + 2c + 2d + 2e$

3. (a) $a \div 1 = a$
(b) $a + a = 2a$
(c) $a - a = 0$
(d) $a \times a = a^2$
(e) $3x + b - 2x = x + b$

4. (a) $16 + a = 5 \times a$
$a = 4$

(b) $\dfrac{42}{b} = 3 \times 7$
$b = 2$

STEP 6-1

1. (a) $40°$ **(d)** $90°$ **(g)** $165°$ **(j)** $165°$
(b) $60°$ **(e)** $115°$ **(h)** $180°$ **(k)** $65°$
(c) $80°$ **(f)** $135°$ **(i)** $15°$ **(l)** $90°$

2. (a) Acute   **(d)** Right   **(g)** Obtuse   **(j)** Obtuse
(b) Acute   **(e)** Obtuse   **(h)** Obtuse   **(k)** Acute
(c) Acute   **(f)** Obtuse   **(i)** Acute   **(l)** Right

3. (See Figure 6-1)

STEP 6-2

1. (Section 6-2 Introduction)

2. (a) Figure 6-2   **(e)** Figure 6-6   **(i)** Figure 6-12
(b) Figure 6-3   **(f)** Figure 6-7   **(j)** Figure 6-13
(c) Figure 6-4   **(g)** Figure 6-10
(d) Figure 6-5   **(h)** Figure 6-11

3. (a) $A = \sqrt{C^2 - B^2}$
$= \sqrt{3^2 - 2^2}$
$= \sqrt{9 - 4}$
$= \sqrt{5}$
$= 2.24$ feet

(b) $B = \sqrt{C^2 - A^2}$
$= \sqrt{160^2 - 80^2}$
$= \sqrt{25{,}600 - 6400}$
$= \sqrt{19{,}200}$
$= 138.56$ km

(c) $C = \sqrt{A^2 + B^2}$
$= \sqrt{112^2 + 25^2}$
$= \sqrt{12{,}544 + 625}$
$= \sqrt{13{,}169}$
$= 114.76$ mm

4. (a) $x = \sqrt{A^2 + B^2}$
$= \sqrt{40^2 + 30^2}$
$= \sqrt{1600 + 900}$
$= \sqrt{2500}$
$= 50$ volts

(b) $x = \sqrt{A^2 + B^2}$
$= \sqrt{75^2 + 26^2}$
$= \sqrt{5625 + 676}$
$= \sqrt{6301}$
$= 79.38$ watts

(c) $x = \sqrt{A^2 + B^2}$
$= \sqrt{93^2 + 36^2}$
$= \sqrt{8649 + 1296}$
$= \sqrt{9945}$
$= 99.72$ mm

$y = \sqrt{A^2 + B^2}$
$= \sqrt{48^2 + 96^2}$
$= \sqrt{2304 + 9216}$
$= \sqrt{11{,}520}$
$= 107.33$ mm

STEP 6-3

1. (a) $O = 35$ mm $\theta = 36°$
$H = \ ?$

SOH or $\sin \theta = \dfrac{O}{H}$

$$\sin 36° = \dfrac{35 \text{ mm}}{H}$$

$$0.59 = \dfrac{35}{H}$$

$$0.59 \times H = \dfrac{35}{H} \times H \qquad (\times H)$$

$$\dfrac{0.59 \times H}{0.59} = \dfrac{35}{0.59} \qquad (\div 0.59)$$

$$H = \dfrac{35}{0.59} = 59.32 \text{ mm}$$

(b) H = 160 km $\theta = 38°$
A = ?

CAH   or $\cos \theta = \dfrac{A}{H}$

$$\cos 38° = \dfrac{A}{160 \text{ km}}$$

$$0.79 = \dfrac{A}{160}$$

$$0.79 \times 160 = \dfrac{A}{\cancel{160}} \times \cancel{160} \qquad (\times 160)$$

$$0.79 \times 160 = A$$

$$A = 126.4 \text{ km}$$

(c) O = 163 cm $\theta = 72°$
A = ?

TOA   or $\tan \theta = \dfrac{O}{A}$

$$\tan 72° = \dfrac{163 \text{ cm}}{A}$$

$$3.08 = \dfrac{163}{A}$$

$$3.08 \times A = \dfrac{163 \times \cancel{A}}{\cancel{A}} \qquad (\times A)$$

$$\dfrac{3.08 \times A}{3.08} = \dfrac{163}{3.08} \qquad (\div 3.08)$$

$$A = \dfrac{163}{3.08} = 52.92 \text{ cm}$$

2. (a) H = 120 miles $\theta = ?$
O = 38 miles

SOH   or $\sin \theta = \dfrac{O}{H}$

$$\sin \theta = \dfrac{38 \text{ miles}}{120 \text{ miles}}$$

$$\sin \theta = 0.32$$

$$\cancel{\text{invsin}} \times \cancel{\sin} \theta = \text{invsin } 0.32 \qquad (\times \text{invsin})$$

$$\theta = \text{invsin } 0.32$$

$$\theta = 18.66°$$

(b) H = 25 feet $\theta = ?$
A = 17 feet

CAH   or $\cos \theta = \dfrac{A}{H}$

$$\cos \theta = \dfrac{17 \text{ feet}}{25 \text{ feet}}$$

$$\cos \theta = 0.68$$

$$\cancel{\text{invcos}} \times \cancel{\cos} \theta = \text{invcos } 0.68$$

$$\theta = \text{invcos } 0.68$$

$$\theta = 47.16°$$

(c) A = 69 cm $\theta = ?$
O = 51 cm

TOA   or $\tan \theta = \dfrac{O}{A}$

$$\tan \theta = \dfrac{51 \text{ cm}}{69 \text{ cm}}$$

$$\tan \theta = 0.74$$

$$\cancel{\text{invtan}} \times \cancel{\tan} \theta = \text{invtan } 0.74$$

$$\theta = \text{invtan } 0.74$$

$$\theta = 36.5°$$

STEP 6-4
1. (Section 6-4 Introduction)
2. (Section 6-4-1)
3. **(a)** Figure 6-22   **(c)** Figure 6-24   **(e)** Figure 6-26
   (b) Figure 6-23   **(d)** Figure 6-25
4. A frustum is the base section of a solid pyramid or cone.

STEP 7-1
1. Exponent
2. **(c)** 4
3. **(a)** 1   **(c)** 1.4   **(e)** 4.02   **(g)** −0.456
   (b) 2   **(d)** 2.176   **(f)** 3.57   **(h)** 0.243
4. **(a)** 56.23   **(c)** 100,000   **(e)** 316,227.76 **(g)** 1.26
   (b) 223.87   **(d)** 10   **(f)** 1.78   **(h)** 1^{10}

STEP 7-2

1.

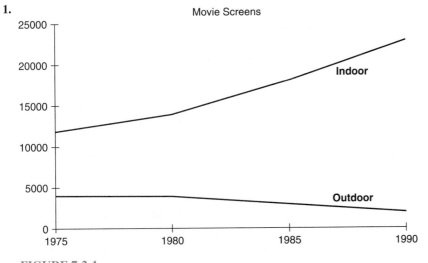

Movie Screens

FIGURE 7-2-1

2.

Tom's Day

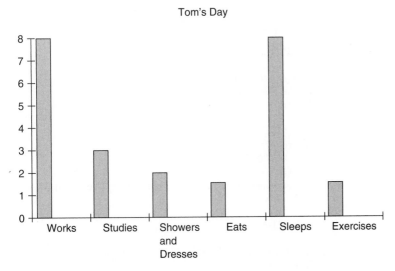

Tom's Day

FIGURE 7-2-2

3. Graphing calculators have all the functions of scientific calculators, and in addition, they can graph functions.

4. (a) Y-Editor: Key is used to enter equation.
   (b) Table: Shows every possible result for all possible input values.
   (c) Graph: Displays the plotted data in graphical form.

STEP 8-1
1. Section 8-1-1
2. Section 8-1-1
3. Section 8-1-2
4. Circuit Simplicity and Accuracy
5. 2
6. Section 18-1-2

STEP 8-2
1. CPU, Memory, Input-Output
2. Analog to Digital Converter
3. Analog, Digital
4. Digital, Analog

STEP 9-1
1. No difference
2. Base 10
3. $2 = 1000, 6 = 100, 3 = 10, 9 = 1$
4. Reset and carry

STEP 9-2
1. 26_{10}
2. 10111_2
3. LSB = least significant bit, MSB = most significant bit
4. 1101110_2

STEP 9-3
1. (a) Base 10   **(b)** Base 2   **(c)** Base 16
2. 21_{10} 15_{16}
3. BF7A
4. Binary equivalent = 100001
   Hexadecimal equivalent = 21

STEP 9-4
1. 8
2. 232_8
3. 4
4. 4267_8

STEP 9-5
1. $0111\ 0110\ 0010\ 1001_{BCD}$
2. 11_{10}
3. Only one digit changes as you step to the next code group.
4. B a b b a g e

STEP 10-1
1. Hardware
2. Software
3. Vacuum tube
4. Diode, transistor

STEP 10-2
1. OR, AND
2. Binary 1
3. AND gate
4. OR
5. Binary 0
6. AND

STEP 10-3
1. NOT gate
2. Binary 0
3. NAND
4. NOR
5. Binary 1
6. NAND

STEP 10-4
1. OR
2. XOR
3. XNOR
4. XNOR

STEP 10-5
1. IEEE/ANSI
2. Instead of using distinctive shapes to represent logic gates, it uses a special dependency notation system to indicate how the output is dependent on the input.

STEP 11-1
1. a. AND   **c.** OR   **e.** NOR
   b. XOR   **d.** NAND   **f.** NAND
2. $\overline{A \cdot B} = \overline{A} + \overline{B}$
3. a. $\overline{AB}$ **b.** $(\overline{A} \cdot \overline{B}) \cdot C$
4. a.

   b.

   c.

   d.

   e.

   f.

STEP 11-2
1. a. 1   **b.** 1   **c.** 0   **d.** A
2. $A(B + C) = (AB) + (AC)$
3. $(AB)C$ or $(AC)B$
4. $A(B + C + D)$

STEP 11-3
1. Product, sum
2. a. $A\overline{B}CD$ **b.** $\overline{A}BC\overline{D}$
3. SOP
4. Fundamental products are ORed and then the circuit is developed.

STEP 11-4
1. 4
2. $y = B(A + \overline{A}) = B(1) = B$
3. Larger
4. With 1s for larger grouping

STEP 12-1

1. **(a)** $1011 + 11101 = 101000$ **(c)** $101 \times 10 = 1010$
   (b) $1010 - 1011 = 11110001$ **(d)** $10111 \div 10 = 1011$

STEP 12-2

1. **(a)** 0101 **(b)** 1001 0101 **(c)** 1101 **(d)** 0111 1001
2. **(a)** -5 **(b)** $+6$ **(c)** -5 **(d)** $+6$

STEP 12-3

1. Adder circuit

STEP 12-4

1. **(a)** 0100 0001 0000 1000
   (b) 0001 0010 1111 0100

STEP 13-1

1. The three blocks are the microprocessor unit (MPU), the memory unit, and the input/output unit. Functions are:

   Microprocessor—logic circuitry recognizes and executes the program stored in memory

   Memory—contains the program of instruction that is fetched and executed by the MPU

   Input/output—allows data to be read in from or sent out to peripheral devices

2. The three buses are address, data, and control. Functions are:

   Address—used by the MPU to select memory locations or input/output ports

   Data—used to transfer data or information between devices

   Control—carries signals used by the processor to coordinate the transfer of data within the microcomputer

3. Hardware—the units and components of the system

   Software—written programs stored as code that are understood by the MPU

STEP 13-2

1. **a.** Machine language—the digital code understood by the MPU

   b. Assembly language—programs written using mnemonics, which are easier to remember

   c. High-level language—an independent language designed to make programming easier that later is compiled into machine language for the MPU

2. A flowchart is a graphic form that describes the operation of a program using oval, rectangular, and diamond-shaped blocks.

3. **a.** Instruction set—contains a list of the machine codes that are recognized by a specific type of microprocessor

   b. Peripherals—input and output devices connected to microprocessor system ports

   c. Programmer—a person who writes a program in any language that will be understood by the MPU

   d. Mnemonic—a short name, or memory aid, used in assembly language programming

   e. Compiler— a program that translates high-level language into machine code

   f. Accumulator—an MPU internal register (register A), used in conjunction with the ALU; a storage location within an MPU chip

   g. Flag register—an internal register within an MPU chip that is a collection of flip-flops used to indicate the results of certain instructions

   h. Op-code—the operation code or "do" portion of an instruction

Answers to Odd-Numbered Problems

APPENDIX

B

Chapter 1
1. c
3. b
5. d
7. c
9. c

(The answers to Communication Skill Questions 11 through 19 can be found in the indicated sections that follow the questions.)

21. (a) $3\frac{3}{4}$ (b) $2\frac{1}{3}$

23. (a) $\dfrac{5}{8} + \dfrac{2}{8} = \dfrac{5+2}{8} = \dfrac{7}{8}$

   (b) $\dfrac{47}{76} + \dfrac{15}{76} + \dfrac{1}{76} = \dfrac{63}{76}$

25. (a) $\dfrac{5}{2} = ?$, 2 into 5 = 2 with 1 remaining

   $\dfrac{5}{2} = 2\dfrac{1}{2}$

   (b) $\dfrac{17}{4} = 4\dfrac{1}{4}$

   (c) $\dfrac{25}{16} = 1\dfrac{9}{16}$

   (d) $\dfrac{37}{3} = 12\dfrac{1}{3}$

27. (a) $\dfrac{4 \div 4}{16 \div 4} = \dfrac{1}{4}$

   (b) $\dfrac{16 \div 2}{18 \div 2} = \dfrac{8}{9}$

   (c) $\dfrac{74 \div 2}{128 \div 2} = \dfrac{37}{64}$

   (d) $\dfrac{28}{45} = \dfrac{28}{45}$

29. (a) $\dfrac{1}{9} \times 4 = \dfrac{1 \times 4}{9 \times 1} = \dfrac{4}{9}$

   (b) $\dfrac{3}{6} \times \dfrac{4}{5} = \dfrac{3 \times 4}{6 \times 5} = \dfrac{12 \div 6}{30 \div 6} = \dfrac{2}{5}$

   (c) $\dfrac{1}{3} \times 4\dfrac{1}{2} = \dfrac{1}{3} \times \dfrac{9}{2} = \dfrac{9}{6} = 1\dfrac{3}{6} = 1\dfrac{1}{2}$

   (d) $2\dfrac{3}{4} \times 4\dfrac{4}{11} = \dfrac{11}{4} \times \dfrac{48}{11} = \dfrac{1}{1} \times \dfrac{12}{1} = 1 \times 12 = 12$

Chapter 2
1. b
3. a
5. b
7. a
9. c

(The answers to Communication Skill Questions 11 through 19 can be found in the indicated sections that follow the questions.)

21. $\begin{aligned}9 \times 10,000 &= 90,000\\ 6 \times 1,000 &= 6,000\\ 2 \times 100 &= 200\\ 3 \times 10 &= 30\\ 7 \times 1 &= \underline{\quad 7}\\ &\quad 96,237\end{aligned}$

23. (a) Units and tens
   (b) Units
   (c) Units
   (d) Units, tens, hundreds, and thousands
25. (a) 2.3
   (b) 0.507
   (c) 9030.04

27. (a) $\dfrac{16}{32} \div \dfrac{2}{2} = \dfrac{1}{2} = 1 \div 2 = 0.5$

   (b) $3\dfrac{8}{9} = 3 + (8 \div 9) = 3 + 0.889 = 3.889$

   (c) $4\dfrac{9}{8} = 4 + \dfrac{8}{8} + \dfrac{1}{8} = 4 + 1 + \dfrac{1}{8} = 5\dfrac{1}{8} = 5 + (1 \div 8)$

   $= 5 + 0.125 = 5.125$

   (d) $195\dfrac{7}{3} = 195 + \dfrac{3}{3} + \dfrac{3}{3} + \dfrac{1}{3} = 197\dfrac{1}{3} = 197 + (1 \div 3)$

   $= 197 + 0.333 = 197.333$

29. (a) $0.777 = \dfrac{7}{9}$

   (b) $0.6149069 = \dfrac{99}{161}$

   (c) $43.125 = 43\dfrac{1}{8}$

31. (a) $\dfrac{20 \text{ ft}}{5 \text{ ft}}$ ← Reduce to lowest terms

   $\dfrac{20 \div 5}{5 \div 5} = \dfrac{4}{1}$

   The ratio of 20 ft to 5 ft is 4 to 1 (4:1).
   (b) Both quantities must be alike, so we must first convert minutes to seconds.

   $2\dfrac{1}{2} \text{ min} = 2\dfrac{1}{2} \times 60 \text{ s} = 150 \text{ s}$

   $\dfrac{150 \text{ s} \div}{30 \text{ s} \div} = \dfrac{5}{1}$

   The ratio of $2\dfrac{1}{2}$ min to 30 s is 5 to 1 (5:1).
33. 10.9
35. (a) 6

   (b) 17.25 or $17\dfrac{1}{4}$ s

   (c) 116 V

328

(d) 505.75, or $505\frac{3}{4}$ Ω

(e) 155 m

Chapter 3
1. c   7. b
3. a   9. a
5. a

(The answers to Communication Skill Questions 11 through 19 can be found in the indicated sections that follow the questions.)

21. **(a)** $+8 - (-5)$
    (b) -0.6×13
    (c) $-22.3 \div (-17)$
    (d) $(+4) \div (-9)$
23. **(a)** $+6 - (+8) = -2$
    (b) $+9 - (-6) = +15$
    (c) $-75 - (+62) = -137$
    (d) $-39 - (-112) = +73$
25. **(a)** $+19 \div (+3) = 6.33$
    (b) $+36 \div (-3) = -12$
    (c) $-80 \div (+5) = -16$
    (d) $-44 \div (-2) = 22$
27. $+15 \div (+5) \times (-3.5) = -10.5$
29. $-6 \div (-4) \div (-3) \times (-15) = 7.5$

Chapter 4
1. c   7. a
3. a   9. b
5. d

(The answers to Communication Skill Questions 11 through 19 can be found in the indicated sections that follow the questions.)

21. **(a)** $9^2 = 9 \times 9 = 81$
    (b) $6^2 = 6 \times 6 = 36$
    (c) $2^2 = 2 \times 2 = 4$
    (d) $0^2 = 0 \times 0 = 0$
    (e) $1^2 = 1 \times 1 = 1$
    (f) $12^2 = 12 \times 12 = 144$
23. **(a)** $9^3 = 9 \times 9 \times 9 = 729$
    (b) $10^4 = 10 \times 10 \times 10 \times 10 = 10,000$
    (c) $4^6 = 4 \times 4 \times 4 \times 4 \times 4 \times 4 = 4096$
    (d) $2.5^3 = 2.5 \times 2.5 \times 2.5 = 15.625$
25. **(a)** $\frac{1}{100} = 10^{-2}$
    (b) $1,000,000,000 = 10^9$
    (c) $\frac{1}{1000} = 10^{-3}$
    (d) $1000 = 10^3$

27.

	Scientific Notation Base of 1 to 10, plus power of ten	Engineering Notation Base greater than 1, plus × 3 power of ten
(a) 475	$= 4.75 \times 10^2$	475.0
(b) 8200	$= 8.2 \times 10^3$	8.2×10^3
(c) 0.07	$= 7 \times 10^{-2}$	70×10^{-3}
(d) 0.00045	$= 4.5 \times 10^{-4}$	450×10^{-6}

29. **(a)** km $\times 0.6$ = mi, 100 km $\times 0.6$ = 60 mi
    (b) $m^2 \times 1.2 = yd^2$, 29 $m^2 \times 1.2 = 34.8$ yd^2
    (c) kg $\times 2.2$ = lb, 67 kg $\times 2.2$ = 147.4 lb
    (d) L $\times 2.1$ = pt, 2 L $\times 2.1$ = 4.2 pt
31. **(a)** $\frac{5}{9} \times (°F - 32) = °C$

$$= \frac{5}{9} \times (32°F - 32) = 0.5555 \times 0 = 0°C$$

$$32°F = 0°C$$

(b) $\frac{5}{9} \times (72°F - 32)$

$$= \frac{5}{9} \times 40 = 22.22°C$$

33. **(a)** meter
    (b) square meter
    (c) gram
    (d) cubic meter
    (e) liter
    (f) degrees Celsius
35. **(a)** 8000 ms = _____ μs
     $= 8000 \times 10^{-3} =$ _____ 10^{-6}
     $= 8000.000. \times 10^{-6} = 8,000,000 \times 10^{-6}$ s
     $= 8000$ ms $= 8,000,000$ μs
    (b) 0.02 MV = _____ kV
     0.02×10^6 V = _____ $\times 10^3$ V
     $0.020. \times 10^3$ V $= 20 \times 10^3$ V
     0.02 MV $= 20$ kV
    (c) 10 km = _____
     10×10^3 m = _____ $\times 100$ m
     $10.000. \times 100$ m $= 10,000$ m
     10 km $= 10,000$ m
    (d) 250 mm = _____ cm
     250×10^{-3} m = _____ $\times 10^{-2}$ m
     $25.0. \times 10^{-3}$ m $= 25 \times 10^{-2}$ m
     250 mm $= 25$ cm

Chapter 5
1. a   7. c
3. d   9. b
5. c

(The answers to Communication Skill Questions 11 through 25 can be found in the indicated sections that follow the questions.)

27. **(a)** $\qquad 4x = 11$
$$4 \times x = 11$$
$$\frac{\cancel{4} \times x}{\cancel{4}} = \frac{11}{4} \qquad (\div 4)$$
$$x = \frac{11}{4}$$
$$x = 2.75$$
(b) $6a + 4a = 70$
$$10a = 70$$
$$10 \times a = 70$$
$$\frac{\cancel{1}0 \times a}{\cancel{1}0} = \frac{70}{10} \qquad (\div 10)$$
$$a = \frac{70}{10}$$
$$a = 7$$
(c) $5b - 4b = \frac{7.5}{1.25}$
$$1b = \frac{7.5}{1.25} \qquad (1b = 1 \times b = b)$$
$$b = \frac{7.5}{1.25}$$
$$b = 6$$
(d) $\frac{2z \times 3z}{4.5} = 2z$ $\qquad \begin{bmatrix} 2z \times 3z = 2 \times z \times 3 \times z \\ = (2 \times 3) \times (z \times z) \\ = 6 \times z^2 \\ = 6z^2 \end{bmatrix}$
$$\frac{6z^2}{4.5} = 2z$$

$$\frac{6z^2}{4\cancel{.5}} \times 4\cancel{.5} = 2z \times 4.5 \qquad (\times \ 4.5)$$

$$6z^2 = 2z \times 4.5$$

$$\frac{\overset{3}{\cancel{6}} \times \cancel{z} \times z}{\cancel{2} \times \cancel{z}} = \frac{(2 \times z) \times 4.5}{2 \times z} \qquad (\div \ 2z)$$

$$3z = 4.5$$

$$\frac{\cancel{3} \times z}{\cancel{3}} = \frac{4.5}{3} \qquad (\div \ 3)$$

$$z = \frac{4.5}{3}$$

$$z = 1.5$$

29. (a) Power (P) = voltage (V) × current (I)
1500 watts = 120 volts × ?
Transpose formula to solve for I.

$$P = V \times I$$

$$\frac{P}{V} = \frac{\cancel{V} \times I}{\cancel{V}} \qquad (\div \ V)$$

$$\frac{P}{V} = I$$

$$I = \frac{P}{V}$$

$$\text{Current } (I) \text{ in amperes} = \frac{\text{power } (P) \text{ in watts}}{\text{voltage } (V) \text{ in volts}}$$

$$I = \frac{1500 \text{ W}}{120 \text{ V}}$$

Current (I) = 12.5 amperes

(b) Voltage (V) in volts = current (I) in amperes × resistance (R) in ohms
120 volts = 12.5 amperes × ?
Transpose formula to solve for R.

$$V = I \times R$$

$$\frac{V}{I} = \frac{\cancel{I} \times R}{\cancel{I}}$$

$$\frac{V}{I} = R$$

$$R = \frac{V}{I}$$

$$\text{Resistance } (R) \text{ in ohms} = \frac{\text{voltage } (V) \text{ in volts}}{\text{current } (I) \text{ in amperes}}$$

$$R = \frac{120 \text{ V}}{12.5 \text{ A}}$$

Resistance (R) = 9.6 ohms

Chapter 6

1. b	**9.** c
3. c	**11.** d
5. d	**13.** a
7. b	**15.** d

(The answers to Communication Skill Questions 17 through 25 can be found in the indicated sections that follow the questions.)

27. (a) $C = \sqrt{A^2 + B^2}$
$= \sqrt{20^2 + 53^2}$
$= \sqrt{400 + 2809}$
$= \sqrt{3209}$
$= 56.65 \text{ mi}$

(b) $C = \sqrt{2^2 + 3^2}$
$= 3.6 \text{ km}$

(c) $C = \sqrt{4^2 + 3^2}$
$= 5 \text{ inches}$

(d) $C = \sqrt{12^2 + 12^2}$
$= 16.97 \text{ mm}$

29. (a) $\sin 0° = 0$
(b) $\sin 30° = 0.5$
(c) $\sin 45° = 0.707$
(d) $\sin 60° = 0.866$
(e) $\sin 90° = 1.0$
(f) $\cos 0° = 1.0$
(g) $\cos 30° = 0.866$
(h) $\cos 45° = 0.707$
(i) $\cos 60° = 0.5$
(j) $\cos 90° = 0$
(k) $\tan 0° = 0$
(l) $\tan 30° = 0.577$
(m) $\tan 45° = 1.0$
(n) $\tan 60° = 1.73$
(o) $\tan 90° = \infty$ (infinity)

31. (a)

Known values	Unknown values
$H = 27$ miles	$O = ?$
$\theta = 37°$	$A = ?$

We must calculate the length of the opposite and adjacent sides. To achieve this we can use either H and θ to calculate O (SOH), or H and θ to calculate A (CAH).

$$\sin \theta = \frac{O}{H}$$

$$\sin 37° = \frac{O}{27}$$

$$0.6 = \frac{O}{27}$$

$$0.6 \times 27 = \frac{O}{\cancel{27}} \times \cancel{27}$$

$$O = 0.6 \times 27$$

$$\text{Opposite} = 16.2 \text{ mi}$$

The next step is to use the Pythagorean theorem to calculate the length of the unknown side now that the length of two sides are known.

$$B = \sqrt{C^2 - A^2}$$
$$= \sqrt{27^2 - 16.2^2}$$
$$= \sqrt{729 - 262.44}$$
$$= \sqrt{466.56}$$
$$\text{Adjacent or } B = 21.6 \text{ mi}$$

(b)

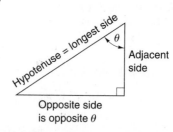

Known values	Unknown values
$O = 29$ cm	$H = ?$
$\theta = 21°$	$A = ?$

We can use either O and θ to calculate H (SOH), or O and θ to calculate A (TOA).

$$\tan \theta = \frac{O}{A}$$

$$\tan 21° = \frac{29 \text{ cm}}{A}$$

$$0.384 = \frac{29}{A}$$

$$0.384 \times A = \frac{29}{\cancel{A}} \times \cancel{A} \ (\times A)$$

$$\frac{\cancel{0.384} \times A}{\cancel{0.384}} = \frac{29}{0.384} \ (\div 0.384)$$

$$A = \frac{29}{0.384}$$

$$A = 75.5 \text{ cm}$$

Now that O and A are known, we can calculate H.

$$C \ (\text{or } H) = \sqrt{A^2 + B^2}$$
$$= \sqrt{75.5^2 - 29^2}$$
$$= \sqrt{5700.25 + 841}$$
$$= \sqrt{6541.25}$$
$$= 80.88 \text{ cm}$$

(c)

Known values	Unknown values
$H = 34$ volts	$\theta = ?$
$A = 18.5$ volts	$O = ?$

Because both H and A are known, we can use cosine to calculate θ (CAH)

$$\cos \theta = \frac{A}{H}$$

$$\cos \theta = \frac{18.5 \text{ V}}{34 \text{ V}}$$

$$\cos \theta = 0.544$$

$$\text{invcos} \times \cos \theta = \text{invcos } 0.544 \quad (\times \text{ invcos})$$

$$\theta = \text{invcos } 0.544$$

$$\theta = 57°$$

To calculate the length of the unknown side, we can use the Pythagorean theorem.

$$B = \sqrt{C^2 - A^2}$$
$$= \sqrt{34^2 - 18.5^2}$$
$$= \sqrt{813.75}$$
$$= 28.5 \text{ V}$$

33. $S = \pi d^2 = \pi \times 2^2 = 12.57 \text{ in}^2$, $V = \frac{1}{6} \pi \times d^3 = 4.189 \text{ in}^3$

35. $L = \frac{1}{2} ps$
$$= \frac{1}{2} \times 24 \times 8 = 96 \text{ in}^2$$

Chapter 7
1. b
3. c
5. a
7. b
9. c

(The answers to Communication Skill Questions 11 through 19 can be found in the indicated sections that follow the questions.)

21. (a) 1   (g) −0.3
     (b) 4   (h) −0.125
     (c) 6.09   (i) −0.9
     (d) 1.7   (j) −0.02
     (e) 0.176   (k) 1.088
     (f) −0.6   (l) 1.88
23. *Step 1:* Find the log of each number.
     Step 2: Change each number to base 10.
     Step 3: Add the base 10 exponents.
     Step 4: Convert the log answer back to a number using antilog.
25. *Step 1:* Find the log of the dividend and divisor.
     Step 2: Change each number to base 10.
     Step 3: Subtract the log of the divisor from the log of the dividend.
     Step 4: Convert the log answer back to a number using antilog.
27. (a) Line graph
     (b) Bar graph
29. (a) Children are able to access educational material.
     (b) **1.** Children are able to access educational material.
         2. Video conferencing
         3. Accessing information on hobbies or interests
         4. Accessing information for work
         5. Sending notes to friends and family

Chapter 8
1. b
3. d
5. c
7. b

(The answers to Communication Skill Questions 9 through 15 can be found in the indicated sections that follow the questions.)

Chapter 9
1. b
3. c
5. d
7. a
9. d

(The answers to Communication Skill Questions 11 through 23 can be found in the indicated sections that follow the questions.)

25. 1001, decimal 9
27. Decimal equivalents are:
     a. 57   **b.** 39   **c.** 06
29. BCD equivalents are:
     a. 0010 0011 0110 0101   **b.** 0010 0100

31. Codes in order are: BCD, gray, excess-3, and ASCII.
33. Stx practice makes perfect etx.

Chapter 10

1. a	**11.** b
3. b	**13.** d
5. b	**15.** c
7. b	**17.** b
9. b	**19.** a

(The answers to Communication Skill Questions 21 through 29 can be found in the indicated sections that follow the questions.)

31. OR gate
33. OR
35. Truth table outputs (Y) should be:

a. 0 (AND)	**c.** 0 (XOR)	**e.** 1 (NOR)
0	1	0
0	1	0
1	0	0
b. 0 (OR)	**d.** 1 (XNOR)	**f.** 0 (NOT)
1	0	1
1	0	
1	1	

Chapter 11

1. c	**11.** c
3. c	**13.** a
5. b	**15.** b
7. c	**17.** c
9. a	**19.** c

(The answers to Communication Skill Questions 21 through 29 can be found in the indicated sections that follow the questions.)

31. a. $Y = (\overline{A} \cdot B) + (B \cdot C) + D$
   b. $Y = (\overline{A} \cdot B) + (\overline{\overline{A} \cdot \overline{B}})$
   c. $Y = (\overline{\overline{A} \cdot \overline{B} \cdot \overline{C}}) + (\overline{A \cdot B \cdot \overline{C}})$
   d. $Y = (LIT + NLT) \cdot (SLW + FST)$

33. a.

   b.

35.

$$Y_2 = \overline{BC} + \overline{A}BC$$

37. $Y_1 = \overline{ABCD} + \overline{AB}C\overline{D} + \overline{A}BC\overline{D} + \overline{A}BCD + A\overline{BC}D$
      $A\overline{B}CD + AB\overline{CD} + AB\overline{C}D + ABC\overline{D} + ABCD$

$$Y_1 = \overline{AB} + AB + AD$$
(50 Inputs to 9 Inputs)

39.

(Question 35, Y_1)

(Question 37, Y_2)

(Question 38, Y_3)

Chapter 12

1. b   **7.** d
3. d   **9.** d
5. a
4. c

(The answers to Communication Skill Questions 11 through 17 can be found in the indicated sections that follow the questions.)

19. a. 1000 0111   **b.** 1100 0010   **c.** 0101 0100
21. a. 0001 0001 0001   **b.** 0001 0101 0101   **c.** 0001 1100
23. $+34 = 0010\ 0010$
     $-83 = 1010\ 1101$
     $+83 = 0101\ 0011$
25. a. $+16$ **b.** -66 **c.** -11
27. a. 1010 0110   **b.** 0111 1111   **c.** 0000 0011

Chapter 13

1. c   **7.** c
3. b   **9.** b
5. a

(The answers to Communication Skill Questions 11 through 19 can be found in the indicated sections that follow the questions.)

21.

23.

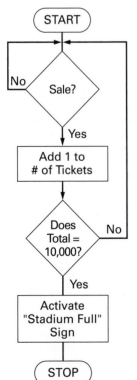

25. Choose value and then follow flow chart.

Index